AF521818

Building Environments: HVAC Systems

First Edition

Alan J. Zajac
Johnson Controls Institute

Johnson Controls, Inc.
Milwaukee, WI 53202

Published by
Johnson Controls, Inc.
Milwaukee, WI 53202

Publisher's Cataloging-in-Publication Data:

Zajac, Alan J., 1953–
Building environments: HVAC systems / Alan J. Zajac. –
1st ed.
p. cm.
Includes index.
ISBN 0-925669-00-8
1. Heating—Control. 2 Ventilation—Control. 3. Air
conditioning—Control. 4. Commercial buildings—
Control. I. Title.
TH7466.5.J64 1997

97-94080

First Edition

10 9 8 7 6 5 4 3 2 1

ISBN 0-925669-00-8

Printed in the United States of America

Table of Contents

Preface

Knowledge of building environments is fundamental to the design, operation and maintenance of today's complex buildings. Knowledge of systems and controls enables building personnel to create a healthy, productive indoor environment. Within this book, the student of HVAC systems will learn how HVAC systems function to create the building environment. The student who invests time in this valuable learning opportunity will be laying a solid foundation for their professional career development.

I have been challenged during my career with the task of developing people: people to engineer, operate, install and service HVAC systems. Like many trainers, I have concluded that the theory and operation of HVAC systems must be presented within a technical, yet easily accessible framework. My challenge was to produce a book in a technical, yet extremely readable format. The result is this one of a kind book, which features top-to-bottom coverage of the entire HVAC system, in a very readable format!

By pooling resources and collaborating with industry experts, we have captured years of knowledge and experience for you in this one volume. Assembled in eighteen chapters are the combined efforts of technicians, engineers, trainers and manufacturers. *Building Environments: HVAC Systems* provides an examination of the physics of HVAC, human comfort, psychrometrics, mechanical system components, control systems and facility management systems.

Whatever your role, whether it's engineering, installation, service, operations or administrative support, this book is for you. *Building Environments: HVAC Systems* is a valuable resource on HVAC systems for both beginners and experts. Experts will find within these pages a handy reference guide. Students of HVAC systems will find technical concepts explained in clear, concise language with supporting illustrations and examples. If you're a student of HVAC systems, consider this book an important first step in your ability to serve others by creating and maintaining healthy, comfortable, productive building environments. Take some advice from those who know, "Knowledge of systems and controls helps you perform in such a way so as to create a quality building environment."

Alan J. Zajac
Author, *Building Environments: HVAC Systems*
Johnson Controls Institute
Johnson Controls, Inc.

About the Cover

Featured on the cover is the Koin Tower, Portland Oregon. Owned and operated by the Louis Dreyfus Property Group, the Koin Tower contains 410,000 square feet of office space, forty four condominiums, retail space, two radio stations, a television station and movie theater.

The building is controlled by a Johnson Controls Metasys® facility management system. Johnson Controls, Inc. is a leading worldwide supplier of facility management and control systems to education, health care, office, government, industrial and retail buildings.

Acknowledgments

Johnson Controls, Inc. would like to acknowledge the combined efforts of all those who contributed their time and expertise in making *Building Environments: HVAC Systems* possible. We thank them and hope that this book will expand their effectiveness in working with HVAC systems. We apologize to those whose names do not appear, because it would be impossible to include them all.

Many technical contributors, editors and expert reviewers helped to create *Building Environments: HVAC Systems*. Their many years of experience and research helped refine the content of this book. Johnson Controls, Inc. would like to acknowledge the contributions of the following individuals: Ray Aschauer, Jodi Delfosse, Kirk Drees, Jeff Gloudeman, Kay Goodwin, Mike Hofschulte, Howie Holland, Dean Johnson, Mary Kaczmarek, Cindy Kasper, Dave Kenney, Jerry Leonard, Eric McAtte, Chuck Miles, Kathy O'Neil, Ken Oakleaf, Jim Peck, Howard Rauch, Lori Reichert, Jackie Schneider, Roy Smith, Dave Sobczak, Gene Strehlow, Ron VanderMeer, Jarrell Wenger, Paul Wichman and Alan Zajac.

Johnson Controls, Inc. would also like to express appreciation to the societies, associations, publishers, manufacturers and building owners who gave permission to reproduce their work. Their willingness to share their knowledge and expertise made *Building Environments: HVAC Systems* possible.

Technical Societies

American Society of Heating, Refrigerating and Air Conditioning Engineers, Inc. (ASHRAE)

Associations

Air Movement and Control Association International, Inc. (AMCA)

Building Owners

Louis Dreyfus Property Group – Koin Tower

Manufacturers

Airguard Industries
Copeland Corporation
Dri-Steem Humidifier Company
Dunham-Bush, Inc.
Dwyer Instruments, Inc.
ITT Fluid Technology Corporation
ITT McDonnell & Miller
Paul Mueller Company
Munters Corporation – Cargocarie Division
The Trane Company
TACO, Inc.
Tecumseh Products Company

Publishers

American Technical Publishers, Inc.
Prentice-Hall Inc.
McGraw Hill
TPC Training Systems, Inc.

1

The Role Of HVAC Systems In Facility Management

A facility's heating, ventilating and air conditioning (HVAC) system maintains the building environment within a desired range of conditions, independent of the outdoor environment.

A well-designed HVAC system enhances personal productivity, makes possible the efficient operation of manufacturing processes and controls, and conserves energy.

HVAC systems provide year-round control of temperature, humidity, circulation, ventilation and purification of the air within a facility. To achieve complete air control, the design of a HVAC system addresses three specific purposes:

- Maintains the comfort and productivity of the facility's occupants,
- Enhances the efficiency of equipment and processes within the facility, and
- Conserves energy used for heating and cooling the entire facility.

Occupant Comfort and Productivity

The HVAC designer strives to create a building environment that is acceptable to most of the people, most of the time. While a comfortable environment will not always improve productivity, an inadequate environment will certainly reduce it.

Creating a comfortable environment begins with identifying the owner's needs. Who will use the facility and how will they use it? Next, the engineer or architect designs the building shell. Only after addressing these initial matters is it time to design the mechanical, electrical and structural details of the HVAC systems.

Equipment and Manufacturing Process Efficiency

Like people, the equipment and processes operating in a facility often work best in a fairly narrow range of temperature and humidity. For example, a cereal manufacturer might be mixing grains and powdered additives with liquids or syrups. To flow well, the grains and additives need moderate temperature and humidity. Before mixing, some of the additives might need refrigeration, others might need heating. Finally, all the piping, valves, sorters, and mixers used to blend the grains and their enhancers are probably controlled by sensitive electronic equipment which has its own environmental needs: freedom from dust plus moderate temperature and humidity. A well-designed HVAC system will optimize the efficiency of both material mixing and process controls.

In most manufacturing and distribution processes, the operating equipment gives off heat which adds to the heat gain in the facility. In the winter this extra heat might be used to heat other portions of the building, but in the summer it adds to the cooling load of the air conditioning equipment. In either season, the architect or engineer relies on the HVAC system for maintaining the interior environment in a range that enhances equipment and process efficiency.

In addition to maintaining the temperature, humidity and air quality within predetermined limits, HVAC systems can support equipment and process efficiency in two other ways:

- They add to equipment and process safety by regulating the building environment. Their control systems can even stop equipment when environmental conditions exceed predetermined, safe ranges.
- They can also perform an operating function by starting and stopping equipment in predetermined sequences.

Energy Conservation

The cost of the energy to warm or cool a building, to operate equipment or to control a process is significant. In addition, the traditional energy sources of wood, coal, natural gas and oil are diminishing. Because they manage building environments, HVAC systems can also perform an important energy conservation function. By intelligently controlling environmental heating and cooling, equipment performance and process operating cycles, the designer ensures that the entire facility is operating efficiently and using a minimum of costly energy resources.

2

Heat, Temperature And Pressure Basics

The first word in HVAC is *heating* and that's where this chapter begins – reviewing the concept of heat – how it transfers from higher temperature regions to lower temperature regions and how it is measured. The chapter concludes with the measurement of pressure. Upon completing this chapter you will be able to:

1. Define heat and its units of measurement.
2. Define the heat transfer processes of conduction, convection and radiation.
3. Apply specific, sensible and latent heat concepts to HVAC systems.
4. Compare the means of measuring temperature using the Fahrenheit, Celsius and Absolute temperature scales.
5. State the difference between atmospheric and gauge pressures and how various gauges measure pressure.

Heat Basics

Heat is a form of energy. Every substance has some heat. Heat transfers (or flows) from sources of higher temperature to places and bodies with lower temperatures. For example, on a cold day, the heat from your warm body continually flows to the cooler air surrounding it.

When heat enters a body or system it adds to the internal energy already present. The amount of heat energy in a body or system is measured in calories (cal) or British thermal units (Btu). A **calorie** is the amount of energy required to raise or lower 1 gram of water 1°C. A **Btu** is the amount of energy necessary to raise or lower 1 pound of water 1°F.

Theoretically, a body or system can have no internal energy whatsoever. The temperature of zero energy is known as **absolute zero**. On the Celsius temperature scale absolute zero is defined at -273°C. To date, scientists have been able to produce laboratory conditions within 1/1000 degree Celsius of absolute zero.

Heat Transfer

Heat continuously moves from places of higher temperature to places of lower temperature. Heat will flow only from hot to cold and will continue to do so until both areas or objects are equal in temperature. The rate of heat transfer will be fastest between objects with the largest temperature differential and be slowest when the two temperatures are nearly equal. **Heat transfer** occurs in three ways:

1. Conduction
2. Convection
3. Radiation

Conduction

Conduction transfer requires physical contact between molecules or atoms. When the molecules of liquids and gases collide or the electrons of solids interact, they transfer energy from molecules or atoms with higher kinetic energy to molecules or atoms with lower kinetic energy levels. For example, your body loses heat to a cold seat at a November football game by conduction. Similarly hot tea conducts heat to a colder teaspoon.

Convection

Convection heat transfer is a combination of two mechanisms; conduction and energy transport due to fluid motion. Heat is conducted through a thin stationary layer of fluid near the solid surface and is then swept away by the moving fluid. Higher flow rates (forced convection) reduce the thickness of the stationary fluid layer and thus give higher heat transfer rates.

Radiation

Regardless of their temperature, all solid materials emit electromagnetic waves or **radiation**. For example, the sun's radiation warms your face as you stand at a window on a winter's day. Stand by the same window at night and your body feels cool because it is radiating heat to the cold window.

Most heat radiation is invisible to the human eye. For example, radiant panels gradually warm a room without any visible effects. If the emitting body is sufficiently hot, however, more of the radiation occurs in the visible part of the electromagnetic spectrum. So, a gas fired infrared heater glows red as it emits radiation in the visible spectrum.

Specific Heat

Specific Heat is a measure of the heat storage capacity of a substance. It relates to the temperature change of a known mass of a substance after a specific amount of heat has been applied. For example, to raise one gram of water one Celsius degree, you have to add one calorie of heat.

Specific heat uses water as a standard of 1.0 and is based upon the fact that it takes 1 **Btu** (British thermal unit) of energy to raise 1 pound of water 1°F. For another example, the specific heat of air is 0.24. This means that 0.24 Btu will raise the temperature of one pound (approximately 13.3 cubic feet) of air 1°F.

Sensible Heat

All matter typically exists in one of three states: it is a solid, a liquid or a gas. Any substance can exist in any one of these states. Heat that changes the temperature of a substance without changing the substance's state is called **sensible heat** which can be measured with a thermometer.

The letter Q is the engineering symbol used to represent heat transfer rate in **Btus per hour (Btu/hr)**. An example of sensible heat change using Q or more precisely Qs (Btu sensible) will follow shortly.

Sensible heat changes are measured with a thermometer. The temperature change is written as **ΔT** and is read as **"delta T"**. The greater the ΔT, the greater the rate of heat transfer.

If you know the weight, temperature and specific heat of a material, you can determine the change in sensible heat. Multiply the sensible change (ΔT) by the mass of the material and by its specific heat. The result is the sensible heat required for the change. For example, to calculate the sensible heat necessary to raise the temperature of one gallon of water from 140°F to 160°F:

Example 2-1

$$Qs = \Delta T \times \text{Weight of water} \times \text{Specific heat}$$

$$Qs = (160 - 140) \times (8.34 \text{ lb/gallon}) \times (1.0)$$

$$Qs = 20 \times 8.34 \times 1.0$$

$$Qs = 166.8 \text{ Btu/gallon}$$

Latent Heat

To change a material from a solid to a liquid or from a liquid to a gas, you have to add heat to the substance. Conversely, to change a gas to a liquid or a liquid to a solid, you have to remove heat. The heat required to change the state of a substance–without changing its temperature–is its **latent heat** or hidden heat.

The conversion of water between its three states requires the addition or removal of latent and sensible heat. Control of latent and sensible heat is at the heart of the HVAC industry. For example:

- It takes 144 Btu of latent heat to change one pound of ice at 32°F to one pound of water at 32°F, or vice versa. This is the **latent heat of fusion.**
- It takes 180 Btu of sensible heat to raise the temperature of one pound of water from 32°F to 212°F.
- It takes 970 Btu of latent heat to change one pound of water (at atmospheric pres-

sure) to steam at 212°F, or vice versa. This is the **latent heat of vaporization.**

The exchange of heat, either sensible or latent, is the basis for most heating and air conditioning processes. These HVAC processes add or remove heat from a medium, such as air or water, at a central point and then distribute this heated or cooled medium to all parts of the structure where it will be used to warm or cool the space.

Heat Content

Every substance contains the heat that was necessary to raise its temperature from absolute zero to its present temperature. This total heat is its **heat content**. Figure 2-1 shows the change in the heat content of one pound of water that has been changed from ice at 32°F to steam at 212°F.

Temperature

Temperature is a measure of the thermal state of matter. Temperature defines matter's tendency to take on or give up heat to other matter it is in contact with. You measure thermal change with a thermometer.

If, upon contact, there is no heat flow, there is no temperature difference. For example, pour a warm drink into an ice-filled glass. Initially, there is a temperature difference between the liquid and the ice, and heat flows from the liquid to the ice. After some time however, the heat in the liquid and the heat in the ice will be equal; the heat flow will stop because their temperatures will be the same.

Measuring Heat Change

Thermometers are used to measure temperature, they do not measure a substances specific physical quantity. They measure only the length of a liquid column, the pressure or volume of a gas, or the electrical voltage in a thermocouple. The scales for these temperature measuring devices vary, depending upon the type of thermometer.

Because water freezes and boils at unvarying temperatures, thermometers are calibrated for these two temperatures. And, because atmospheric pressure does affect the tempera-

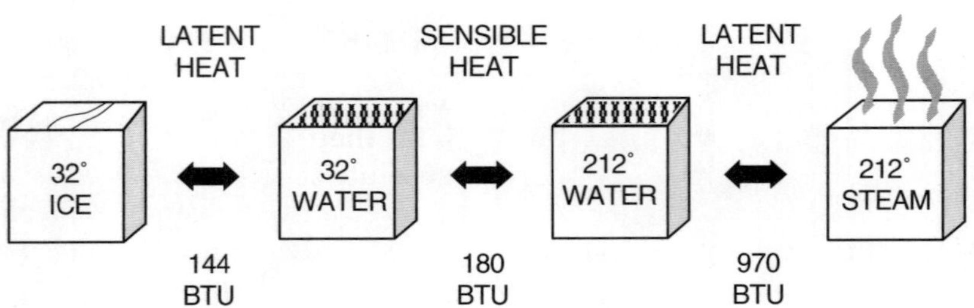

Fig. 2-1 Heat Content of One Pound of Water

ture at which water boils, thermometers are also standardized to the atmospheric pressure at sea level.

The two most common temperature scales are:

1. Fahrenheit Scale. The **Fahrenheit** scale is probably the most common throughout the United States. It places absolute zero at -460°F, water's freezing point at 32°F and water's boiling temperature at 212°F.
2. Celsius Scale. The **Celsius** scale is also known as the **Centigrade** scale and is the most commonly used scientific thermometer throughout the world. On the Celsius scale absolute zero is -273°C, water freezes at 0°C and boils at 100°C.

Several liquid filled laboratory thermometers are shown in Figure 2-2. Note that two of the scales, the Rankine (based on the Fahrenheit scale) and the Kelvin scales (based on the Celsius scale), use absolute zero as their starting points and lowest temperatures.

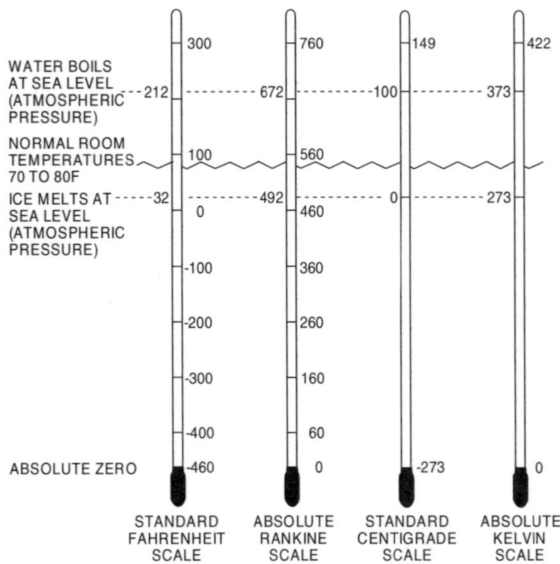

Fig. 2-2 Typical Laboratory Thermometers

Converting between °F and °C

Two equations (Example 2-2) make it easy to convert between Fahrenheit temperatures and Celsius temperatures.

> ***Example 2-2***
>
> °F = 9/5 (°C) + 32° and °C = 5/9 (°F - 32°)
>
> For example, to convert 77°F to Celsius degrees:
>
> °F = 77
>
> °C = 5/9 (77° - 32°)
>
> °C = 5/9 (45°) = 25°C

Absolute Temperature Scales

The Rankine and Kelvin temperature scales base their zero-points on the temperature of absolute zero (-460°F or -273°C). As Figure 2-2 shows, both scales are graduated differently. Water boils at 672° on the Rankine scale and 373° on the Kelvin scale.

Dry and Wet Bulb Thermometers

The ordinary thermometer is a **dry bulb thermometer**. It measures only the *sensible heat content* of the air. Now, if you cover the bulb of an ordinary thermometer with a wetted wick, you have created a wet bulb thermometer. See Figure 2-3. Instead of measuring sensible heat, the **wet bulb thermometer** measures the *latent heat content* of the air.

As the water evaporates from the wick of the wet bulb thermometer the latent heat absorbed by the vaporizing water lowers the temperature of both the wick and thermometer bulb.

The drier the surrounding air, the greater the potential evaporation rate and the greater the difference between dry and wet bulb readings.

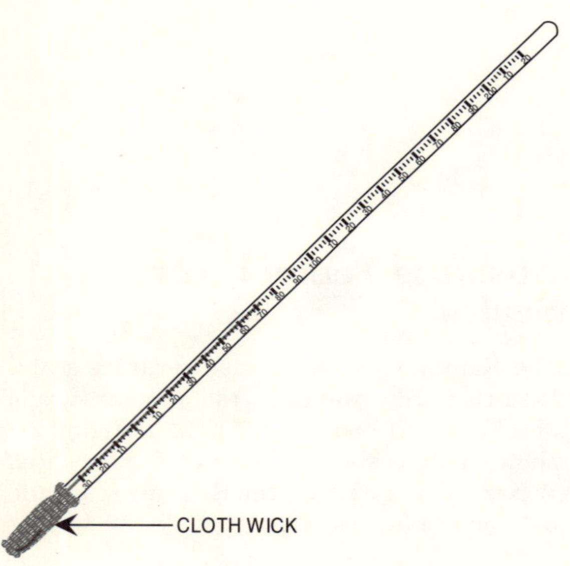

Fig. 2-3 Wet Bulb Thermometer

Conversely, the wetter the surrounding air, the slower the evaporation rate and the smaller the difference between dry and wet bulb thermometer readings. Finally, when the air is saturated (the relative humidity is 100%), evaporation stops and the dry and wet bulb readings will be identical.

Pressure

Everything on the surface of the earth is influenced by the pressure of the air, or atmosphere, above it. This pressure is **atmospheric pressure**. Atmospheric pressure decreases as you move above sea level, for example, when climbing a mountain or flying in a plane. To account for elevation changes, all pressure reading instruments are standardized to the atmospheric pressure reading at sea level.

In addition to atmospheric pressure, certain processes may increase or decrease the pressure on an object or fluid. For example, an air compressor pushes air into a smaller volume and greatly increases its pressure within the compressor's tank. Conversely, a vacuum pump removes air molecules from a vessel and dramatically decreases the vessel's internal pressure.

Liquids also experience a variety of pressures. For example, the water in an air conditioning system is controlled by three pressures:

1. Atmospheric pressure,
2. Elevation pressure determined by the water's weight and its height above a specified level, and
3. Mechanical pressure developed by the system's pump(s).

Measuring Pressure

All pressure measuring devices must react to two pressures. The first is a background or reference pressure, which most of the time is simply atmospheric pressure. The second pressure is the result of an elevation change or a mechanical process.

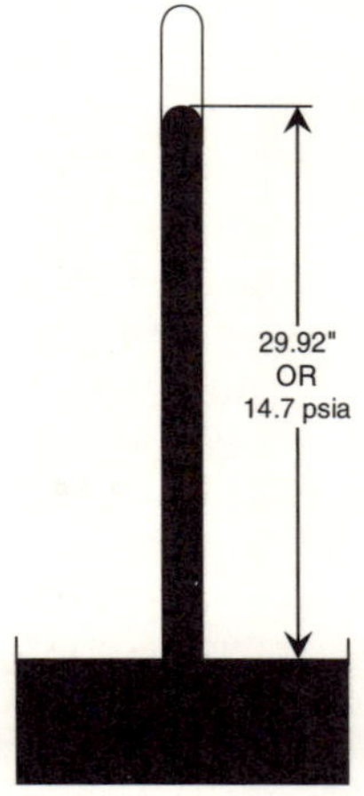

Fig. 2-4 Mercury Barometer

A **barometer** measures atmospheric pressure. A simple mercury barometer may be made from a glass tube slightly more than 30 inches in length. Seal the tube at one end, fill it with mercury and then invert the tube in a pan of mercury. See Figure 2-4. The mercury will drop in the tube until the weight of the atmosphere on the surface of the mercury in the pan just supports the weight of the mercury column. At sea level, and under certain average climatic conditions, the height of mercury in the tube will be 29.92 inches.

A mercury **manometer**, Figure 2-5, can measure an induced pressure with respect to atmospheric pressure. To make a manometer place mercury in a glass, U-shaped tube. With both ends of the tube open to the atmosphere the mercury will stand at the same level in both sides of the tube. Next, mount a scale on one side of the tube with its zero point at the mercury level.

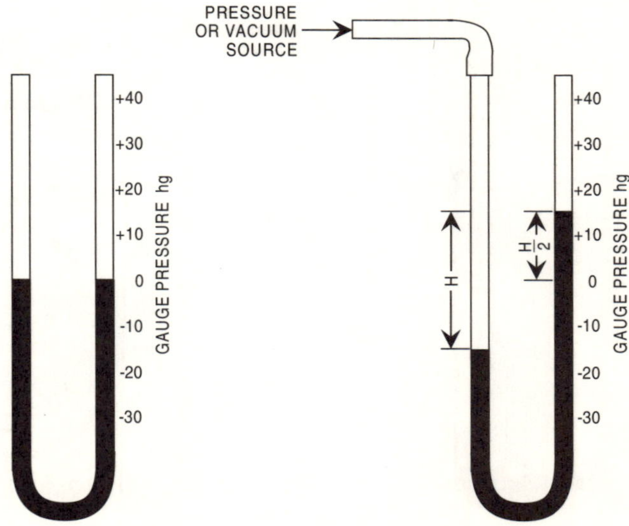

Fig. 2-5 Mercury Manometer at Rest (Left) & Pressurized (Right)

The induced pressure change causes a change in the height of the two mercury columns. The pressure difference is read on the scale. The manometer's scale is usually calibrated in inches so that pressure is read as **inches of mercury, in. Hg**. The HVAC industry commonly uses "in. Hg" to express vacuums that occur within systems.

Since the mercury in one arm of the manometer tube falls as the level in the other arm rises, and since the pressure is read as the difference in levels, each 1/2 inch on a manometer scale represents a 1-inch difference in pressure. For example, a 2 in. Hg reading results from a 1 in. rise on the scale of one arm and a 1 in. drop on the scale of the other arm.

Because water is 13-1/2 times lighter than mercury, a water manometer is easier to read than a mercury manometer since it is 13-1/2 times more sensitive to pressure changes. The water manometer measures pressure in inches of water and a 1 in. change is abbreviated as 1". H_2O or 1" **WG (water gauge)**. Because the pressure for most HVAC distribution ducts is greater than atmospheric pressure, these pressures are usually measured with manometers filled with water or red gauge oil.

Types of Manometers

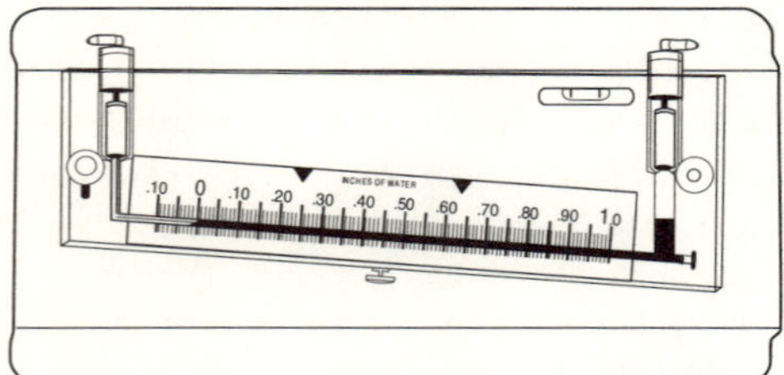

Fig. 2-6a Inclined Manometer

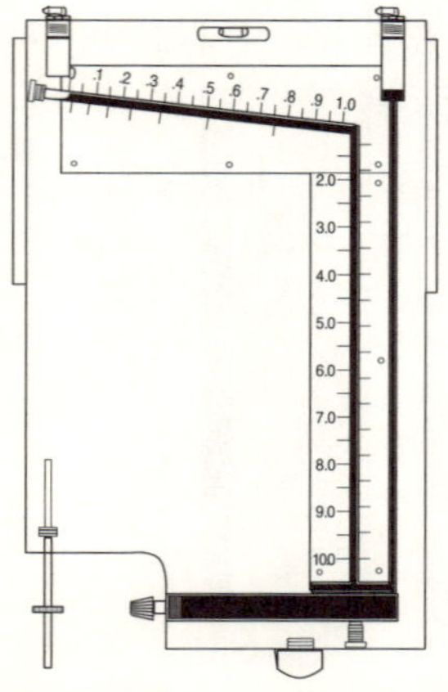

Fig. 2-6b Inclined Vertical Manometer

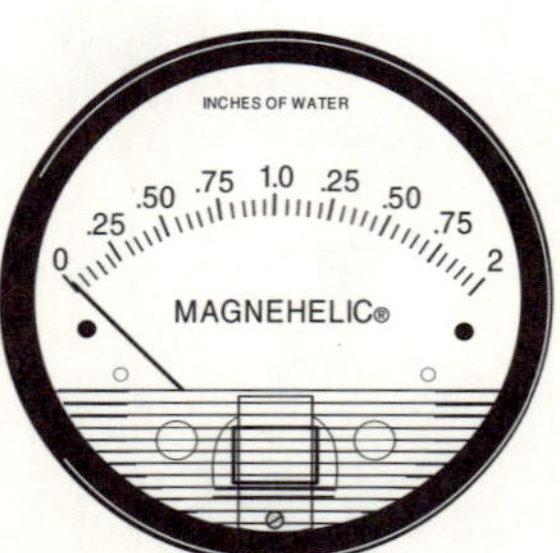

Fig. 2-6c Magnehelic® Manometer

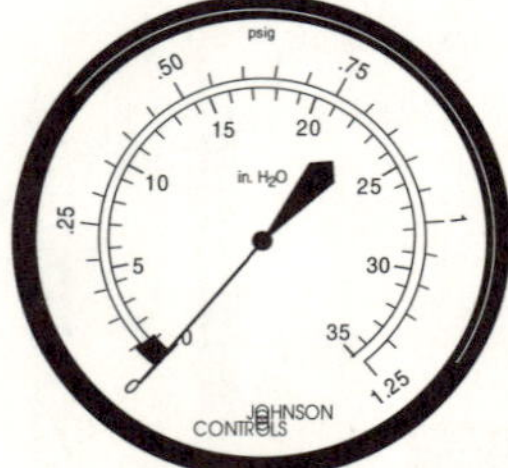

Fig. 2-6d Inches-of-Water Gauge

Types of Manometers

Figure 2-6a shows an **inclined manometer** that reads 0" WG; it spans a small range of pressure (less than 1" WG) over a longer tube and makes it possible to read very small differences in pressure. Figure 2-6b shows an inclined vertical manometer which is often used for accurately measuring low differential pressure as well as higher readings such as 10" WG. Most inclined manometers use a red gauge oil in place of water to make reading the gauge easier. Gauge oil is lighter than water and is therefore more sensitive to pressure changes. Because of their precision, inclined manometers are often used to calibrate other instruments.

Figure 2-6c is a **magnehelic mechanical manometer** and like the inclined gauge, it does not contain water and is also very accurate. Figure 2-6d is a **standard air pressure gauge** used for adjusting control instruments. It is calibrated in **inches of water and pounds per square inch gauge** or **psig**.

Figures 2-7 and 2-8 show two other methods of measuring pressure. Figure 2-7 is a **Bourdon (spring) tube gauge**. Figure 2-8 is a **metal diaphragm gauge**. These gauges are commonly used in commercial applications. Because of their mechanical components they are not as reliable as a manometer or barometer. However the spring tube and metal diaphragm gauges are much more suitable for measuring the higher than atmospheric pressures frequently encountered in commercial applications.

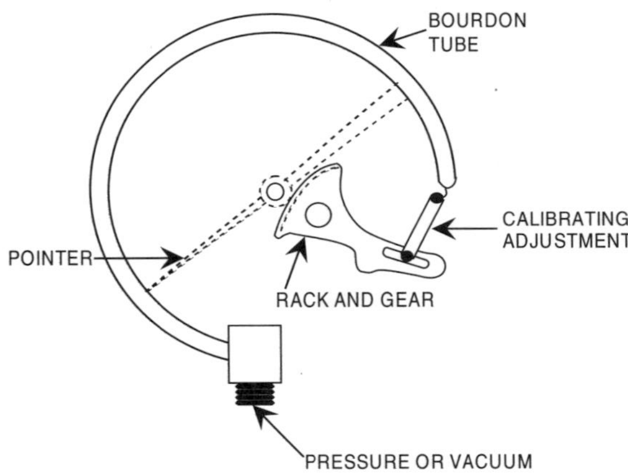

Fig. 2-7 Bourdon(Spring) Tube Gauge

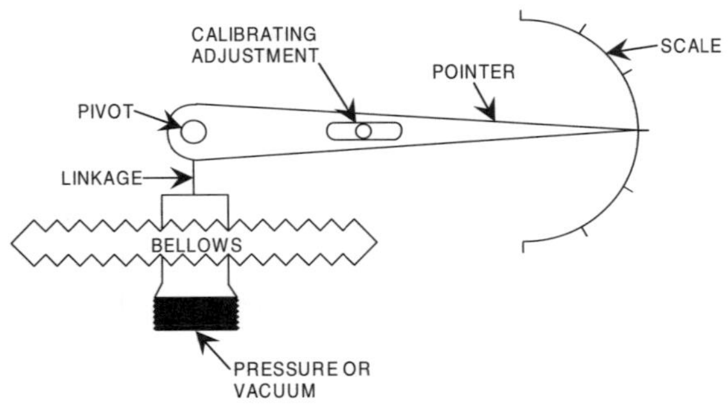

Fig. 2-8 Metal Diaphragm Gauge

Gauge and Absolute Pressure

An ordinary pressure gauge has one side of its measuring element exposed to the medium under pressure and the other side exposed to the atmosphere. Since atmospheric pressure varies with altitude and climatic conditions, gauge pressure readings will not represent a precise value unless the atmospheric pressure is also known. Typically, gauge pressures or pressures above atmosphere are read as **pounds per square inch gauge or psig**. Psig gauges do not take atmospheric pressure into account on its scale. Once atmosphere =

0 psig, some gauges may read both atmospheric pressure and the medium under pressure. These gauges measure **pounds per square inch absolute or psia**.

Summary

The second chapter has been an overview of the basic principles of heat, temperature and pressure that apply to the field of heating, ventilating and air conditioning. You should remember the following points.

- The amount of heat in a substance is expressed as either calories or British thermal units (Btu).
- Heat flows, or transfers, from places of higher temperature to places of lower temperature by three methods: conduction, convection and radiation.
- Conduction occurs by physical molecular/atomic contact.
- Convection is a heat transfer by the combination of conduction and fluid motion near the solid surface.
- Like the sun warming the interior of your car on a winter day, heat also transfers by radiation.
- Specific heat is the energy change in a substance per degree change in temperature.
- Water is the standard measure of specific heat loss or gain. It takes one Btu to raise one pound of water 1°F.
- A sensible heat gain or loss is one that affects only the temperature of a substance. Sensible heat changes are measured with a thermometer and do not affect the state of the substance.
- To change the state of a substance latent heat must be added to change it from a solid to a liquid or a liquid to a gas or vice versa. Latent heat is hidden heat because it cannot be measured with a thermometer. Changing water between its three states is an example of the need to add or subtract latent heat.
- The latent heat of fusion is the latent heat necessary to change frozen water to liquid or at 32°F [°C] vice versa.
- The latent heat of vaporization is the latent heat necessary to change liquid water to water vapor.
- An ordinary thermometer is a dry bulb thermometer and it measures only sensible heat. A wet bulb thermometer measures latent heat.
- The air, water and other liquids of a HVAC system experience three forms of pressure: atmospheric, elevational and mechanical.
- A barometer measures atmospheric pressure. A manometer measures a second pressure with respect to atmospheric pressure.
- There are many other forms of manometers and gauges used in the HVAC industry: the inclined and inclined vertical manometers, Bourdon spring tube and metal diaphragm gauges.

3

Managing Human Comfort

Perhaps the greatest challenge before the HVAC engineer and designer is to create and manage a system that makes *people* comfortable. Upon studying this chapter you will be able to:

1. List the factors affecting people's comfort.
2. Explain the role of metabolism in thermal comfort.
3. List the three primary means of losing metabolic heat.
4. Compare the effects that dry bulb temperature, humidity, mean radiant temperature and air velocity have on metabolic heat transfer.
5. Describe how the body reacts to thermal extremes.
6. Understand how metabolism and activity affect thermal design.
7. Cite the ASHRAE standard for and definition of thermal comfort.
8. List five considerations when estimating metabolic heat transfer rates.
9. Identify the ergonomics the designer can manipulate to achieve thermal comfort.

Defining Human Comfort

The primary job of a heating, ventilating and air conditioning system is usually to make people feel comfortable. The system helps people enjoy their surroundings and it improves their efficiency. To meet people's needs, the HVAC system will effectively manage the temperature, humidity and air flow and, to some extent, the air quality of the interior environment.

People's interaction with their environment is very complex, however, and defining comfort to everyone's satisfaction is nearly impossible. First, comfort is a subjective feeling and is impossible to measure. Second, one person's comfort may be another person's discomfort. The ideal temperature for one person may be too warm or too cool for another. Third, of the many variables that affect comfort — thermal, visual, acoustical and air quality — HVAC systems can control only temperature, humidity and, to some extent, air quality and noise. Tobacco smoke, for example, can be only partially eliminated by the HVAC system. And, assuming that the HVAC system operates quietly, the system should not have any effect on visual or acoustical conditions.

Figure 3-1 summarizes some of the variables that affect people's comfort. It's a complex picture made even more complex because several of these variables depend on each other and cannot be controlled separately.

Allowing for all these individual variables would easily overwhelm the HVAC system designer. Fortunately, of all the conditions that affect personal comfort, probably none is more influential than temperature. And controlling temperature is something HVAC systems do very well.

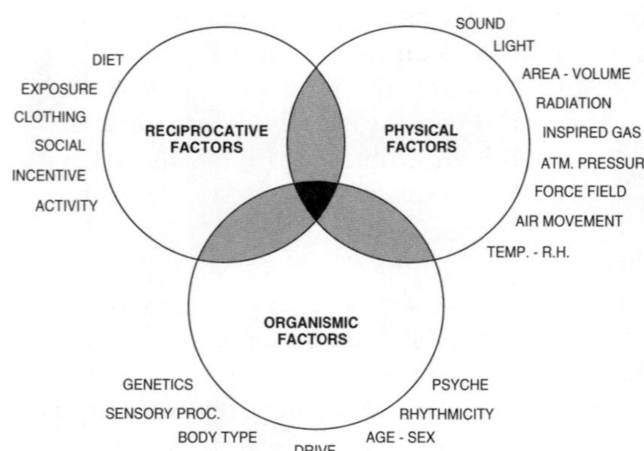

Fig. 3-1 Components of Comfort

Metabolism and Heat Loss from People

To manage thermal comfort with a HVAC system, the designer must allow for heat loss from your body. To do this, it is first necessary to consider the individual as a heat producing system. The food you eat is metabolically converted into heat energy. This process of **metabolism** is quite inefficient — only about 20% of the energy you produce is used for tissue building, work and movement. Eighty percent of your metabolic energy is given up to your environment as waste heat. For the designer, metabolic heat can be a major heat source in meeting rooms, auditoriums and the like.

The HVAC system makes you feel comfortable because it helps you maintain your natural cooling processes. Your body cools itself in three ways:

1. **Evaporation** or **latent heat loss**
2. **Radiation** or **radiant heat loss**
3. **Convection** or **sensible heat loss**

Your metabolic heat gain can be expressed as the heat loss equation: M = E + R + C .

where: M = metabolic heat produced
E = evaporative heat loss or gain
R = radiant heat loss or gain
C = convective heat loss or gain

This equation depicts the concept that to maintain your body core temperature of 98.6°F, your body must lose metabolic heat at the same rate as it takes on heat from the environment. The equation neglects conductive heat losses which are generally very small. It also neglects the relatively small heat storage capacity of your body. Notice too that radiant and convective heat losses can be positive or negative. You can give up or take on heat through both of these transfer methods.

Evaporative Heat Loss

Your body constantly loses heat through evaporation. For example, your body uses about 100 Btu per hour just keeping your lungs moist. It does this by evaporating moisture into the air you breathe as the air flows down through your mouth, nose and trachea.

When you exercise, you produce mechanical heat that your body regulates through perspiration. During hot weather, when you are also taking on heat from your surroundings, you are cooled by perspiration because your body supplies the necessary latent heat to evaporate the moisture.

Radiant Heat Loss

You gain or lose heat by radiation depending upon the difference in temperature between your skin and clothing and the mean radiant temperature. The **mean radiant temperature**, MRT, is an average of the temperature of all the surfaces in a direct line of sight of your body.

- If the mean radiant temperature is lower than your body temperature, you radiate heat to the surfaces surrounding you.
- If the mean radiant temperature is higher than your body temperature, the surfaces surrounding you radiate heat toward you.

All surfaces radiate heat to some extent, the amount depends upon the temperature of the surface. In a normal environment, your surface temperature (your exposed skin and clothing) is about 80°F. This temperature is normally higher than the surrounding temperatures in the room, so most of the time you radiate heat toward the walls, ceiling, floor, windows, etc.

Consider yourself as a vertical cylinder, the surface of which is a uniform temperature. The surface temperature of this cylinder is your mean radiant temperature. To determine your total radiation loss, or gain, you must know the temperatures of all the surfaces that are in a direct line of sight of your cylinder.

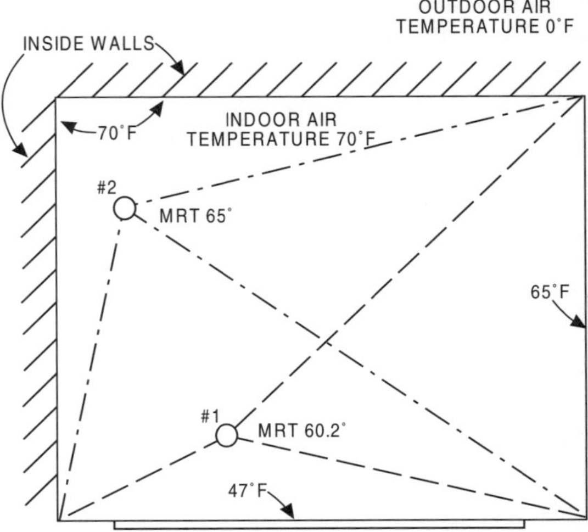

Fig. 3-2 Variations in Mean Radiant Temperature throughout a Room

Figure 3-2 shows a typical case of how the mean radiant temperature that you feel might vary on a cold day. It neglects radiant losses or gains from the ceiling, floor and furniture. As you moved about the room, the MRT could vary up to 5°F or more from one position to another. Such variations are typical, even with modern, well-insulated buildings.

As you can see in the example, the mean radiant temperature depends upon your position within the room. As you move about, the angular relationships between you and your surroundings change. As a result the MRT also changes.

Convective Heat Loss

You can gain or lose heat from your surroundings. If the air temperature is less than your skin temperature, you lose heat by convection. If the air temperature is higher than your skin temperature, you gain heat by convection. At normal room temperatures, you continuously lose sensible heat to the environment. As the room temperature approaches 80°F, the normal temperature of your skin and clothing, the convective heat loss stops and you begin to feel warm. At temperatures above 80°F you gain heat from your surroundings.

Factors Affecting Metabolic Heat Loss or Gain

The effectiveness of evaporation, radiation or conduction heat loss depends upon environmental conditions. These conditions are summarized below:

Evaporation losses depend upon:

1. The skin surface temperature of the person, assuming that the dry bulb air temperature is 80°F or below.
2. The relative humidity, but only slightly, if the air temperature is 80°F or below. It is much more important if the dry bulb is above 80°F.
3. Dry bulb air temperature.
4. The velocity of the air. Higher velocities increase the evaporation rate and therefore the heat loss.

Radiation losses depend upon:

1. The mean radiant temperature (MRT).
2. The average temperature of exposed skin and clothing.
3. Surface emissivities, or how well the surfaces radiate and absorb radiant energy.

Convection losses depend upon:

1. The dry bulb air temperature.
2. The air velocity.
3. The average temperature of exposed skin and clothing.

This summary shows that there are four thermal parameters that affect metabolic heat transfer relationships. These are:

1. Dry bulb air temperature
2. Humidity
3. Mean radiant temperature
4. Air velocity

It also shows that in typical indoor conditions, your body continually gives up its metabolic heat through the combination of evaporation, radiation and convection. Therefore, the metabolic heat generated by a building full of people is a significant source the HVAC designer must consider.

Physiological Reactions to Thermal Extremes

Engineers and scientists have studied the human body to better understand how it reacts to thermal extremes. Figure 3-3 summarizes some of these reactions. In the extreme, the body always reacts to protect the most vital organs. For example, during severe cooling the body sacrifices the less essential hands, feet and skin while protecting the head and internal organs. Likewise, in extreme heat, the body stops sweating to preserve enough moisture for the vital internal organs to prevent them from dehydrating.

In a Cold Environment	In a Hot Environment
1. The body's surface temperature declines as the body loses heat faster than it is being produced or taken on from the environment.	1. The body's surface temperature increases because the body is unable to lose metabolic and environmental heat fast enough.
2. As the core temperature starts to drop, the body reduces blood flow to its surface. The reduced flow cools the skin further, reducing convection and radiation losses.	2. As the core temperature rises, the body increases blood flow to the surface. The flow increases skin temperature and an increase in convection and radiation losses.
3. Declining skin temperature also reduces evaporation heat loss from the skin.	3. Perspiration increases, increasing evaporation heat loss.
4. As body temperature continues to fall, involuntary shivering begins — increasing mechanical heat generation.	4. Body temperature continues to rise. To conserve blood and water for the internal organs, the body reduces blood flow to the skin.
5. If temperature continues to fall, extremities may become severely frostbitten and the victim may succumb to hypothermia.	5. If temperature continues to rise, the victim may succumb to heat stroke and risks permanent brain injury.

Fig. 3-3 Range of Physiological Reactions to Thermal Extremes

Designing for Metabolism and Sedentary Activity

Figure 3-4 shows the average metabolic rates for people in various activities. Notice that the range from sleeping to maximum exertion involves about a 20-fold increase in hourly heat production

Activity	Body Heat Production Btu per Hour
Sleeping	255
Awake, Prone	300
Seated at Rest	380
Sedentary	400
Standing at Ease	430
Walking, 2 mph	780
Walking, 4 mph	1400
Maximum Exertion	300 - 4800

Fig. 3-4 Metabolic Heat Produced by Various Activities

If several people were engaged in all these activities at the same time in the same room, no heating/cooling system could keep all of them comfortable — their range of heat production would be too great. To design a practical system then, designers must select some "average" activity and tailor the system to it.

For office occupancies, designers typically choose **sedentary** activity — an activity between seated at rest and standing at ease — as an average condition for healthy people to be served by the system. Sedentary activity is also considered the most important case because it permits people to think about their comfort or discomfort.

For design purposes, a seated person doing light work is a good example. This average person generates about 380 Btu/hr. By comparison, a 100 watt light bulb generates about 340 Btu/hr.

Defining Thermal Comfort

Allowing for all the personal, metabolic and thermal variations that might occur in a building would easily overwhelm the HVAC system designer. Because of this, the designer strives to create an energy-balanced environment for people engaged in average activities. This idea is expressed in the following definition:

> **Thermal comfort** exists when a person is surrounded by an environment whose temperature and relative humidity permit the person to lose, without conscious effort, metabolic heat at the same rate he or she produces it.

The ASHRAE Thermal Comfort Standard

The American Society of Heating, Refrigeration and Air Conditioning Engineers (ASHRAE) has also struggled with defining comfort. Recognizing that comfort has both physiological and psychological aspects, ASHRAE defined thermal comfort in its **Thermal Environmental Conditions for Human Occupancy Standard 55-1992** in the following way:

> "that condition of mind which expresses satisfaction with the thermal environment; it requires subjective evaluation."

This positive feeling of comfort is based on the interior environment being neither too warm nor too cold. To translate this range of comfort into numbers the designer can use, Figure 3-5 is reprinted from the **Standard 55-1992**. It show the range of temperatures and humidities that people generally find comfortable when they are dressed in seasonal clothing and engaged in sedentary activities.

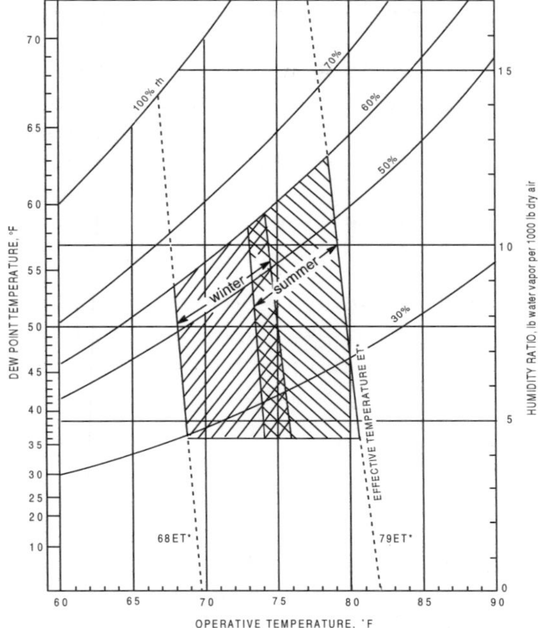

Fig. 3-5 Operative Temperature and Humidity Ranges for People Engaged in Light, Sedentary Activity

Reprinted with the permission of ASHRAE

Figure 3-6 details the acceptable ranges for the two thermal parameters outlined in ASHRAE Standard 55-1992:

Within the occupied zone, from 4 inches above the floor to 67 inches above the floor, and 2 feet or more from walls and fixed heating or air conditioning equipment, the conditions shown in Figure 3-6 should exist. Figure 3-6 is based upon a 10% dissatisfaction criterion.

Estimating Metabolic Heat Transfer Rates

Many researchers have studied human metabolic heat losses and have collected physiological data on the processes. Some generalizations may be made about the results of these studies. For the HVAC designer the metabolic losses in sedentary adults may be summarized in the following way:

For a sedentary person in an environment with a 75°F mean radiant temperature, 40% relative humidity, and air velocity of 45 feet per minute, or less,

1. The evaporative transfer rate is about 100 Btu/hr,
2. The radiation transfer rate is about 150 Btu/hr, and
3. The convective transfer rate is about 150 Btu/hr.

The relative humidity is not very important to evaporative heat losses in a normal interior

Variable	Range
Dry Bulb Temperature	68 to 79°F @ 50% RH
Short Term Cycles (15 min or less)	2°F or less swing
Relative Humidity	30% to 60% (36°F Dew Point)

Fig. 3-6
Range of Variables in ASHRAE Comfort Standard 55-1992

space because evaporation is controlled largely by the skin temperature of the person. In other words, the skin allows only so much water through for evaporation. In the temperature range specified, the amount of this moisture is always below the amount that can be taken up by the surrounding air.

It is erroneous to think that the mean radiant temperature and the air temperature are always nearly equal. They can vary widely within the same room. Winter measurements invariably find the mean radiant temperature to be below the air temperature. In the summer, air conditioned spaces often have a mean radiant temperature that is above the air temperature.

Some conclusions can now be drawn about estimating transfer of metabolic heat.

1. Air temperature and mean radiant temperature are about equally important to comfort.
2. Relative humidity does not adversely affect comfort so long as it is between 30% and 60%.
3. Air temperature and mean radiant temperature should be designed and programmed together for best results. Conditioned air is often introduced into the space near outside walls and windows to minimize the cold wall effect.
4. Air velocity is not important if it is less than 50 ft/min; almost all systems are well below this figure.
5. Short cycles of temperature and humidity changes are sources of discomfort and should be avoided.

Designing for Thermal Comfort

To produce and maintain the desired thermal environment the designer must consider a number of factors. These factors are:

1. Weather conditions and geographical locations
2. Building use and occupancy
3. Architectural details
4. Heating and/or cooling sources
5. Heating and/or cooling distribution systems
6. HVAC system controls

Weather, facility location, building use and occupant activities cannot be controlled by the designer. Therefore, the designer must use design conditions to predict the effect of these variables on the interior environment. Most of these design conditions can be found in HVAC literature and technical guides.

The architectural and structural aspects of wall construction, insulation, window sizes and types, floor plan and sun orientation all affect heating and cooling loads. These details also affect how fast a thermal exchange occurs within the structure and to some extent control the HVAC system's ability to provide uniform conditions in response.

Finally, the heating-cooling source, the distribution means and the HVAC system controls are usually within the system designer's province. By carefully selecting these elements, the designer can usually achieve the thermal comfort objective of making most of the people comfortable while performing sedentary activities.

Summary

In this chapter you learned the importance of evaluating and accommodating human comfort in the selection and control of HVAC systems. The important points to remember include:

- Comfort is far more than just the right temperature and humidity. For example, sound, light, air quality, genetics, body type and gender also affect comfort. One person's comfort may easily be another person's discomfort.
- Our body's metabolism continually produces more internal heat than we need. To maintain the best body temperature, we give up the excess heat through evaporation, radiation and convection. In spaces filled with people this cast off metabolic heat becomes a heat gain that the HVAC system must accommodate.
- Dry bulb temperature, humidity, mean radiant temperature and air velocity affect all three forms of heat loss.
- The body has specific mechanisms to deal with the extremes of cold and hot air temperatures.
- Designing for human comfort is modeled on sedentary activity. Based on research, ASHRAE has designated a range of temperatures and humidity ratios that most people find comfortable when they are engaged in sedentary activity and dressed in clothing of the season.

4

Determining The Loads On The HVAC System

Every building has unique HVAC needs. Even identically designed and constructed buildings have different occupants and equipment and are used in different ways. To allow for these differences, the design engineer always begins with a load evaluation of the building's heating and cooling needs. During your study of this chapter you will be able to:

1. Explain how the three forms of heat transfer apply to buildings.
2. Describe how the location of a facility affects HVAC load calculations.
3. Describe how the site orientation of the building influences HVAC load calculations.
4. Evaluate the weather conditions affecting the facility.
5. State which indoor conditions influence HVAC load calculations.
6. Identify which building characteristics affect HVAC loads.
7. Explain how to determine a facility's heating loads.
8. Explain how to determine a facility's cooling loads.

Buildings are complex, involving many architectural, structural, mechanical, electrical and control systems. These systems are often hidden from view and taken for granted by the building tenants. Yet, after a cold, windy walk from the parking lot, tenants expect the building to be warm and comfortable.

The engineer is responsible for making the many systems work together effectively. Typically, the responsibility for calculating sensible and latent loads and designing the building's HVAC system rests with the mechanical design engineer. Often a licensed professional engineer, the engineer will follow uniform building codes and state administrative codes for the design.

Heat Transfer Coefficients

Ultimately, all load calculations represent a mathematical model of the three modes of heat transfer: conduction, convection and radiation. Heat passage through roofs, doors, glass and wall sections involves all transfer modes. For example, conduction occurs as heat energy travels through the wall, and convection results from the air circulating through the wall. Radiation or solar heat gain is generally only a concern for the cooling or heat gain calculation. Typically, for walls, the conduction component represents the larger portion of the total heat transfer.

To some extent, every material resists the passage of heat energy through itself. The rate of this resistance varies with the specific properties of each material. This resistance is an insulating effect represented by a **thermal resistance value**, or **R-Value**. The higher the R-Value, the more inhibited is the transfer of heat energy.

For load calculations, conduction is best defined through the use of an overall heat transmittance factor known as a **U-Factor**. The U-Factor or overall **heat transmission coefficient**, describes the *rate* of heat energy moving through an area of a building section for each degree of temperature difference between the air on the warm side and the air on the cool side of the section.

There is a reciprocal relationship between an R-Value and a U-Factor as shown below:

$$U = \frac{1}{R} \quad or \quad R = \frac{1}{U}$$

where: U = Overall Heat Transfer Coefficient, in Btu/hr/sq ft/°F

R = Thermal Resistance

When multiple layers of materials are involved, the resistances of R-Values can be determined and then totaled as shown below:

Multi-layer surface: (in series)

$$U = \frac{1}{R_T} \quad or$$

$$U = \frac{1}{R_1 + R_2 + R_3 + R_4 + \ldots}$$

In some instances, different construction methods and materials may be used along the same surface in a building. For example, the structural portion of a wall may have concrete block while the exterior cosmetic portion has face brick.

A description of typical wall and roof sections is given in the *ASHRAE Fundamentals Handbook*. In addition, a complete list of resistance values can be found in the same handbook.

Example 4-1 illustrates the method for determining a U-Factor.

Example 4-1: Determine the heat transmission coefficient for the building wall section shown below.

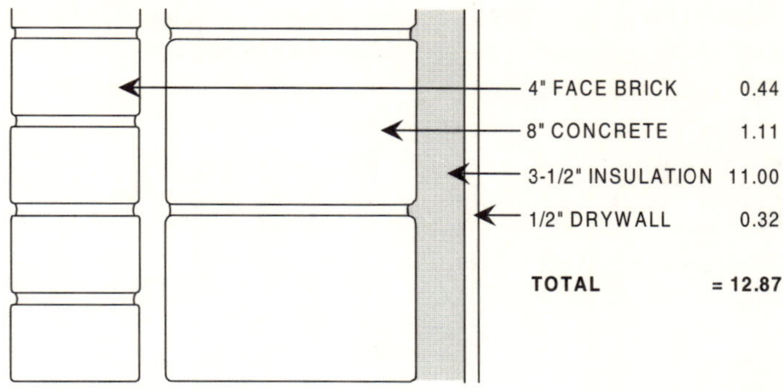

Solution: Look up the transmission values for the several wall materials. Then complete the calculation as follows:

$$U\text{–}Factor = \frac{1}{R_T} = \frac{1}{12.87} = 0.077\ Btu/hr/sq\ ft/°F$$

Establishing HVAC Design Conditions

Perhaps the most important part of a heat transfer calculation is a thorough evaluation of the outdoor and indoor **design conditions**. These conditions help to determine the *worst case* heating or cooling requirement. Designing for worst case conditions ensures that the building will operate as desired during the coldest and hottest days of the year.

In addition to designing for worst case weather extremes, most consulting engineers also add as much as a 40% additional capacity to *pick up* the building from a cold start or add a safety margin into their design loads. It is important that the controls manufacturer does not eliminate this safety margin. Instead, the design of controls should throttle back the HVAC equipment to its most cost efficient operating condition, while continuing to maintain occupant comfort.

Effect of Facility Location

Weather, including precipitation, temperature, humidity, wind and solar energy, has a significant effect on HVAC load calculations. To analyze weather conditions, the designer needs to know where the building will be. The information the designer needs and the sources for the information are summarized below:

Location: City, State, Country

Latitude: Degrees (See *ASHRAE Fundamentals Handbook*)

Elevation: Feet Above Sea Level (See *ASHRAE Fundamentals Handbook*)

Effect of Site Orientation

The direction a building faces can also have a significant effect on load calculations. For example, glass on a southern exposure experiences high solar heat gains, while a northern exposure experiences much less. Solar heat gain is another name for radiant heat transfer, the third transfer method. The direction of each exposed building surface will be labeled on the building plans such as north, south, east, west, northeast, northwest, southeast or southwest.

Selecting Design Weather Conditions

Outdoor design conditions must be defined for both winter and summer. Weather data has been collected and statistically analyzed to aid the designer in selecting the appropriate outdoor design conditions. In most cases, a 10 or 15 year period is used to determine the frequency of occurrence of extreme temperatures and humidities. From this data, the designer decides whether to design for weather extremes or to be less conservative and design for conditions more near average.

Generally, a one or two percent occurrence value is used. For an example, the 2.5% occurrence for Milwaukee, Wisconsin is -4°F. The State of Wisconsin Administrative code, however, requires a minimum design temperature of -10°F. Typically, the 1% occurrence is used for critical facilities, such as hospitals, nursing homes and research facilities. The 2% design occurrence is used for non-critical facilities such as commercial buildings.

There are sophisticated methods of organizing detailed historical weather data. In some cases this data has been collected in ten minute intervals going back over eighty years. **Bin data**, is one method used to organize weather data. The bin method organizes this data by the number hours a given condition, such as temperature, may occur in a given period of time.

Recommendations for Indoor Conditions

The indoor design conditions may be selected according to the comfort requirements of the building occupants. Thermal comfort information may be found in ASHRAE Standard 55-1981, "Thermal Environmental Conditions for Human Occupancy." The final decision is often dictated by building codes which usually follow ASHRAE Standard 90A-1980 "Energy Conservation in New Building Design." For suggested indoor design temperatures for specific applications, see the *ASHRAE Applications Handbook*.

Envelope Losses

Calculating for **envelope losses**, heat transmission through conduction, through walls, glass, doors and roofs can be performed with the calculation $Q = U \times A \times \Delta T$ $(ti\text{–}to)$
where:

Q = Btu/hr

U = U-Factor (overall heat transmission coefficient)

A = Area in square feet

ΔT (ti-to) = difference between heating design indoor air conditions (ti, dry bulb temperature indoors °F) and heating design outdoor air conditions (to, dry bulb temperatures outdoors °F).

Example 4-2: Determine the heat loss through the wall described in Example 4-1. The wall is 30 feet wide by 12 feet high. The heating design indoor air is 70°F and heating design outdoor air is 5°F.

Answer: The 30 by 12 foot wall will lose 1801.8 Btu/hr through the wall.

Solution:

$Q = U \bullet A \bullet \Delta T$ (ti-to)

$Q = .077 \bullet (30 \bullet 12) \bullet (70 - 5)$

$Q = .077 \bullet 360 \bullet 65$

$Q = 1801.8\ Btu/hr$

Determining Building Characteristics

A building survey will determine the variety of heat loads which affect occupancy comfort. Architectural drawings can be used to accurately identify the structural and functional aspects of the building. Some of the more important building characteristics to evaluate are:

- Physical dimensions of spaces to be conditioned (length × width × height).
- Materials used in construction of the building envelope.
- Directional orientations of heat transmitting surfaces (interior and exterior).
- Dimensions of heat transmitting surfaces (interior and exterior).
- Occupancy levels (peak and average).
- Occupancy scheduling (time of day, seasonal, etc.).
- Use of the space being conditioned.
- Interior heat sources (motors, equipment, lighting, etc.).
- Ventilation requirements of specifications and codes.

All of these building characteristics contribute to the heating or cooling load. Determining whether a characteristic affects heating or cooling, or both, can be done by asking yourself, "Does this load exist at the time of the worst case heating (or cooling) condition?"

Determining Heating Loads

The **peak heating load**, determined from the **heat loss calculation**, is used to select the building's heating equipment. Because this load usually occurs at night, during unoccupied hours, internal heat sources are not considered a part of the heating equipment. Note that some industrial building applications may be an exception to this rule. Loads to consider for heating include:

1. Transmission through walls, glass, ceilings, roofs, doors and floors exposed to the out-of-doors.
2. Transmission through partitions to unheated spaces.
3. Infiltration of outdoor air.
4. Ventilation. This depends on occupant needs and the make-up air requirements.
5. Transmission through below-grade walls and basement floors to the ground.

Building heating loads are further described in the *ASHRAE Fundamentals Handbook*.

Determining Cooling Loads

The **peak cooling load**, commonly referred to as the **heat gain calculation**, is used to select the building's cooling equipment. Because this load usually occurs during occupied hours, internal heat sources are considered a part of the cooling load. Cooling loads include:

1. Transmission through walls, glass, ceilings, roofs, doors and floors exposed to the out-of-doors.
2. Transmission through partitions from unconditioned spaces.
3. Solar radiation impinging on walls, glass, roof and doors.
4. Ventilation requirements.
5. Latent and sensible heat losses from people.
6. Lighting and ballasts.
7. Appliances and equipment within the conditioned space.
8. Duct and motor gains from the cooling system itself.
9. Infiltration of outdoor air.

Building cooling loads are further described in the *ASHRAE Fundamentals Handbook*.

Analysis of heating and cooling loads can be tedious work. Fortunately, it can be made easier if you use one of the many computer programs available from heating and cooling equipment manufacturers.

Degree Days

Degree days were developed as a guide to estimate heating or cooling needs. When future needs are known they are then used to estimate the cost of heating or cooling during that particular time period. The number of heating degree days for any specific date is found by averaging that day's highest and lowest temperatures, then subtracting that result from 65°F. Cooling degree days are derived in much the same way, the only exception being that the difference is added to 65°F. Degree days acknowledges 65°F as the heating/cooling crossover assumption.

> ***Example 4-3:*** Given that the high for January 26 was 34°F and the low was 0°F, find the number of heating degree days.
>
> ***Answer:*** The average temperature was 17°F (34 + 0 ÷ 2 = 17).
>
> ***Solution:*** 65 - 17 = 53 heating degree days.

The National Weather Service compiles and reports the heating/cooling degrees to date. It also maintains previous year's degree day records to be used for heating and cooling cost estimating.

Summary

In this chapter you learned the importance of evaluating a building's materials, location, use and equipment to determine its heating and cooling loads. The important points to remember are:

- Conduction, convection and radiation modes of heat transfer apply to buildings as well as people.
- A measure of a material's resistance to heat transfer, and therefore an expression of its insulating value, is the resistance or R-Value.
- Conversely, an expression of the ease with which a material transfers heat is its heat transmission coefficient, or U-Factor.
- The R-Value and U-Factor are mathematically inverse expressions of one another.
- A facility's location determines the weather extremes it must withstand. The facility's site orientation determines the radiant solar heat gain it will experience and the effect the prevailing winds will have on infiltration.
- You should consult local and state building codes for the required design weather conditions. Usually, only critical facilities such as hospitals, nursing homes and research buildings have HVAC systems designed to withstand the worst weather.
- You can estimate peak heating and cooling loads after considering characteristics such as building dimensions and materials, site orientation, occupancy type and levels, use and equipment and ventilation requirements.

5

Psychrometrics: The Properties Of Air

To condition air or water with a variety of equipment you have to understand psychrometry and the psychrometric chart. By understanding these concepts you will understand the application of many of the physical principles used throughout the industry of heating, ventilating and air conditioning. After studying this chapter you will be able to:

1. Define psychrometry, the psychrometric chart and psychrometrics.
2. Describe the composition of air.
3. Explain Boyle's, Charles' and Dalton's Laws.
4. Read a psychrometric chart and locate different conditions on it.
5. Use a psychrometric chart to determine the effect of sensible and latent heat changes on initial air conditions.

Changing the Condition of Air

Imagine that you were chosen to decide what today's weather should be. What temperature would you choose? What would be the humidity? The ideal wind speed?

Questions like these are answered every day by HVAC system designers because their systems control the *indoor* weather—the temperature, humidity and air velocity inside a building. By controlling a building's HVAC equipment, they are determining the weather for everyone inside the building.

Look over the psychrometric chart in Figure 5-1.

At first glance the chart may seem a bit complex. After completing Chapters 5 and 6 however, you will understand all the components of the psychrometric chart, and you will be able to use it to predict the temperature and moisture content of the indoor weather. The psychrometric chart will in fact enable you to be an indoor meteorologist!

What is Psychrometry and Psychrometrics?

Psychrometry is the study of air, temperature and water vapor relationships. For more than a century, scientists and engineers have measured and examined these relationships. They found that temperature, relative humidity, enthalpy, moisture content and dew point temperature are related to each other. They also precisely determined the values of these parameters throughout the normal range of HVAC operations. To assist engineers and designers, these relationships and values have been graphically combined in what is known as the **psychrometric chart**. Because all of these relationships are related mathematically, the engineer only needs to know two of them for a given environmental condition, the rest can be read from the chart.

Strictly speaking, psychrometry and the psychrometric chart are separate and distinct concepts. However, throughout much of the HVAC industry, both concepts are combined under the general term of **psychrometrics**. Regardless of the semantics, *every HVAC process, and virtually every item of HVAC equipment and control, has a purpose which can be analyzed and determined with psychrometrics.*

The Composition of Air

The atmosphere is a mechanical mixture of many components. Chemical reactions such as the formation of smog, ozone or acid rain are not a part of this discussion. The air that you breathe and HVAC systems manage is a mixture of:

1. Dry air,
2. Water vapor, and
3. Impurities, such as smoke, dust, pollen, bacteria and noxious gases.

Pure dry air is made up of several gases: about 77% nitrogen, 22% oxygen, 0.04% carbon dioxide and 0.96% other gases, such as argon, neon, helium and krypton. HVAC systems can add and remove heat from the dry air component of a building's air, but they do not affect the proportion of gases making up the dry air.

Water vapor in the atmosphere comes from river, lake and ocean evaporation and plant transpiration. This vapor is actually a low pressure, low temperature steam. In the laboratory, water vapor behaves predictably when changes are made to its temperature, pressure

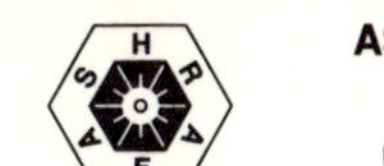

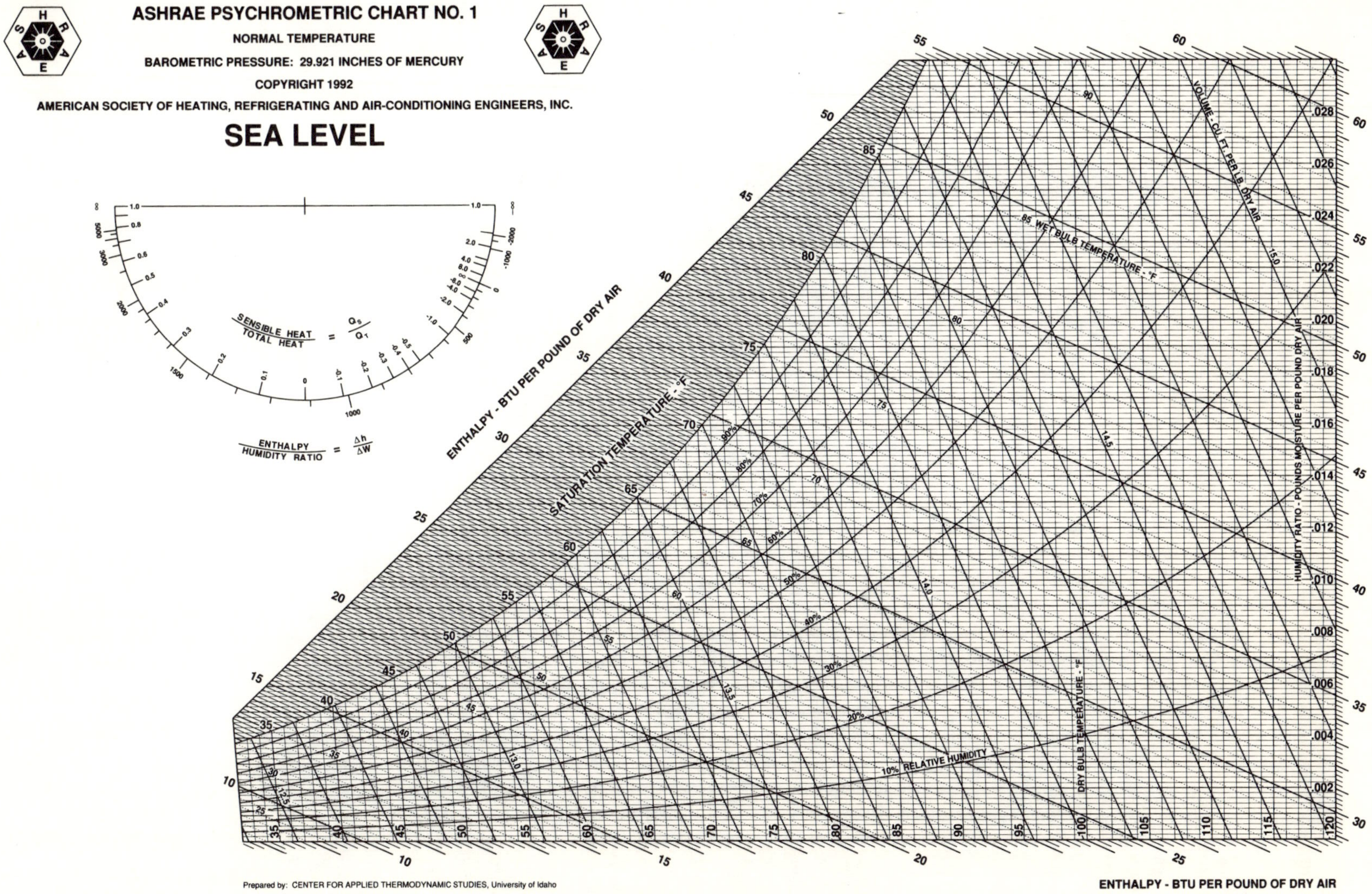

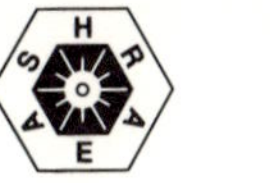

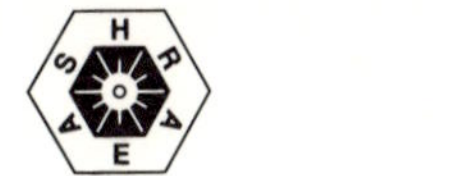

ASHRAE PSYCHROMETRIC CHART NO. 1
NORMAL TEMPERATURE
BAROMETRIC PRESSURE: 29.921 INCHES OF MERCURY
COPYRIGHT 1992
AMERICAN SOCIETY OF HEATING, REFRIGERATING AND AIR-CONDITIONING ENGINEERS, INC.
SEA LEVEL
SENSIBLE HEAT / TOTAL HEAT = Qs / QT
ENTHALPY / HUMIDITY RATIO = Δh / ΔW
ENTHALPY - BTU PER POUND OF DRY AIR
SATURATION TEMPERATURE - °F
WET BULB TEMPERATURE - °F
VOLUME - CU. FT. PER LB. DRY AIR
HUMIDITY RATIO - POUNDS MOISTURE PER POUND DRY AIR
DRY BULB TEMPERATURE - °F
10% RELATIVE HUMIDITY
Prepared by: CENTER FOR APPLIED THERMODYNAMIC STUDIES, University of Idaho
ENTHALPY - BTU PER POUND OF DRY AIR

Fig. 5-1 ASHRAE Psychrometric Chart No. 1 *Reprinted with the permission of ASHRAE*

or volume. An important HVAC function is to add or remove water vapor from building air.

Airborne impurities are either particles or gases. Several different types of HVAC filters successfully remove the larger particulate impurities. These particles are also quite easily removed by the condensation that occurs inside air conditioners and chillers. Removing the extremely small particles and the gaseous impurities is a far more difficult process and requires specialized equipment.

Predicting Air's Behavior

Over the years, scientists have developed several gas laws to explain the behavior of air and water vapor when they are heated, cooled and pressurized.

Seventeenth century scientist Robert Boyle found that when he held temperature constant, a gas decreased in volume as he pressurized it. Conversely, the gas expanded when he decreased the pressure. This behavior of a gas is known as **Boyle's Law**.

Charles' Law is named after an early nineteenth century French chemist, Jacques Charles, who observed that as he held pressure constant and heated a gas, it expanded in volume. Conversely, it shrank when he cooled it.

In 1802, John Dalton found that when he brought gases together in a common space, they spread throughout the space. The several gases didn't stratify as some liquids might; instead, they behaved as if the other gases were not present. He found that the pressure of the gaseous mixture was equal to the sum of the pressures of the individual gases. This law is known as **Dalton's Law of Partial Pressures**.

Drawing on the work of these and many other scientists, in 1911 Dr. Willis Carrier, presented a paper entitled, "Rational Psychrometric Formulae." With it he explained how wet and dry bulb temperatures, dew points and relative humidity affect the heat energy of air. Because of his presentation and subsequent work, he is credited with founding the air conditioning industry.

Water Vapor in Air

Remember the latent heat of vaporization? It is the amount of heat that must be added to water to change it from its liquid state to water vapor. Similarly, it is the heat that must be removed from water vapor to make it condense back to a liquid. Controlling the amount of water vapor in the air is an extremely important HVAC function, second only to temperature control.

When the water vapor content of the air is high, you feel warmer for two reasons:

1. For every pound of water vapor there is an additional 970 Btus of heat in the air, the latent heat of vaporization.
2. Water vapor retards your body's evaporation rate and the slowed rate makes you feel warmer. The higher the air's water vapor content, the slower your evaporation rate.

Conversely, when the water vapor content of the air is low, you feel cooler for two reasons:

1. For every pound of water vapor removed from the air, there is a reduction of 970 Btus in the air's heat, the loss of the latent heat of vaporization.
2. When the air's water vapor content is low, your body's evaporation rate is enhanced and you feel cooler also.

Defining Humidity

Air, as referred to in the HVAC industry, is a mixture of dry air and water vapor. The airborne moisture is called **humidity**. Quantitatively, humidity is measured and described in two ways

The **specific humidity** is the weight of the water vapor in a pound of dry air. It is expressed as **lb/lb dry air** or **grains/lb dry air**. There are 7000 grains in one pound in the English measurement system. Specific humidity is also referred to as the **humidity ratio** (w) since it expresses the ratio of the weight of water vapor to the weight of dry air. Scientists have used Dalton's Law to calculate specific humidities for a wide range of conditions.

Saturated air contains 100% of the moisture it can hold, while perfectly dry air contains no water vapor at all. **Relative humidity** expresses the *percentage of saturation* at a specific dry bulb temperature. Again, using Dalton's Law, relative humidity is the ratio of the actual partial pressure of water vapor in the air compared to the saturation partial pressure, as measured at a specific temperature.

Relative humidity is always *relative* (to the dry bulb temperature). For example, assume the air contains a known amount of specific humidity. At a low dry bulb temperature, this specific humidity translates into a high relative humidity. Conversely, at high dry bulb temperatures, the same amount of specific humidity translates into a low relative humidity. Warm air is able to support more water vapor than cool air.

When discussing humidity a third concept is also important: dew point. As a mixture of dry air and water vapor is cooled the air becomes increasingly saturated. At a dry bulb temperature known as the **dew point**, the air has a relative humidity of 100%; it is saturated. At this temperature dew will form on colder surfaces such as a lawn, a car's windshield or the coils in an air conditioner. The dew point temperature varies as the amount of moisture in the air varies; the higher the dry bulb temperature, the more moisture the air can hold.

Measuring Humidity

Measuring partial pressures to determine humidity is a precise laboratory procedure. Fortunately, engineers and maintenance personnel can more conveniently determine humidity using a device called the **sling psychrometer**, and its accompanying psychrometric table or chart. A typical psychrometer is pictured in Figure 5-2.

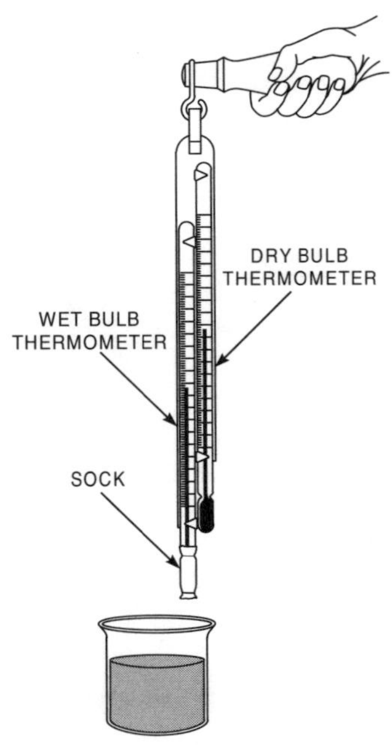

Fig. 5-2 Sling Psychrometer

The two thermometers of a sling psychrometer are identical, one merely has a wick placed over its bulb. Although the air passing over both bulbs is at the same temperature, the wet bulb thermometer will nearly always have a lower reading. Its reading is lower because of the evaporative cooling effect of the water evaporating from the saturated sock.

To use the sling psychrometer, first saturate the wick with distilled water at room temperature. Next, rotate the psychrometer about the handle at a constant rate of about four revolutions per second. Because it enhances evaporative cooling, this rotation will depress the wet bulb temperature below the dry bulb temperature. Continue to rotate the psychrometer until you obtain a constant reading on the wet bulb scale. This wet bulb reading is the lowest temperature, at which water vapor can be evaporated into the surrounding air.

The temperature difference between the two thermometers depends on the amount of moisture already in the air. If the air is moist, the evaporation rate is slower, little latent heat is lost from the wetted bulb, and the wet bulb reading will be only slightly lower than the dry bulb reading. In fact, if the humidity of the surrounding air is 100%, no evaporation can occur, and the two temperature readings will be identical. On the other hand, if the air is dry, evaporation occurs easily, latent heat is readily lost through evaporation, and the wet bulb reading will be significantly lower than the dry bulb reading.

Once you know the dry and wet bulb readings, you can use the psychrometric table supplied with the instrument to determine the relative humidity. A typical table is shown in Figure 5-3.

Percent Relative Humidity

Difference between dry and wet bulb readings

DB temp. °F	1	2	3	4	5	6	7	8	9	10	11	12	13	14	15	16	17	18	19	20	21	22	23	24	25	26	27	28	29	30
32	90	79	69	60	50	41	31	22	13	4																				
36	91	82	73	65	56	48	39	31	23	14	6																			
40	92	84	76	68	61	53	46	38	31	23	16	9	2																	
44	93	85	78	71	64	57	51	44	37	31	24	18	12	5																
48	93	87	80	73	67	60	54	48	42	36	31	25	19	14	8	3														
52	94	88	81	75	69	63	58	52	46	41	36	30	25	20	15	10	6	0												
56	94	88	82	77	71	66	61	55	50	45	40	35	31	26	21	17	12	8	4											
60	94	89	84	78	73	68	63	58	53	49	44	40	35	31	27	22	18	14	6	2										
64	95	90	85	79	75	70	66	61	56	52	48	43	39	35	31	27	23	20	16	12	9									
68	95	90	85	81	76	72	67	63	59	55	51	47	43	39	35	31	28	24	21	17	14									
72	95	91	86	82	78	73	69	65	61	57	53	49	46	42	39	35	32	28	25	22	19									
76	96	91	87	83	78	74	70	67	63	59	55	52	48	45	42	38	35	32	29	26	23									
80	96	91	87	83	79	76	72	68	64	61	57	54	51	47	44	41	38	35	32	29	27	24	21	18	16	13	11	8	6	4
84	96	92	88	84	80	77	73	70	66	63	59	56	53	50	47	44	41	38	35	32	30	27	25	22	20	17	15	12	10	8
88	96	92	88	85	81	78	74	71	67	64	61	58	55	52	49	46	43	41	38	35	33	30	28	25	23	21	18	16	14	12
92	96	92	89	85	82	78	75	72	69	65	62	59	57	54	51	48	45	43	40	38	35	33	30	28	26	24	22	19	17	15
96	96	93	89	86	82	79	76	73	70	67	74	61	58	55	53	50	47	45	42	40	37	35	33	31	29	26	24	22	20	18
100	96	93	90	86	83	80	77	74	71	68	65	62	59	57	54	52	49	47	44	42	40	37	35	33	31	29	27	25	23	21
104	97	93	90	87	84	80	77	74	72	69	66	63	61	58	56	53	51	48	46	44	41	39	37	35	33	31	29	27	25	24
108	97	93	90	87	84	81	78	75	72	70	67	64	62	59	57	54	52	50	47	45	43	41	39	37	35	33	31	29	28	26

Fig. 5-3 Psychrometric Table for a Sling Psychrometer

Modern Air Conditioning Practice, 3rd Edition. by Harris, Norman C., ©1983

Adapted by permission of McGraw-Hill

Example 5-1: On a warm and humid day in Florida you read 92°F on the dry bulb thermometer and 89°F on the wet bulb thermometer of a sling psychrometer. Using Figure 5-3, what is the relative humidity?

Solution: To use the table you need the difference between the two readings. In this example that is 3°. Read down the left column and locate the dry bulb reading of 92°F and then read across to the temperature difference. The relative humidity is 89%, or quite humid.

Example 5-2: On a warm and sunny day in Arizona you read 90°F on the dry bulb scale and 60°F on the wet bulb scale of the same psychrometer. What is the relative humidity?

Solution: The dry bulb temperature is between table values so estimate a relative humidity reading of 13.5% for a temperature difference of 30°. This relative humidity is quite dry.

There are several other devices for simultaneously measuring wet and dry bulb temperatures. Each device must be calibrated by the manufacturer, and its accompanying psychrometric table must be standardized under laboratory conditions. Because the rate at which air passes over the evaporative surface is crucial to accurate readings, you must carefully follow the manufacturer's suggestions when using any device.

Defining Other Psychrometric Properties of Air

Several air properties depend on the *volume* of the air. For example a pound of air with a specific humidity of 0.004 pounds of moisture takes up 12.5 cu ft of volume at 35°F and has a relative humidity of 93%. At 112°F this same pound of air takes up 14.5 cu ft of volume and has a relative humidity of about 8%. This is a 16% increase in volume and a 91% decrease in relative humidity, yet the weight of the air remains the same.

This example demonstrates that the amount of space occupied by air can greatly affect the operation of air distribution equipment, such as fans and air compressors. The volume of air per unit of dry air is the **specific volume**, expressed in **cu ft/lb dry air**. Specific volume also varies with atmospheric pressure and relative humidity. For precise results, specific volumes must also be measured in the laboratory. However, for design purposes the engineer may accurately use one of the many tables or charts that summarize this data.

For the engineer selecting and controlling cooling equipment, knowing the total heat content or enthalpy of the air is very important. **Enthalpy** is the sum of the sensible and latent heats and is expressed in **Btu/lb dry air**. For the purpose of quantifying the total heat of air, 0°F is used as a baseline.

To be very precise, enthalpy is the sum of the sensible and latent heats of *both* the air and its water vapor.

Because the value is very small, the sensible heat of water vapor is typically ignored. For HVAC purposes, enthalpy is the sensible heat of the air plus the latent heat of water vapor.

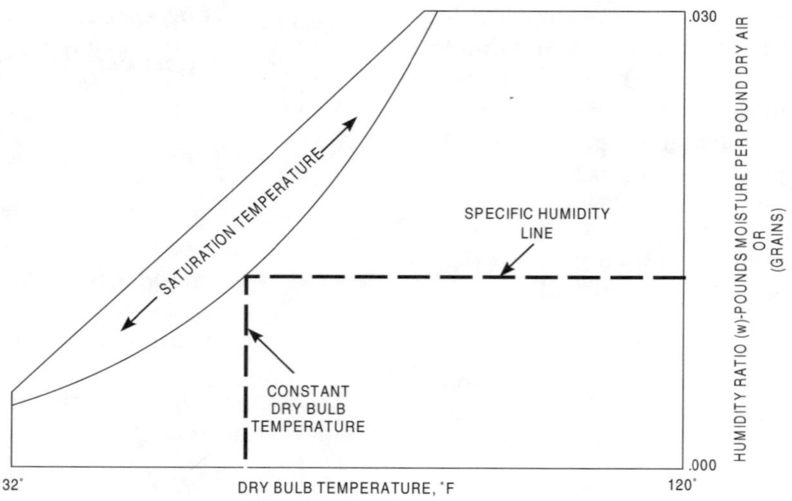

Fig. 5-4 Outline of Psychrometric Chart

Introduction to the Psychrometric Chart

Measuring the properties of air and water vapor mixtures requires sophistication found only in a laboratory. Fortunately, because these properties are also well-studied and predictable, scientists have summarized them in tables of *moist air properties*. To make using this information easier, the values of moist air tables can be plotted in **psychrometric charts**.

Please note that all psychrometric relationships depend upon atmospheric pressure. *Psychrometric charts are usually standardized on sea level pressure. If you are working at higher elevations (Denver for example) be sure to use a chart specifically prepared for your elevation.*

Constructing the Psychrometric Chart

After looking over the many lines and curves of the psychrometric chart, lean back and using the solid lines, note the overall outline of the chart. It is displayed in Figure 5-4. The outline of the ASHRAE chart is created by first placing a scale of dry bulb temperatures along the bottom of the chart. These temperatures cover the range of typical HVAC applications (32-120°F). Next, a scale of moisture measurements grains or humidity ratio forms the right side of the chart. Lines of *constant dry* bulb temperature extend vertically upward from the dry bulb scale and lines of humidity ratios or specific humidity extend horizontally right to left from the humidity scale. In the ASHRAE Chart No. 1, the dry bulb temperatures range from 32 to 120°F, and the humidity ratios range from .000 to .030 lb/lb dry air.

Next, the upwardly curving line along the left side of the chart is created by plotting the dry bulb temperature for each saturation temperature of the humidity ratio. The saturation value of the humidity ratio occurs when the wet bulb and dry bulb temperatures are equal. This curve represents the saturation temperature of moist air, which is also the line of 100% relative humidity, which is also the line of dew point temperatures.

Note that while the ASHRAE Chart No. 1 uses English units, metric charts are also common. Some charts, as you will see in Chapter 6, may also include additional information such as sensible heat factor.

Examining the Psychrometric Chart

One of the handiest features of the psychrometric chart is that *if you know any two properties of air, you can read the other properties from the chart.* Let's look at how useful this chart is.

Determining Dry Bulb Temperature Using Figure 5-1, locate the dry bulb temperatures along the bottom scale of the chart. Notice that it is easy to read the scale to the nearest degree.

Determining Wet Bulb Temperature Using Figure 5-1, the wet bulb temperatures extend upward from the bottom scale at about a 45° angle until they intersect the saturation curve. It is also easy to read the wet bulb temperatures to the nearest degree.

Determining Relative Humidity Relative humidity, you will recall, is the actual amount of vapor in the air compared to the maximum amount that could be present. Look for the lines of less than 100% relative humidity on Figure 5-1. They appear in curved line increments of 10% beginning in the lower left of the chart. Notice that the relative humidity lines curve upward as the dry bulb temperatures increase. This is because warm air can hold more water vapor and consequently more easily support higher moisture content than can cooler air.

When you know the value of any two air properties and have located them on the psychrometric chart, you have established the air's condition.

Determining Dew Point The dew point (**DP**) is the temperature at which the moisture content of the air is 100%. By knowing any other two air properties you can easily determine the **DP**.

Determining Wet Bulb As discussed earlier, wet bulb (WB) is helpful when determining the latent heat content of the air. Since the reading is taken at saturation, the point is plotted on the saturation line. Plot a WB temperature of 61°F on the saturation line and extend this line downwards at a 45° angle. At the intersection of the wet bulb and dry bulb readings relative humidity can be determined.

Determining Enthalpy Enthalpy is very useful in estimating the amount of heat in Btus added to or removed from the air during a given HVAC process. Enthalpy in Btus per lb dry air is read on the enthalpy scale at the upper left hand corner of the chart.

Determining how much heat is removed from the air by a specific HVAC process is also easy to do with the psychrometric chart.

Determining Specific Volume Specific volume is the amount of volume one pound of dry air occupies for a given condition. The space varies by the temperature and barometric pressure in accordance with Boyle's and

Charles' Laws. Knowing the specific volume is particularly useful for evaluating fan performance and selecting motor sizes for high and low temperature applications.

The specific volume lines on the psychrometric chart slant upward to the left from the bottom of the chart. On Figure 5-1, notice that the specific volume of a pound of dry air ranges from about 12.4 cubic feet at 32°F to over 15.0 cu ft for temperatures over 110°F. This approximately 20% difference in volume can be significant when selecting equipment that must handle air of widely varying temperatures.

When rating the capacity of various air handling units, manufacturers do not know all the variations of temperature and pressure that their equipment might be subjected to. To help the manufacturer and the engineer compare equipment capacities, the cubic feet per minute (CFM) rating of equipment is often expressed at *standard* air conditions. **Standard air** (SCFM) is defined as having:

- A specific volume of 13.3 cu ft/lb dry air,
- A temperature of 70°F, both at
- An atmospheric pressure of 29.92 in Hg.

Note that in the ASHRAE comfort zone, Figure 3-5, the specific volume is about 13.6 cu ft/lb of dry air. For HVAC calculations that are primarily concerned with human comfort ranges, using standard air is also sufficiently accurate.

The following examples will build your skills in the use of the psychrometric chart.

Example 5-3: Assume that using a sling psychrometer you obtained a wet bulb temperature (**WB**) of 61°F and a dry bulb (**DB**) temperature of 75°F. What is the relative humidity (**RH**)? Look back at Figure 5-1.

Solution: Look back at Figure 5-1. Locate the **DB** line of 75°F and read upward to where it intersects the 61°F **WB** line. The wet bulb lines extend diagonally downward from the saturation curve and you have to approximate the **WB** reading between the 60 and 65°F lines. After finding the intersection of the **DB** and **WB** lines, read between the curving 40 and 50% **RH** lines. The relative humidity is 42%. Figure 5-5 provides a skeletal solution.

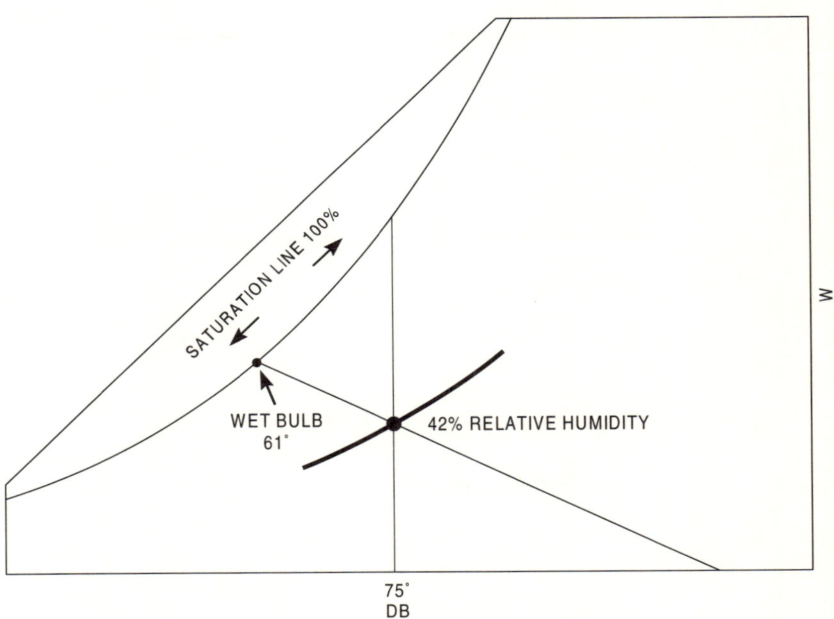

Fig. 5-5 Solution for Example 5-3

Example 5-4: Given a **DB** temperature of 75°F and humidity ratio (**w**) of .008 lb/lb dry air, what is the relative humidity, **RH**?

Solution: See Figure 5-6. Locate the **DB** line of 75°F and read upward to where it intersects the .008 **w** line extended from the right scale. This intersection lies between the relative humidity curves of 40% and 50%. Again, locate a relative humidity value of 42%.

To confirm this value, extend the 75° **DB** line upward until it intersects the saturation temperature line, or the 100% **RH** line. At saturation the corresponding humidity ratio is .019. The given **w** of .008 divided by the saturation **w** of .019 = .42, or 42%. $\frac{.008}{.019} = .42, \quad .42 \times 100\% = 42\%$ This is very close to the reading we made from the chart and is accurate enough for HVAC matters.

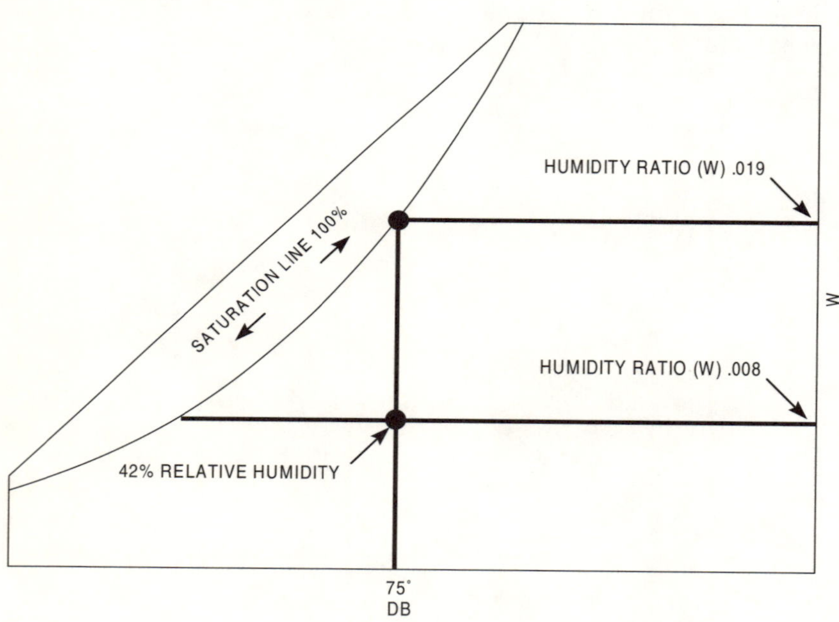

Fig. 5-6 Solution for Example 5-4

Example 5-5: Using the previous example again, suppose air with an initial condition of 75°F **DB** and 42% **RH** is cooled by decreasing its sensible heat. See Figure 5-7. While maintaining the same humidity ratio but gradually decreasing the **DB** temperature, note that the relative humidity steadily increases. The relative humidity will continue to increase until the 100% **RH** line or a saturation temperature of 51° is reached. This point is called the dew point. It is found on the dew point line, which is another name for the saturation line.

Also note that if the air is cooled even more, the **DP** temperature continues declining and the moisture content of the air starts to fall. When the moisture content of the air (specific humidity) decreases below that of the entering air, the process is called **dehumidification**. Dehumidification occurs when the air is cooled below its dew point, causing the water vapor to condense on surfaces colder than the dew point. For example, when the evening temperature falls below the air's dew point temperature, dew forms on the grass and other cool objects. The cooler the evening, the greater the amount of moisture that leaves the atmosphere by way of condensation.

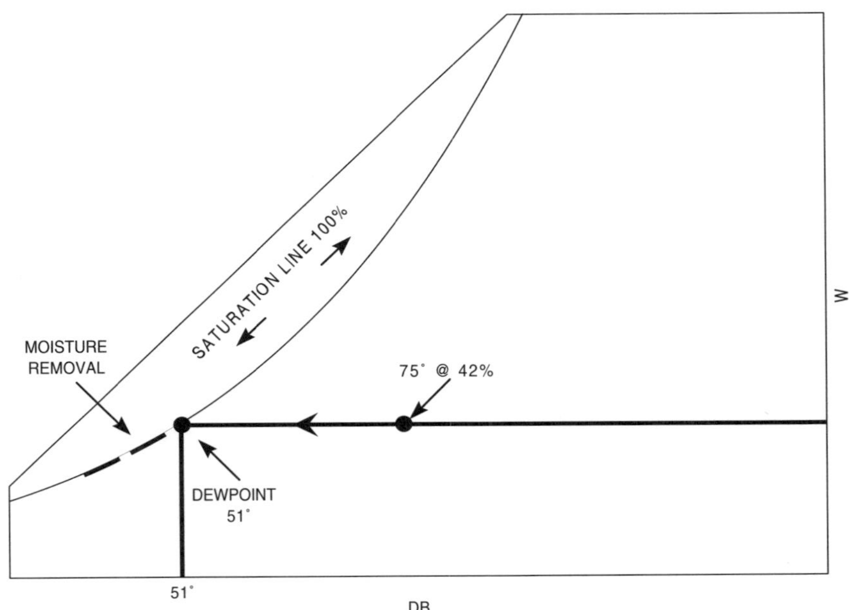

Fig. 5-7 Solution for Example 5-5

Example 5-6: Given a **DB** temperature of 75°F and a **WB** temperature of 61°F, what is the enthalpy (**h**) of the air at these temperatures?

Solution: See Figure 5-8. Read 75°F upward on the **DB** line until it intersects the 61°F **WB** line at 42% **RH**. Now follow the 61°F **WB** toward the saturation line. Past the saturation line, start drawing an enthalpy line outward to the enthalpy scale. Enthalpy, **h**, is about 27 Btus/lb dry air.

Note: There is a subtle deviation between the wet bulb and enthalpy lines in the psychrometric chart. This deviation will not be discussed in this text.

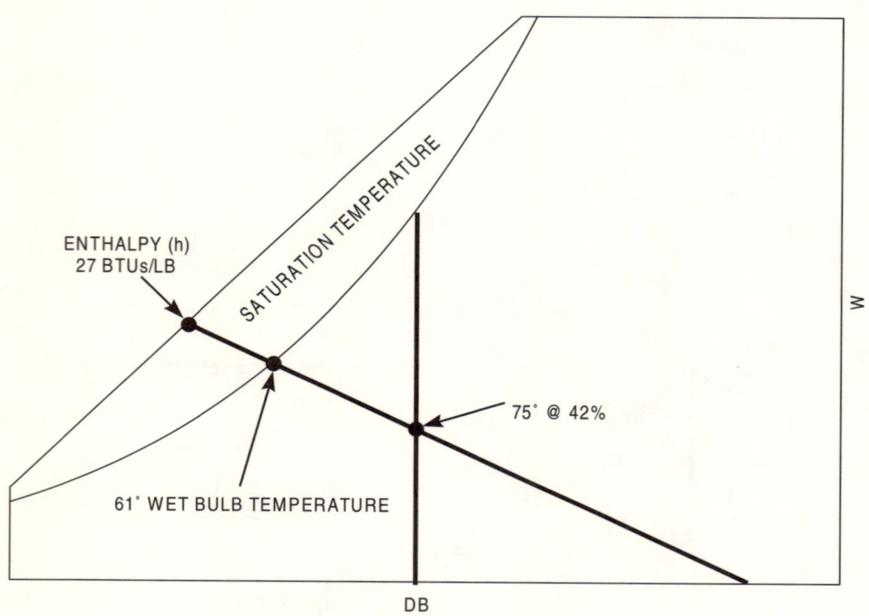

Fig. 5-8 Solution for Example 5-6

Example 5-7: Given the condition of 75° **DB**, **w** = .008, and 27 Btu/lb **h**, how much heat is removed if the air is cooled by a sensible HVAC process to 55°F with a **RH** of 88%?

Solution: See Figure 5-9. From the initial condition trace the **DB** change along the humidity ratio line of .008 lb/lb dry air to the new **DB** of 55°F. From this new condition read upward toward the enthalpy scale. The new **h** is about 22 Btu/lb. The heat content change is 27 - 22 = 5 Btu/lb dry air.

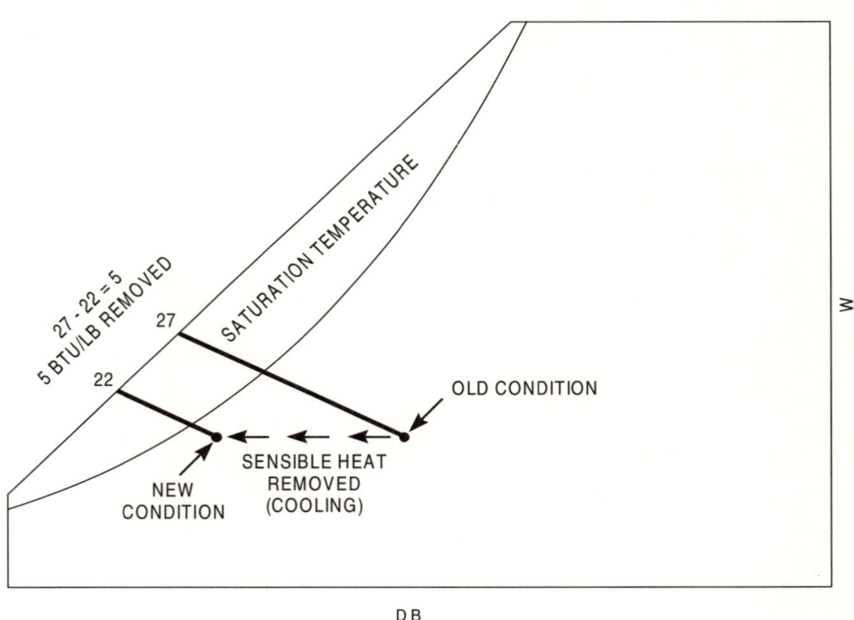

Fig. 5-9 Solution for Example 5-7

Example 5-8: Locate the condition of 75°F **DB**, 42% **RH**. What is the specific volume of the air for this condition?

Solution: See Figure 5-10. After plotting the initial condition, estimate the specific volume at 13.65 cu ft/lb of dry air.

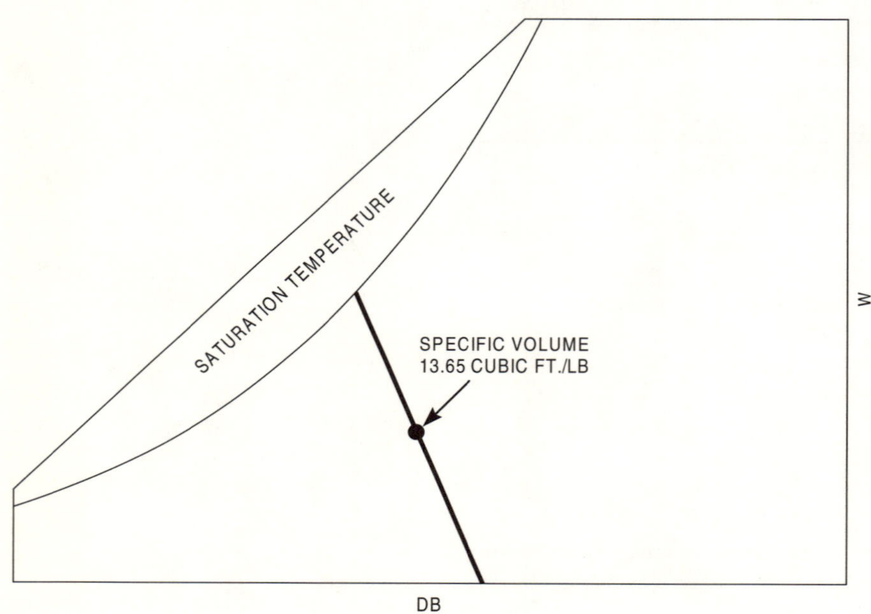

Fig. 5-10 Solution for Example 5-8

Summary

Now you should be familiar with the psychrometric chart. You should be able to confidently locate an initial condition on the psychrometric chart when you know only two air properties. From that condition you should be able to read values for all the other air properties. Finally, given a change in one or more of the values, you should be able to locate the new condition and read all the new air properties.

In addition, after studying chapter 5, you should remember that:

- Psychrometrics is a general term that refers to the use of both psychrometry (the study of air temperature and water vapor relationships) and the psychrometric chart.
- Three laws of science explain our understanding of how air behaves to changing pressure, temperature and gaseous mixture. Boyle's Law says that a gas shrinks in volume when it is pressurized under constant temperature. Conversely, it expands when the pressure is reduced. Charles' Law holds that, under constant pressure, a gas expands when heated and it contracts when cooled. Dalton's Law of Partial Pressures explains that when gases are mixed together they spread throughout the space and the pressure of the resulting mixture is the sum of the partial pressures of the separate gases.
- Airborne water vapor adds heat to the air (the latent heat of vaporization) and its presence retards the evaporation of more water into the air.
- Water vapor is called humidity. Specific humidity is the absolute weight of water vapor present in a pound of dry air. Specific humidity is also a measure of the humidity ratio, the ratio of the weight of water vapor to the weight of dry air.
- Relative humidity expresses the percentage of the actual amount of water vapor to the potential amount of water vapor at saturation.
- The dry bulb temperature of the air when it is saturated is the dew point temperature.
- The sling psychrometer is a convenient device for determining the wet bulb temperature and the relative humidity.
- The specific volume of dry air varies with the temperature, humidity and atmospheric pressure. The specific volume should be estimated when selecting or evaluating air distribution equipment.
- Enthalpy is the sum of the sensible and latent heat of the air.
- Psychrometric charts are usually standardized for sea level pressures. Others have been prepared for higher elevations.

6

Plotting HVAC Processes On The Psychrometric Chart

The best way to understand HVAC processes is to plot the processes on a psychrometric chart. Familiarity with the chart will also lead you to a better understanding of HVAC equipment and controls. After studying this chapter you will be able to:

1. Plot sensible and latent heating processes on the psychrometric chart.
2. Recognize the directions the eight HVAC processes follow on the psychrometric chart.
3. Explain the sensible heating and cooling processes, their equipment and their effect on the air's temperature, humidity and energy.
4. Describe the heating and humidifying process, its equipment and its effect on the air's temperature, humidity and energy.
5. Explain the evaporative cooling process, its equipment and its effect on the air's temperature, humidity and energy.
6. Define the cooling and dehumidifying process, its equipment and its effect on the air's temperature, humidity and energy.
7. Locate the condition points of an air mixing process on the psychrometric chart.
8. Calculate the changes in sensible heat and air flow in an all air heating/cooling system.

9. Determine the sensible heat factor and use it to evaluate the heating/cooling requirements of an all air system.

10. Recognize how small the zone of human comfort is when plotted on the psychrometric chart.

The previous chapter presented the basics of air and water vapor relationships as they apply to HVAC processes. You learned about the psychrometric chart and the information plotted and measured on the chart. Now, you will learn how the psychrometric chart is used to plan HVAC processes and select the equipment used in those processes.

Your furnace affects the dry bulb temperature and the relative humidity, but it does not affect the quantity of moisture in the air. This process is a **sensible heat change**. By adding only sensible heat to your home, the moisture content, latent heat, and dew point remain constant. You can also think of this process as a constant moisture content, constant latent heat or constant dew point process. See Example 6-1 and Figures 6-1 and 6-2.

Example 6-1: You live in the Midwest and it's winter. You've just returned from a skiing trip. Your house feels cold. When you arrived the temperature is 57°F and the relative humidity is 40%. So, you hurry over to thermostat on the wall and turn it up to 70°F. Soon, the furnace is blowing warm air and you feel comfortable again. If your furnace doesn't have a humidifier, turning up the thermostat adds only sensible heat to your home. Locate these conditions on Figures 6-1 and 6-2. This point rests on the **w** line of .004 lb/lb dry air. Also at this condition the enthalpy, **h**, is about 18.0 btu/lb dry air.

CONTINUED

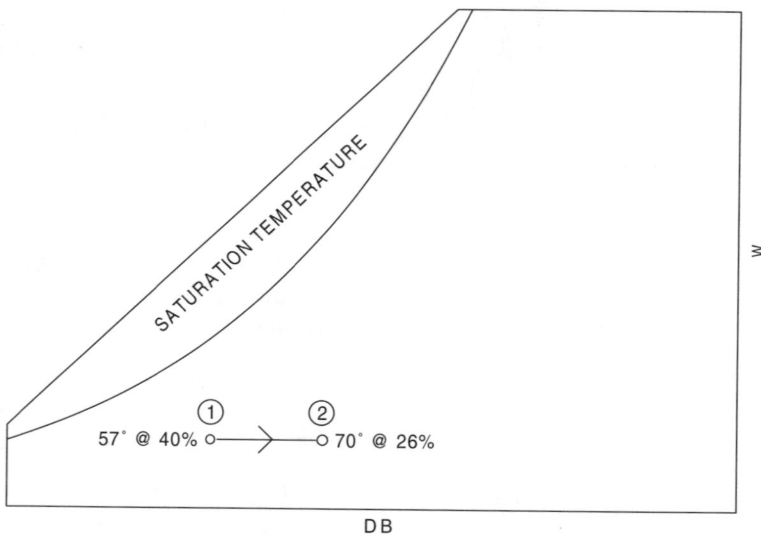

Fig. 6-1 Sensible Heat Change

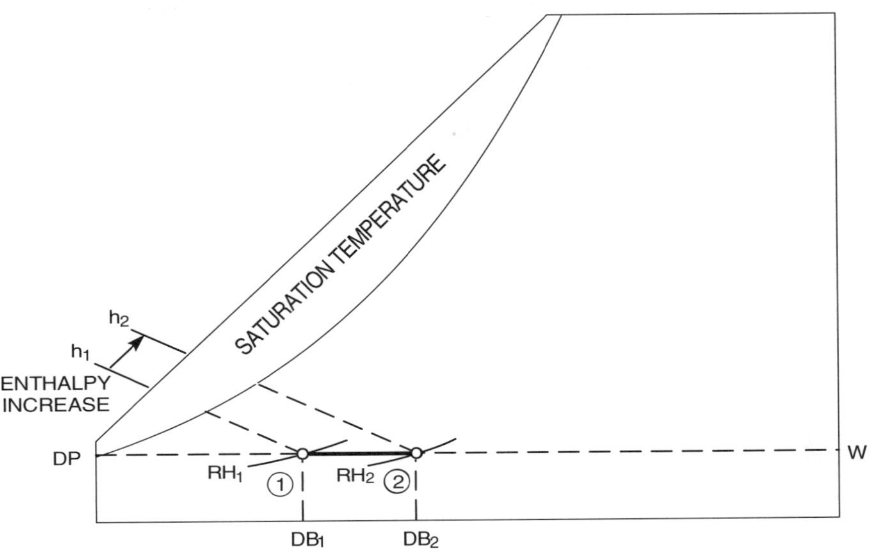

Fig. 6-2 Solution for Example 6-1

> ***Example 6-1 (CONTINUED)***
> Thirty minutes or so after adjusting the thermostat, you notice that the temperature is 70°F and the furnace has turned off. Since your furnace produces only sensible heat, you can follow the change horizontally along the humidity ratio or **w** = .004 line on the psychrometric chart. Trace it to where it intersects the 70°F dry bulb reading. Notice the other environmental data at this condition. The **RH** has dropped to about 26% and the enthalpy, **h**, has increased to 21.4 btu/lb dry air. The air feels dryer (although it contains the same amount of water vapor) and now contains more heat. See Figure 6-2.

Sketching the Eight HVAC Processes

The preceding example suggests that HVAC processes follow general directions on the psychrometric chart. Indeed, they do. *Any change in the condition of air can be represented by a straight line connecting the initial condition with the final condition.* For example, the condition points of sensible heating and cooling processes fall along horizontal lines on the chart. Likewise, the condition points of exclusively moisture content changes fall along vertical lines. You can see the general directions of the eight HVAC processes in Figure 6-3. An explanation of each process follows the figure.

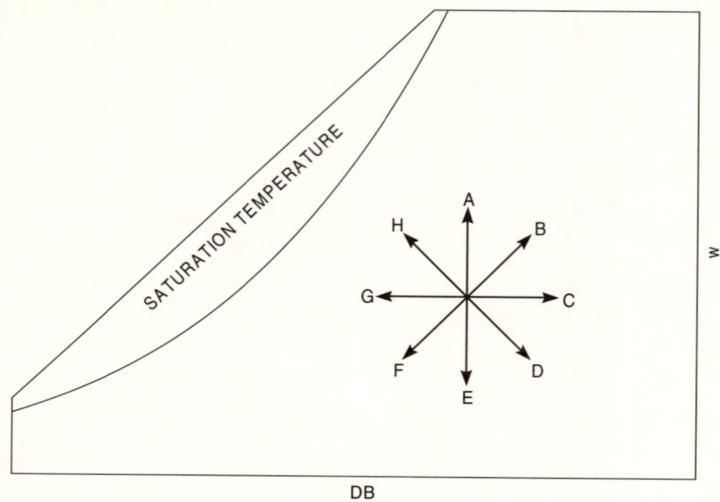

A = Humidifying only
B = Heating and humidifying
C = Sensible heating only
D = Chemical dehumidifying
E = Dehumidifying only
F = Cooling and dehumidifying
G = Sensible cooling only
H = Evaporative cooling only

Fig. 6-3 Psychrometric Representations of HVAC Processes

The *humidifying only* line A shows the effect of adding water vapor to the air. The moisture content of the air increases while the DB temperature remains the same. This process adds latent heat but no sensible heat to the air.

Heating and humidifying, line B, shows the result of adding sensible heat and humidity. Depending on the slope of line B, as the **DB, w** and **h** rise, the **RH** may increase, change very little or even decrease.

The *sensible heating only* line C follows the change described in Example 6-1. **DB** and **h** increase. **RH** decreases, while **w** remains constant.

Desiccant materials absorb moisture from the air and are used for *chemical dehumidifying only*, line D. The **w** and **RH** drop while the enthalpy remains constant. The process is adiabatic (*a dee uh bat ik*). In an **adiabatic process** the air is cooled or heated without a loss or gain in heat energy. Since this method of dehumidifying is slow and quite expensive, it is used primarily for sensitive applications.

The *dehumidifying only* process, line E, exists in theory only because it is extremely difficult to achieve by mechanical means. In this process the dry bulb temperature remains steady while water vapor is stripped from the air. Notice that the declining **w** means that the enthalpy is falling also.

Line F, *cooling and dehumidifying*, is the process line for what lay people call air conditioning. The process removes both enthalpy and water vapor from the air. As a result, the **DB, RH**, and **WB** fall. Although the cooling and dehumidifying process is displayed as a straight line, the process actually follows a gradual, sloping line that *approximates* a straight line. More on this later.

A *sensible cooling only* process, line G, does not remove any water vapor from the air. The moisture content remains the same, the **DB** and **h** fall, and the **RH** increases.

Evaporative cooling only, line H, is also an adiabatic or constant enthalpy process. In the process, air of higher sensible heat is directed through or over water of lower sensible heat. This causes some of the water to evaporate, converting sensible heat into latent heat. Because of the energy conversion, the air's **DB** decreases and the **RH** rises.

Caution: Although it is convenient to think of HVAC processes as occurring in specific directions on the psychrometric chart, all of the processes do not follow perfectly straight lines. Note also that although these processes were plotted near the center of the chart, in fact their initial condition could be anywhere on the chart.

Next, let's study several of these processes in more detail. The illustration to the left of each psychrometric chart depicts the air as it enters and leaves the HVAC equipment.

Sensible Heating and Cooling

There are several forms of sensible heating devices:

- The forced air gas or oil fired home furnace.
- Radiant hot water or steam systems. In both processes water is heated in a boiler and then circulated to radiators where its heat radiates into the surroundings. Because both systems are also closed systems, no water or steam escapes into the room air.
- Electric resistance coils. The coils are used as portable units, or they may be placed in the air ducts, floors or walls of all-electric buildings.

In Figure 6-4, the air passing through an electric resistance coil gains only sensible heat.

There are two primary sensible cooling devices:

- Coils filled with chilled water
- Coils filled with refrigerant

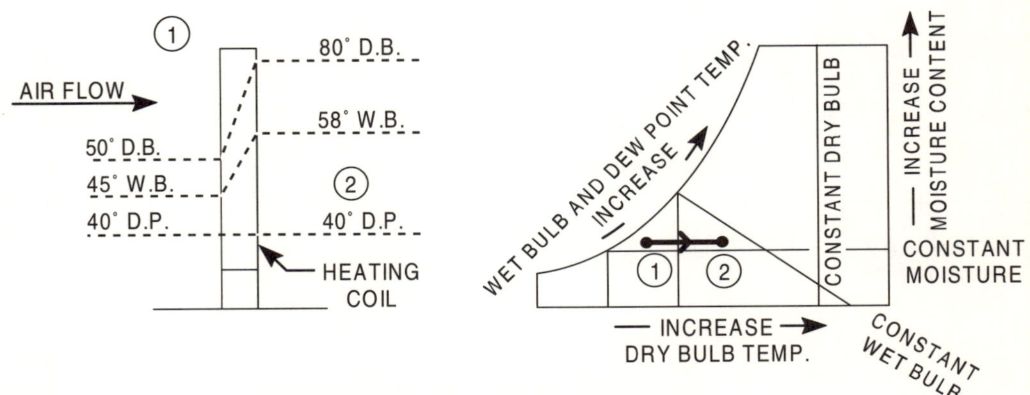

Fig. 6-4 Sensible Heating

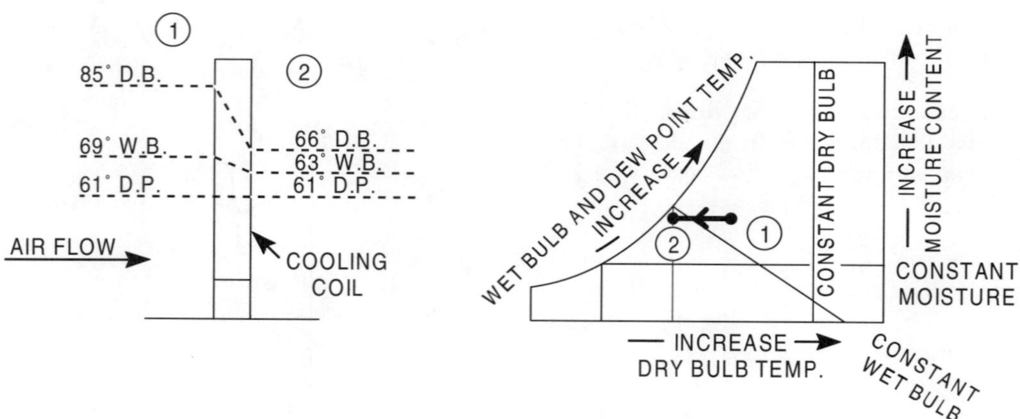

Fig. 6-5 Sensible Cooling

Sensible cooling coils are maintained at a temperature below the dry bulb temperature of the air. As the air passes through the coils, it gives up its sensible heat to the colder coils. Sensible cooling also lowers the wet bulb temperature and enthalpy. In Figure 6-5, notice how dramatically sensible cooling can also raise the relative humidity.

Heating and Humidifying

Heating and humidifying is typically a two-step process. First, the air is heated and then it is humidified. The heating devices are the same as those listed for the sensible heating processes. The humidifying devices are generally of three types:

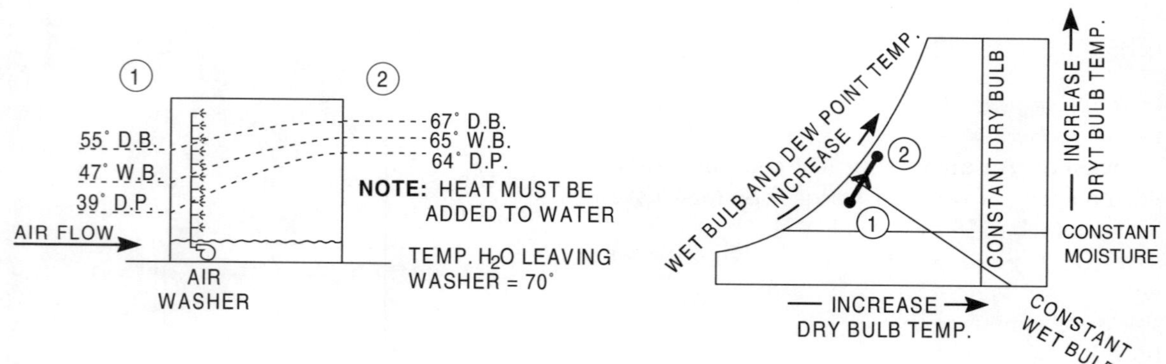

Fig. 6-6 Air Washer Humidifier

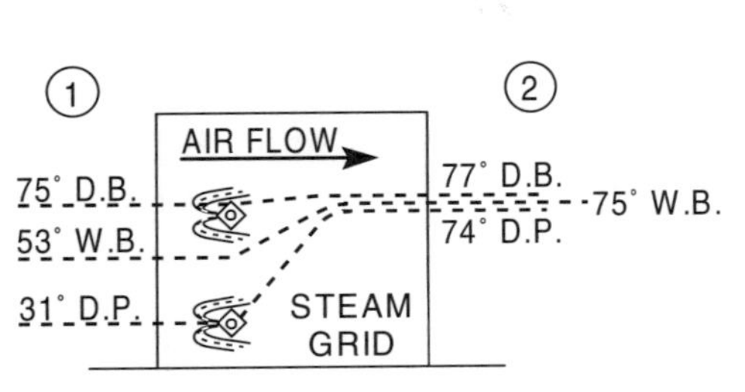

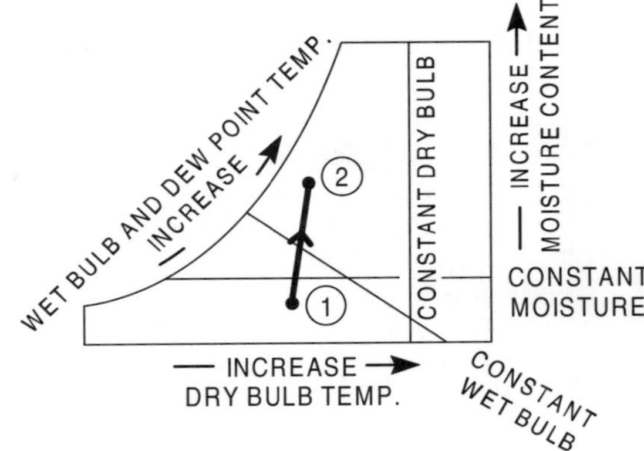

Fig. 6-7 Steam Grid Humidification

- Evaporative humidifiers fitted to furnaces
- Spray chambers or air washers used in industrial applications
- Steam grid humidifiers used in commercial applications

In Example 6-1, after your furnace had warmed the house, the relative humidity would be about 26%. At that level, the air and your skin feel uncomfortably dry. The cure for the problem is to add a humidifier to the furnace. Water in the humidifier evaporates into the warm air stream passing over or through the humidifier.

In some industrial systems, humidity is added to the air by directing the air flow through an air washer. An air washer is a chamber filled with a spray of warmed water. The mist is so fine that the chamber actually becomes an **atomizer**. Figure 6-6 depicts humidifying with an air washer and shows the process on the psychrometric chart. Notice that the dry bulb heat gain is the result of using warmed water in the spray chamber.

Steam grid humidifiers are often used in commercial applications. Air passes through a grid of steam nozzles that releases the steam directly into the air. The addition of moisture to the air is very rapid. Figure 6-7 depicts humidifying with a steam grid. Notice that while the humidity content, relative humidity and enthalpy all increase, the dry bulb temperature rises very little. This is because a steam grid radiates a small amount of heat. See Chapter 15 for more information on humidifiers.

Evaporative Cooling

Remember how the reading on a wet bulb thermometer is usually lower than the dry bulb? The difference is due to the evaporation from the wet bulb's wick. For water to evaporate, it needs heat–the latent heat of vaporization–to change it from a liquid to a gas. Depending on the condition of the surrounding air, evaporative cooling on the wet bulb thermometer takes place in two ways.

1. In the first way, the sensible heat of the surrounding air is *higher* than that of

the wick. Some of the air's sensible heat will be converted into the latent heat of vaporization, changing the state of the water on the wick to water vapor in the air. If you could measure the sensible heat of the air, you would notice a slight decline as a result of the vaporization. In this view, evaporation lowers the sensible heat of the air and serves as the basis of cooling the air.

2. In the second way, the sensible heat and latent heat of the air are *lower* than that of the water on the wick. Now, the wetted wick can give up its energy to the surrounding air in two ways: as a sensible heat exchange and as a latent heat exchange. If you could measure the sensible heat of the air, you would notice a slight increase as it passed over the wick. Similarly, you would notice a slight increase in the latent heat of the air as it took on latent heat from the wetted wick. In this view, evaporation lowers the sensible and latent heat of the wetted wick.

So, depending on your perspective and the relative amounts of sensible and latent in the air, evaporative cooling can either lower the sensible heat of the air or lower the sensible and latent heat of the water on the wick.

There are two primary evaporative cooling devices:

- Air washers
- Cooling towers

Air Washers

In an air washing device, air with high sensible heat is directed through a fine mist of water known as a spray chamber. See Figure 6-8. As the air passes through the spray the sensible temperature of the air declines, while the moisture content and relative humidity rise. Air washers are also constant enthalpy devices. They are adiabatic. They do this by converting sensible heat in the air to latent heat. The conversion lowers the air's dry bulb temperature and raises its moisture content and relative humidity, while keeping the wet bulb temperature constant.

Theoretically, the air leaving a 100% efficient evaporative cooling unit is saturated. Much like the air immediately surrounding the wick

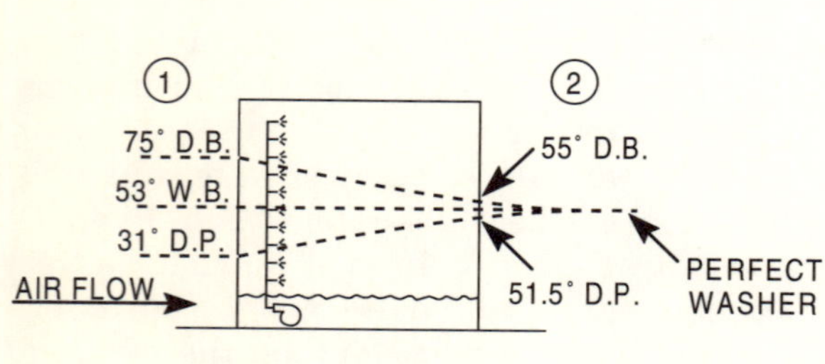

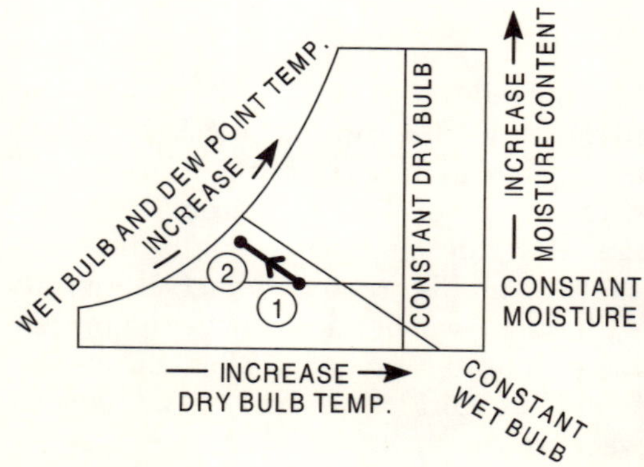

Fig. 6-8 Evaporative Cooling Process

of a wet bulb thermometer becomes saturated. When an air washer achieves 100% saturation, it is called a **perfect washer** because it achieves **adiabatic saturation**. Although achieving perfect adiabatic saturation is impossible, some industrial processes do produce very high levels of efficiency at the cost of building very elaborate systems.

Despite their energy efficiency, evaporative cooling devices are not universally popular. For one reason, they require that the incoming air be quite dry. This makes the process more practical in drier climates, such as the Southwest United States where lower wet bulb temperatures are commonplace, rather than the coastal regions. Secondly, the systems also require regular maintenance to prevent algae growth and the buildup of water minerals in the recirculating water supply. Because the unit's water supply easily develops a bacterial growth, evaporative coolers are often called **swamp coolers**.

Cooling Towers

In many air conditioning processes, the high sensible and latent heat of the air in the conditioned space is eventually transferred to the water flowing in the cooling device. The water, in turn, must give up its increased sensible and latent heat so that it can again absorb heat from the incoming air. This is the second view of evaporative cooling described previously. This heat transfer takes place in a **cooling tower**: a device that facilitates a heat transfer from warmed water to the cooler air of the atmosphere.

Most of this heat transfer is due to vaporization. In fact, a small percentage of the warm water entering a cooling tower actually vaporizes. This small amount accounts for about a 10°F drop in the temperature of the water that passes through the tower. The rest of the heat transfer is due to the conductive heat transfer from the warmer water to the cooler air. Because cooling towers rely on the atmosphere, most are located out-of-doors.

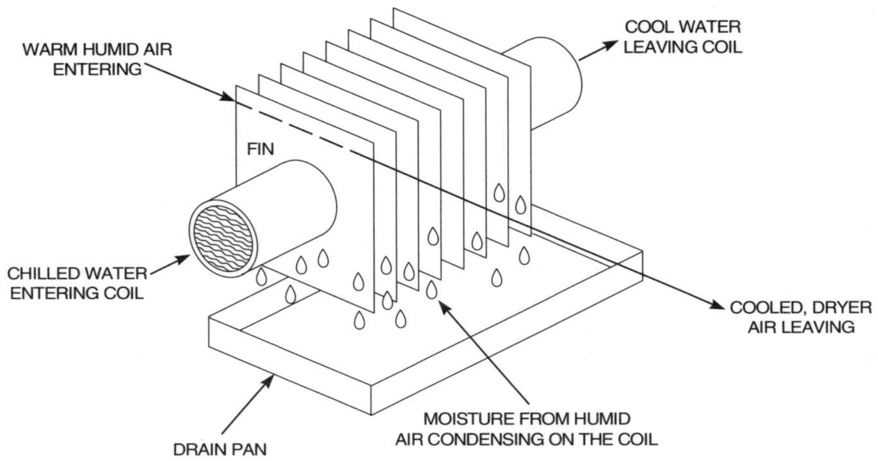

Fig. 6-9 Chilled Water Cooling Coil Operation

Modern Air Conditioning Practice, 3rd Edition. by Harris, Norman C., ©1983

Adapted by permission of McGraw-Hill

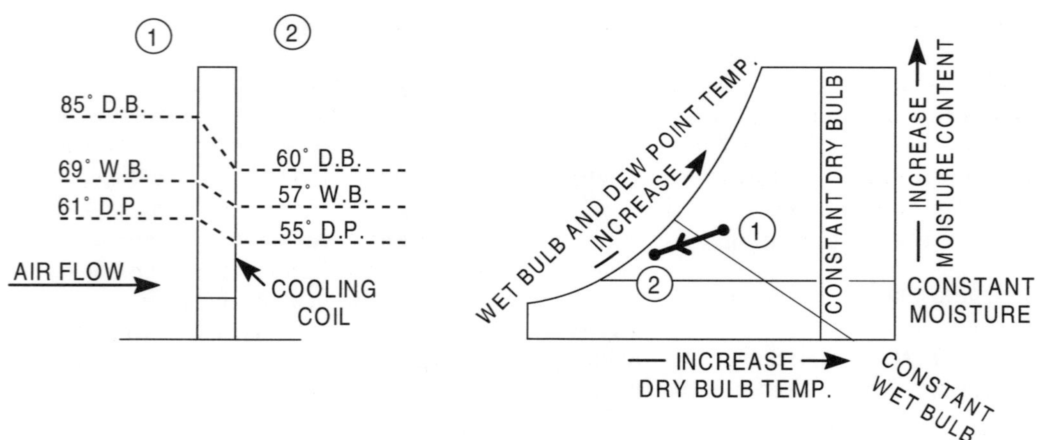

Fig. 6-10 Cooling and Dehumidifying

Cooling and Dehumidifying

In most parts of the country, the summer's air and the interior of buildings are high in both sensible and latent heat. The popular way to remove these heats is to cool and dehumidify the air. Cooling and dehumidifying is achieved by passing air through:

- a **chilled water (CW)** coil, or
- a refrigerant filled coil, called a **direct expansion (DX)** coil.

Figure 6-9 shows a chilled water coil. The coil water absorbs sensible heat from the warm air flowing past the fins. If the surface temperature of the coil and fins is below the dew point of the air, some of the moisture in the air condenses on the coil, removing some latent heat. The temperature of the coil necessary for this condensation is known as the **apparatus dew point**, or **ADP**.

This cooling coil operation is heart of the cooling/dehumidifying process presented in Figure 6-10.

Although this process looks straightforward, it is in fact quite complex. The complexity occurs because not all the air passing through the coils will completely contact the fins of the coils. Some air will always slip through, only partially cooled and dehumidified. The problem is called **bypass**. You can greatly reduce the bypass problem if you are willing to spend more for equipment and operating costs. Manufacturers test their coils for bypass and assign a **bypass factor** to each model. Bypassing will be covered in more detail in the chapter describing heat exchangers.

Lines AB and BC in Figure 6-11 depict the *theoretical* operation of the cooling/dehumidifying process. As the entering air (A) contacts the cooling coil's fins they absorb its sensible heat, cooling the air to its dew point (B). Since the apparatus dew point is well below the air's dew point, the process theoretically continues from (B) downward to (C), the **ADP**. Because of the bypassing air, however, the *actual* air conditioning process occurs along the curved, dotted line, AC.

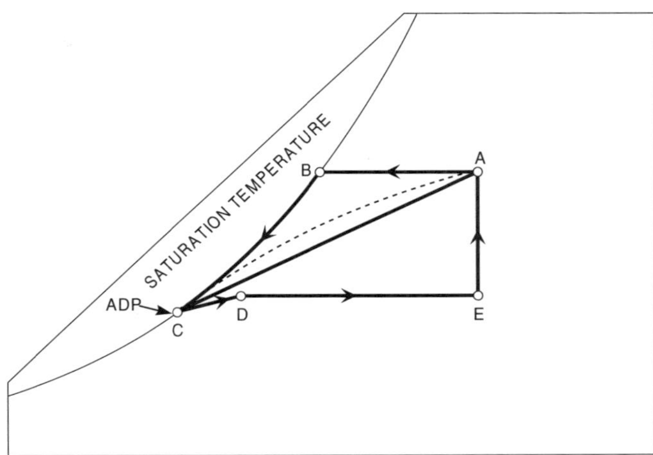

Fig. 6-11 Cooling/Dehumidifying Process

The line AC is the coil process line. The **coil process line** defines the efficiency of the cooling coil which is the sensible, latent and total heat removal capacity of the unit. It is a complex curve and depends on:

- The number of fins per inch of coil length
- The coil surface area
- The velocity of the air passing through the coil
- The chilled water or refrigerant temperature

Manufacturers rate their cooling/ dehumidifying equipment and also show their coil process lines as straight lines on the psychrometric chart.

For practical purposes, it is accurate enough to be able to locate the end points of the cooling/dehumidifying process on the chart. Straight lines can then be used to depict the direction of process change and to aid in equipment selection.

Returning to Figure 6-11, the cooling/dehumidifying process will never reach condition (C) due to the air bypassing the coils. Instead, the air leaving the coils will be at some condition, (D). This temperature is known as the **air-off-the-coil** temperature. Because of bypassing, an air-off-the-coil temperature of 54°F would be common for a unit with an **ADP** of 48°F. Notice too, that (D) is not located on the saturation curve. Because of bypassing, the air-off-the-coil has a RH of about 90%. Lastly, the sensible heat change in this cooling/dehumidifying process is represented by line DE and the latent heat change is represented by line EA.

Air Mixing

As you begin to study the topic of air mixing, you need to visualize the air flow path and components of an **all-air** heating/cooling system. A typical system appears in Figure 6-12.

Looking along the top of Figure 6-12, air returning from the room to the conditioner is **return air, RA**. Part of the **RA** is exhausted

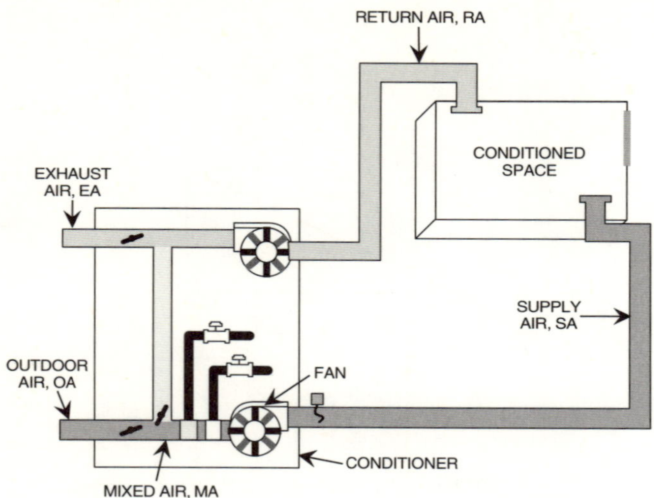

Fig. 6-12 All Air Heating/Cooling System

outside as **exhaust air, EA** to maintain air quality. To make up for the exhaust air, **outdoor air, OA** is added to the return air. The mixture of **RA** and **OA** is called **mixed air, MA**. The mixed air typically passes through a filter before entering the conditioning unit. Next, depending upon the season, the **MA** is heated, cooled, humidified or dehumidified. The conditioned air, now called **supply air, SA**, is then distributed throughout the facility by a fan and duct system.

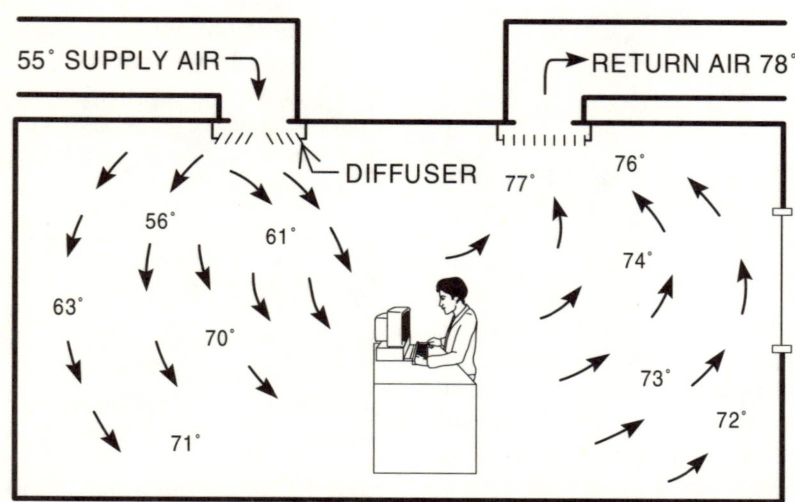

Fig. 6-13 Air Diffusion and Mixing Pattern in a Typical Room - Cooling

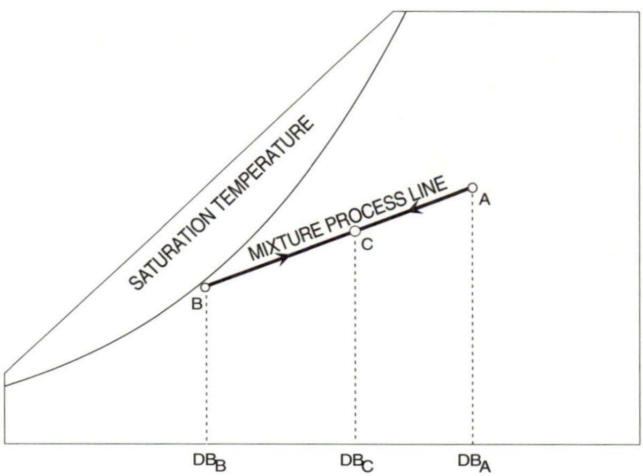

Fig. 6-14 Mixture Process Line

As the **SA** enters each room or area, the diffuser distributes it throughout the space. The diffuser creates a mixing pattern with the **SA** and the air already present in the space. See Figure 6-13. As the supply air enters each space, an equal volume of return air flows out of the space.

The temperature, humidity, dew point and enthalpy of the mixed air, **MA**, are proportional to their values in the two air streams, **RA** and **OA,** and to the volumes of **RA** and **OA**. These values are proportional because the air mixing process conserves both the mass and energy. For example, the mass of $MA = w_{RA} + w_{OA}$ (where w represents weight). This is the law of conservation of mass. Similarly, the mixing process conserves energy and $h_{MA} = h_{RA} + h_{OA}$. The concept of proportional air mixing applies to any number of air streams.

Since the effect of mixing air streams is proportional, the process will also plot as a straight line on the psychrometric chart. The result will lie on the line joining the conditions of the two air flows. See Figure 6-14. The condition of the mixture will always lie somewhere along this **mixture process** line. If, for example, during the summer **RA** represented by condition (A) is mixed with a nearly equal amount of the **OA** of condition (B), the mixture will be some condition (C) along the line joining (A) and (B).

$$DB_{ma} = \frac{CFM_{oa}t_{oa} + CFM_{ra}t_{ra}}{CFM_{oa} + CFM_{ra}}$$

Example 6-2: During the summer the return air from a facility measures 80°F **DB**. If 8000 cfm of this return air is mixed with 2000 cfm of outside air measuring 92°F **DB**, what will be the **DB** temperature of the mixture as it enters the heating/cooling unit depicted in Figure 6-12?

Solution: Substitute the information into the above equation.

$$DB_C = \frac{2000\ (92) + 8000\ (80)}{2000 + 8000} = \frac{824{,}000}{10{,}000} = 82.4°F$$

You can see from the equation that the dry bulb temperature of the mixture is a simple ratio of the air stream quantities. For example, the location of point C in Figure 6-14 depends on the ratio of the air flowing from the two sources. If half of the air is of condition A and half is of condition B, point C will be half way between points A and B. Similarly, if one-third of the air mixture was made up of condition B air and two-thirds condition A, point C would be at the third point nearest A.

Because mass and energy are conserved during mixing, the **DB** temperature of the mixture is also proportional to the amounts (expressed as cubic feet per minute) and dry bulb temperatures of the two air flows. This relationship is summarized in the equation at the bottom of this page.

Estimating the Heating and Cooling in an All Air System

When you are evaluating an all air system, you must know or assume several factors:

1. The design conditions of the building or space
2. The condition of the air entering the space
3. The amount of air distributed to the space
4. The capacity of the heating/cooling equipment being used

This list is also presented in the typical order in which you would determine these factors.

You determine the room or space heating or cooling needs by estimating the probable heat loss or heat gain. The concept of these procedures was discussed in Chapter 4. By putting together the room condition requirements and the design weather conditions, you can then estimate the heating, cooling and air handling capacity of the equipment.

To add sensible heat to a space, the supply air must be warmer than the space temperature. Similarly, to add latent heat, the supply air must have a higher moisture content than the space's moisture content. Assuming that the entering air is distributed evenly throughout the space, the conservation of mass and energy apply again and the air entering and the air already present will mix proportionally.

You can determine the amount of heat that an air supply gives up to heat a space during the winter or that it absorbs as it cools the space during the summer with the formula shown at top of next page.

$$DB_C = \frac{CFM_A}{CFM_A + CFM_B} DB_A + \frac{CFM_B}{CFM_A + CFM_B} DB_B$$

where: DB_C = *DB temperature of the mixture at point C (Figure* 6–14) *on the chart*

CFM_A, CFM_B = *flow rates, in cfm, of airstreams A* and *B (Figure* 6–14)

DB_A, DB_B = *DB temperatures, in °F, of airstreams A* and *B (Figure* 6–14)

$$q_S = W \cdot c \cdot \Delta T$$

where: q_S = *quantity of sensible heat in Btu per hour*

W = *weight of substance, in this case the weight of air*

c = *specific heat of air, in Btu/lb/°F*

ΔT = *sensible temperature difference (change) in substance*

The above formula is based on the concept of specific heat, discussed in Chapter 2. You can convert this formula to a flow rate of standard air in cubic feet per minute, **CFM**, by making the following substitutions: **W**, the weight of air is about 0.075 lb/cu ft, the specific heat of air, **c**, is 0.24 Btu/lb. and there are 60 minutes per hour.

$$0.075\ (0.024)\ 60 = 1.08$$

$$q_S = 1.08 \times CFM \times \Delta T$$

As a rule of thumb, when working with all air systems, you can assume a temperature change of about 20°F. You will remember from the cooling and dehumidifying discussion earlier in this chapter that an air-off-the-coil temperature of 54-55°F is common. Furthermore, an average room design temperature during the warm weather months is about 75°F. Thus, a temperature change ΔT °of 20°F is common.

> ***Example 6-3:*** You estimate the warm weather heat gain of a room at 160,000 Btu/hour. You also assume a 20°F difference between **SA** and **RA**. What volume of air in CFM, at 55°F, do you need in this space to maintain a 75°F temperature?
>
> ***Solution (using reworked equation):***
>
> $$CFM = q_S \,/\, (1.08\ \Delta T)$$
>
> $$= 160{,}000 \,/\, (1.08 \times 20) = 7407\ CFM$$
>
> *You can use this same approach during heating operations to determine the amount of sensible heat that the supply air will release to the space.*

Tracing and Labeling Complex Processes on the Psychrometric Chart

Before going on, let's review two points about using the psychrometric chart.

1. *Process paths do not always follow straight lines.* Drawing a HVAC process as a straight line on the psychrometric chart usually describes the actual path of the process. However, for some processes a straight line represents the *results* of process change, not the path itself.

   For example, adding both sensible and latent heat to a space may be plotted as separate straight lines or as one line when both processes occur simultaneously. Imagine a well-insulated interior room heated with electric resistance. When you turn on the heater and add sensible heat to the room, the process would plot as a straight horizontal line, AB, extending to the right on Figure 6-15. Now, if you move in a portable humidifier to add humidity to the room,

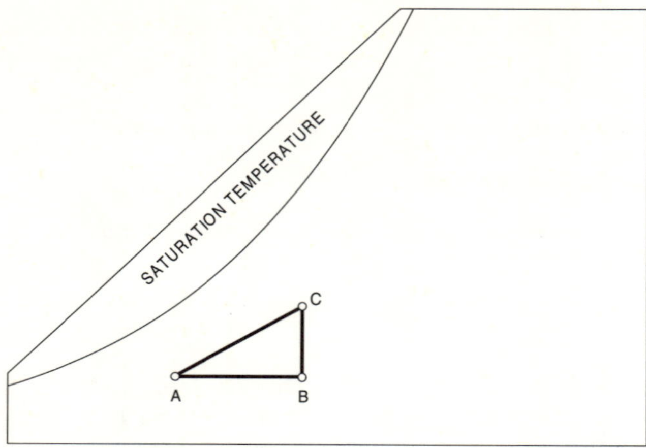

Fig. 6-15 Adding Sensible and Latent Heats

the addition of latent heat would plot as a vertical line, BC, on Figure 6-15. Or, you could turn on the heater and humidifier simultaneously and plot both processes as a line sloping upward to the right, AC.

Air mixing is another example where the straight line drawn between two condition points accurately represents the process path.

Cooling/dehumidifying, on the other hand, follows a straight line process and then a curved line process along the saturation line. First, the cooling apparatus lowers the sensible temperature of the air to the saturation temperature, line AB in Figure 6-16. However, the

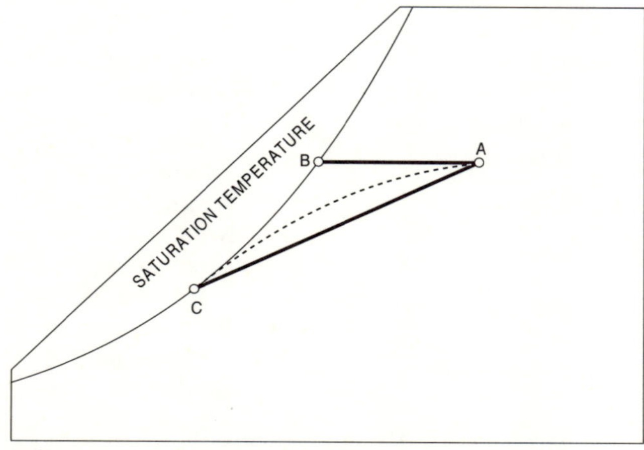

Fig. 6-16 The Steps of Cooling/Dehumidifying

temperature of the cooling apparatus is usually well below the air's dew point temperature. So, the air continues to cool, while the moisture in the air condenses within the cooling apparatus, line BC in Figure 6-16. Eventually the air's temperature (assuming *no* bypassing) will reach the temperature of the cooling equipment, the apparatus dew point, condition C. Because plotting both lines of the cooling/dehumidifying process doesn't provide you with much useful information, you should just draw the *results* of the process–a straight line extending from condition A to condition C in Figure 6-16.

2. *Process paths may be described with several names*. Depending on why you drew it, the line connecting two condition points may be described with several names. For example, Figure 6-16 displays the cooling/ dehumidifying process and could be labeled as the cooling/dehumidifying process line. Line AC of Figure 6-11 was also labeled the **coil process line** in an earlier section because it shows the effect of the cooling coil operation, or coil process. Line AC is also known as the **total heat line** because it depicts the change in total heat from condition A to condition C, as read on the enthalpy scale. Finally, AC is also labeled the **sensible heat factor line**.

Using the Sensible Heat Factor for Equipment Selection

To maintain facility design conditions during warm weather in many regions of the country, you must remove both sensible and latent heat. The ratio between sensible and latent heat gives rise to another concept: the **sensible heat factor** (SHF). Its mathematical form is:

$$SHF = \frac{q_S}{q_S + q_L}$$

where: q_S = *sensible heat gain in Btu's per hour*

q_L = *latent heat gain in Btu's per hour*

This equation tells you what proportion of a total heat change belongs to the change in sensible heat. The values for q_S and q_L are typically determined from a heating/cooling load analysis like those presented in Chapter 4.

The sensible heat factor is also useful for determining an apparatus dew point on the psychrometric chart. *If you draw a line between the design condition point and the saturation curve at the slope of the sensible heat factor, the line will intersect the saturation curve at the apparatus dew point of the equipment that satisfies the design conditions.*

For example, locate the semicircular scale in the upper left corner of ASHRAE Psychrometric Chart No. 1. The inner scale of the semicircle is the sensible heat factor. After calculating an SHF, locate the ratio on the inner scale. Next, draw a line connecting the SHF and the center mark of the semicircle. See Figure 6-17. This line depicts the slope of the sensible heat factor. Now, transfer a line

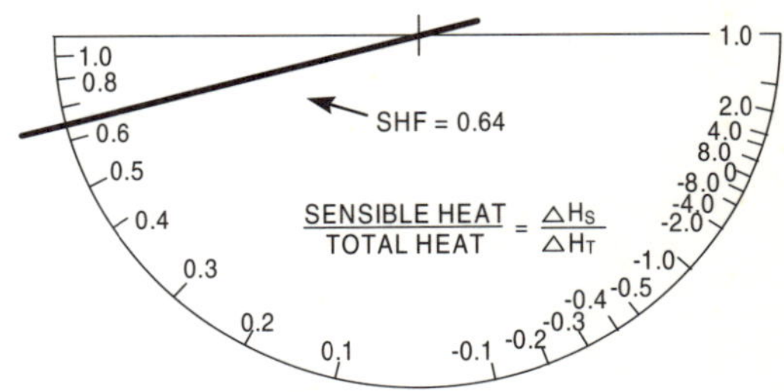

Fig. 6-17 Plotting the Sensible Heat Factor

of this same slope to the psychrometric chart and let it connect the design condition point and the saturation curve. The point where the line intersects the dew point curve is the apparatus dew point.

Example 6-4: You estimate a space will have a total heat gain of 40,000 Btu/hr and of that 14,500 Btu of that will be latent heat. The design condition for the space is 80°F and 50% **RH**. What is the apparatus dew point of the equipment necessary to meet the design condition?

Solution: First, calculate the SHF.

$$SHF = q_S/(q_S + q_L) = (40{,}000 - 14{,}500)/40{,}000 =$$

$$= 25{,}500/40{,}000 = 0.6375 = 0.64$$

Next, locate this ratio on the SHF scale. Transfer the SHF slope to the psychrometric chart with the line passing through the room design condition.

Finally, read where the SHF line intersects the saturation curve at 48°F. This is the required apparatus dew point. See Figure 6-18.

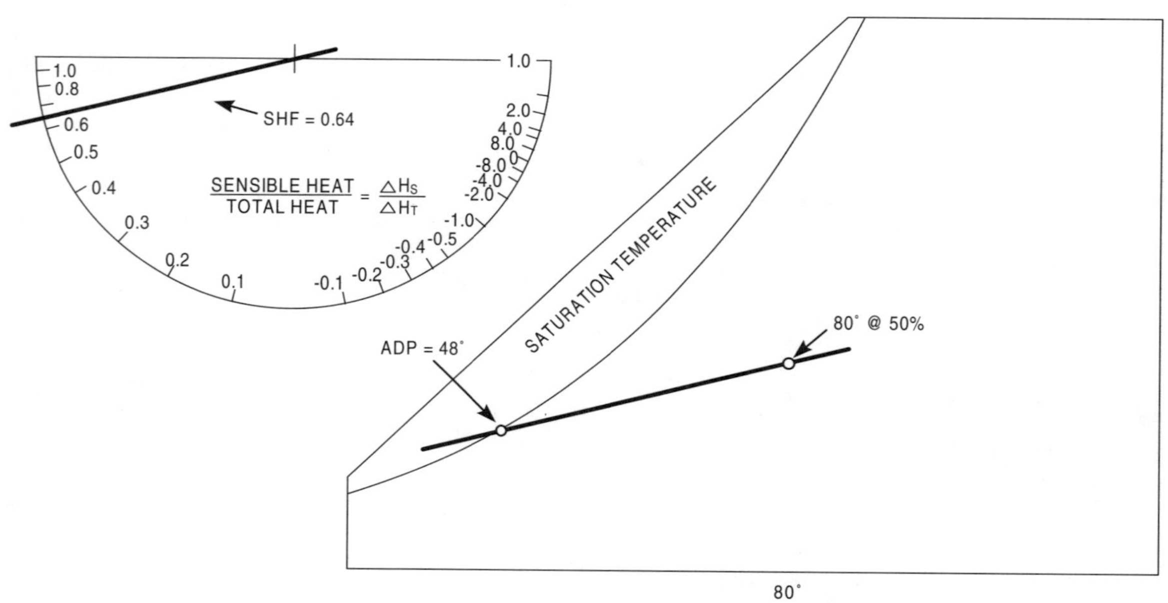

Fig. 6-18 Plotting the Sensible Heat Factor Line

Now, let's work through several examples that show how to use the concepts of air mixing and the sensible heat factor to evaluate equipment needs.

Examples 6-5, 6, 7:

You are evaluating an air handling unit for a small office building that is estimated to have a total sensible heat gain of 234,175 Btu/hr and a total latent heat gain of 41,325 Btu/hr. The room design condition is 75°F at 50% RH and the estimated return air condition is 77°F and 48% RH. Fifteen percent of the air volume will come from the outside air. The design climate condition is 95°F with a 80°F WB.

Example 6-5: What is the mixed air (**MA**) temperature entering the cooling equipment?

Solution: Begin by plotting all the information you can on the psychrometric chart. You can plot the desired room condition, the return air condition and the design outdoor air temperature (**OA**). As the building's heat gain increases, its sensible and latent heat will be carried away by the return air (2) which is then always warmer than the desired space condition (1).

Since some **RA** will be lost as exhaust air, some makeup or ventilation **OA** must be added to the system to replenish it. You can also draw a line between the **RA** condition, (2), and the **OA** condition, (3). **MA**, the mixture of **OA** and **RA** will lie somewhere along this line. In our case, for a mixture of 15% **OA** and 85% **RA**, the mixed air temperature is 80°F, condition (4). The 80° temperature is 15% of the distance between the **RA** and **OA** condition. Since the **RA** is the greater percentage, 85%, the temperature will be closer to the **RA** condition. So far, the information will look like that on Figure 6-19.

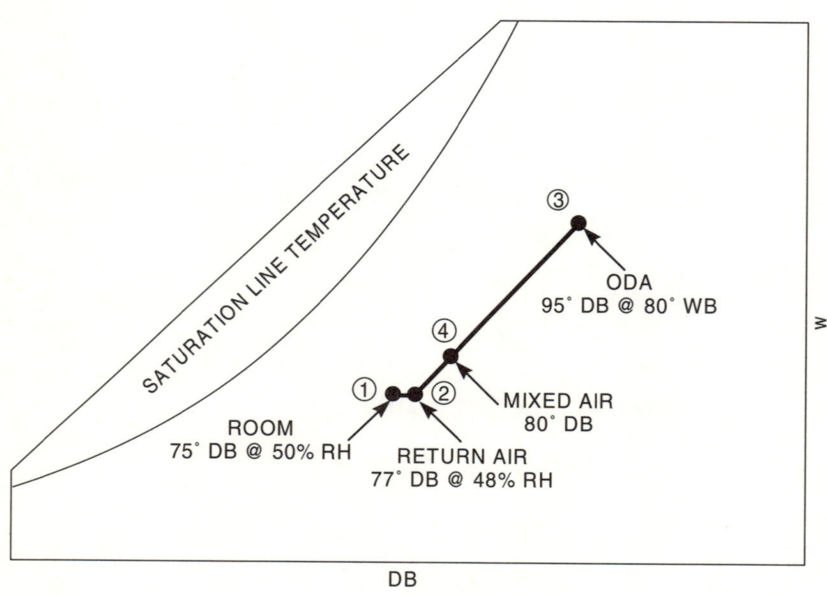

Fig. 6-19 Plotting Design Temperature, RA, OA *and* MA

Example 6-6: What is the Sensible Heat Factor?

Solution:
SHF = 234,175 / (234,175 + 41,325) = 0.85

Now, you can locate the **SHF** on its scale and then transfer the slope downward to find the **ADP** condition 5. Project the slope of the sensible heat factor from the room air condition downward to the left until it intersects the saturation curve. The **SHF** line crosses the saturation curve at about 53°F. The information will look like that on Figure 6-20.

You will notice an air-off-the-coil temperature, condition 6, of 55°F in Figure 6-20. This temperature is the result of the coil bypass factor discussed earlier. The air-off-the-coil is also produced by the temperature controls having a 55° setting.

Example 6-7: What is the required CFM capacity of the air handling unit?

Solution:

$$qs = 1.08 \bullet CFM \bullet \Delta T \quad or \quad CFM = qs \;/\; 1.08 \bullet \Delta T$$

where ΔT = *temperature difference between the desired room air and the air-off-the-coil temperature.*

$$CFM = 234{,}175/1.08 \times (75\text{–}55°) = 234{,}175/1.08 \times 20° = 10{,}841\ CFM$$

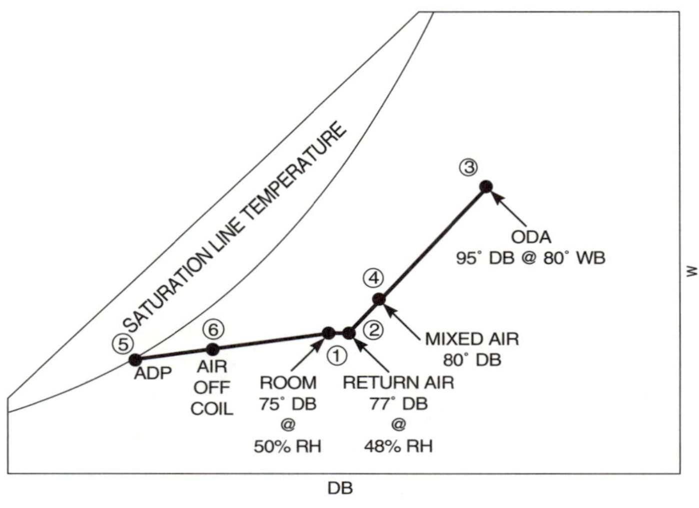

Fig. 6-20 Transferring the Sensible Heat Factor Line

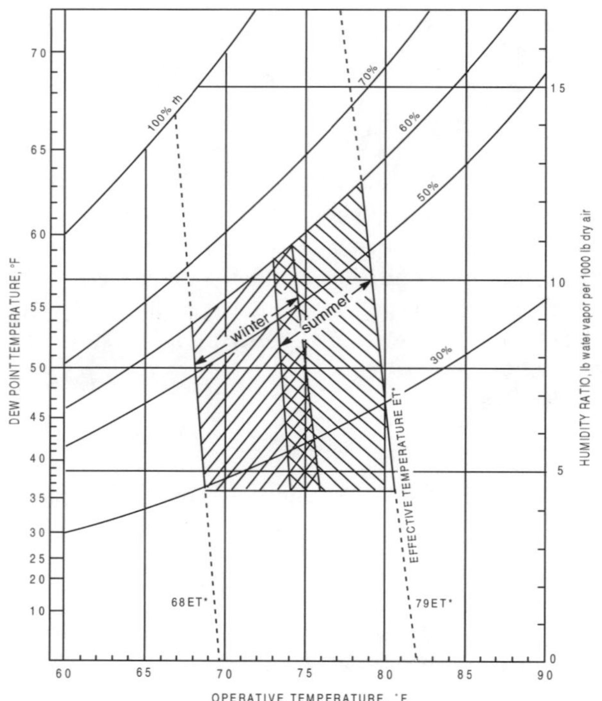

Fig. 6-21
Winter and Summer ASHRAE Comfort Zones
Reprinted with the permission of ASHRAE

Operating the HVAC System for Human Comfort

To put the examples of this chapter into a less detailed perspective, Figure 6-21 shows the winter and summer ASHRAE comfort zones superimposed on the psychrometric chart. The zone recommendations are based on research and assume that people are engaged in light, sedentary activities and wearing clothing typical of the season. Notice that even in the winter, when people's seasonal clothing will be heavier, ASHRAE recommends temperatures above 67°F.

One of the bigger challenges facing the engineer is to achieve human comfort goals through the HVAC system design. For example, most chilled water air conditioning coil temperatures are about 48°F to ensure sufficient cooling. Because of bypassing, the temperature of the air leaving a cooling unit is often between 55 and 60°F. When the right cfm of air is mixed with room air, this conditioned air will adequately remove the heat gain in the space. However, anyone sitting in the direct path of the chilled air entering the room may indeed feel uncomfortable, 60°F is pretty cool.

The psychrometric chart is the single most useful tool for evaluating heating and cooling systems.

While looking over this chart, also notice that:

1. The wintertime heating objectives expressed on the chart are different from the summertime cooling objectives. These comfort zones also correspond to good energy management. Over-heating in the winter and over-cooling in the summer consume extra energy and do not add to occupant comfort.
2. Because people wear heavier clothes in the winter, they can tolerate lower temperatures than in the summer. Similarly, because they wear lighter clothes in the summer, they can tolerate warmer summer temperatures.
3. In the intermediate seasons of spring and fall, the winter temperatures are nearly too warm because of the clothing people wear. Then, during the summer people find these same temperatures too cool because they are wearing much lighter clothing.
4. The range of comfortable relative humidity also depends upon the season. People generally like slightly higher humidities in the winter than the minimum of the winter comfort range. Conversely, in the summer they generally like slightly lower humidities than the maximum of the summer range.

Summary

In Chapter 6 you learned how to plot the several HVAC processes on the psychrometric chart. Whenever you are selecting, evaluating or controlling HVAC equipment, plotting the HVAC processes will always give you greater insight into system operation. In addition, you should remember that:

- Any change in the condition of air can be shown by drawing a straight line between the initial and final conditions.
- The eight HVAC processes lie along general directions on the psychrometric chart although not every process follows a single straight line between its initial and final conditions.
- In an adiabatic process, the air is cooled or heated without a loss or gain in heat energy during the process.
- The two main sensible cooling devices are coils filled with chilled water or refrigerant.
- Heating and humidifying is typically a two-step process: first the air is sensibly heated and then it is humidified.
- The two main commercial and industrial evaporative cooling devices are air washers and cooling towers.
- The **DB** temperature of air passing through an air washer declines while its moisture content rises. The **WB** of the air does not rise since the process is adiabatic.
- The chilled water or refrigerant absorbs the entering air's sensible heat and latent heat of condensation as the cooling coil dehumidifies the air.
- The apparatus dew point is the temperature of the coil necessary for condensation.

- The temperature of the air leaving a cooling/dehumidifying device is called the air-off-the-coil temperature and is higher than the apparatus dew point.
- The effect of mixing air streams is proportional and the condition of the mixture will lie along the mixture process line joining the two conditions.
- To determine the amount of heat that an air supply gives up to heat a space or that it absorbs to cool a space, use the formula, $q_s = W \cdot c \cdot \Delta T$.

7 HVAC System Types

So far you've covered the basics of heat transfer, human comfort, HVAC load evaluation and the psychrometrics of HVAC processes. For the remainder of this text you will focus on the equipment and systems that perform HVAC processes. This chapter begins with ASHRAE's methods of describing and classifying HVAC systems. Next, you will learn the details of the three principal heating, ventilating and air conditioning systems: all-air, all-water and air-water. After studying this chapter you will be able to:

1. Recognize the components of air conditioning systems.
2. Understand how ASHRAE describes HVAC systems.
3. Identify the components and functions of the basic central system, including its primary and secondary systems.
4. Distinguish between heating/cooling zones and rooms.
5. Understand how an all-air system operates and identify its major advantages and disadvantages.
6. Understand the differences between all-air single-path and dual-path systems.
7. Recognize the three variations of the single-path system.
8. Recognize the four variations of the dual-path system.

9. Describe a 100% outdoor air system and its major advantage.
10. Understand how an all-water system operates and list its advantages and disadvantages.
11. Identify the six different piping arrangements that distinguish all-water systems.
12. Identify and explain how the following terminal units work: radiators, convectors, baseboard, fin-tube, unit heaters, radiant panels, unit ventilators and induction units.
13. Understand how an air-water system operates and cite its several advantages and disadvantages.

Components of Air Conditioning Systems

If you examined many HVAC systems, you would find elements common to nearly all of them. These components include:

1. *A heating device* that transfers heat to air or water or creates steam. Heating devices include furnaces, boilers, electric resistance coils, heat pumps and solar heaters.
2. *A cooling device* that removes heat from air, water or refrigerant gas. Cooling devices include chillers, cooling towers and evaporative coolers.
3. *A distribution system*, made up of ducts or pipes, or both, that carries the air, water or steam to the conditioned space.
4. *Equipment* that moves the air, water or steam to the conditioned space.
5. *Heat transfer devices* that transfer the heat between the heated medium and the conditioned space. **Diffusion devices** mix the incoming supply air with the air already in the conditioned space. **Terminal units** promote the heat transfer between the heated or chilled water and the air.
6. *Operational equipment and features* that include valves, dampers, safety devices, automatic controls, sound and vibration attenuators and thermal insulation.
7. *Specialty devices* including humidification/dehumidification equipment, air filtration devices and water treatment systems.

Classifying HVAC Systems

After you had examined many HVAC systems, you would have a difficult time describing all the combinations of systems and equipment you had seen. This problem plagues even the experts: finding the right words to describe a system without listing every major element.

To help with this descriptive problem, ASHRAE has developed a vocabulary for classifying HVAC systems. The first step in the ASHRAE classification is to label the system by its *cooling* medium. Nearly every system uses air, water or a combination of air and water to cool the conditioned space. Therefore, the three major HVAC categories are:

- all-air systems,
- all-water systems and
- air-water systems.

For all-air systems, ASHRAE further describes the conditioning equipment, the distribution arrangement, the number of zones served by the system, and other operational features. For all-water systems, ASHRAE primarily describes piping and zoning capabilities. Air-water systems rely on the primary and secondary equipment of both air and water systems. Their descriptions also depend on those systems.

Details of ASHRAE classifications follow in each system description. In time you will become familiar with the entire ASHRAE system, but for now you will learn only the most common categories.

The Basic Central System

Site orientation, architectural design, building materials, and occupancy use all help make a building unique. To meet its HVAC needs, the designer can put together dozens of systems. Most buildings larger than a few thousand square feet are unique enough that no single packaged HVAC system would be flexible enough. For this reason, designers use a **central system** for most large buildings. First, the designer selects every component of the system. Then, the designer places the system *centrally* in a basement, service area or penthouse to reduce space and distribution requirements. Since the designer is selecting and placing every piece, the central system also becomes a **built-up system**.

For every central system, the designer also develops ideas and estimates for two other sub-systems. The **primary system** converts energy from electricity or fuel and is the source of hot or chilled water. The **secondary system**, or **distribution system**, delivers the heating or cooling throughout the building. Construction costs; fuel costs; and the cooling, heating and ventilating requirements usually have the most influence on primary system choices. Architectural constraints on the distribution space, system noise and the esthetics of terminal units often are the most important concerns in choosing the secondary system.

Describing a basic central air-conditioning system is a good beginning for understanding how HVAC systems function. For descriptive purposes, the **basic central system** is an all-air system. The central system can also be a built-up system–the designer is free to specify whatever equipment and manufacturer meets the design needs. See Figure 7-1.

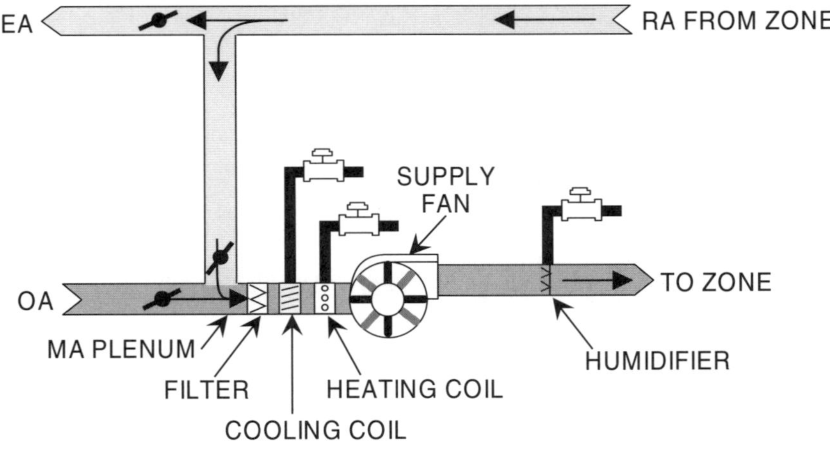

Fig. 7-1 Basic Central Air-Conditioning System

Look at the upper right corner of the figure. This is the return air section of the central system. Air carrying the heat gain flows back from the conditioned spaces in return air ducts, pulled along by the slight suction, or negative pressure, of the supply fan. Some of the **return air**, **RA**, can be exhausted outside, **exhaust air**, **EA,** as it flows into the central system equipment.

Next, is the outdoor air section of the system. Here, an amount of **outdoor air**, **OA,** is drawn in by the suction of the supply fan to replace the exhaust air. The **RA** and **OA** mix in the mixed air section or plenum of the central system. Baffles in the air chamber ensure a thorough air mix.

From the plenum, the **mixed air**, **MA**, carrying many airborne particles, is drawn through the filter section.

From the filter section, the **MA** is then **drawn through** the cooling or heating sections, containing the cooling or heating apparatus. Air leaving the cooling/heating equipment is drawn into and accelerated by the supply fan. The supply fan then pushes the air through the humidification equipment, if there is any.

The intake or negative side of the supply fan develops a negative pressure or suction. The outlet side of the supply fan develops a positive pressure. Pressures on the order of -2" WG in the plenum section and the +3" WG after the fan section are common.

Now look at Figure 7-2. Many smaller central systems do not need a return fan, but most large ones do. The return fan helps exhaust the conditioned space. It also provides a positive return air flow and exhausting of the **RA**.

Packaged and Unitary HVAC Systems

In contrast to the built-up nature of the central system, the designer could chose a **packaged or unitary system**. These self-contained HVAC systems are installed in a selected location and connected to the distribution system. Most units of this type are all-air systems. Typical of unitary units are window air condi-

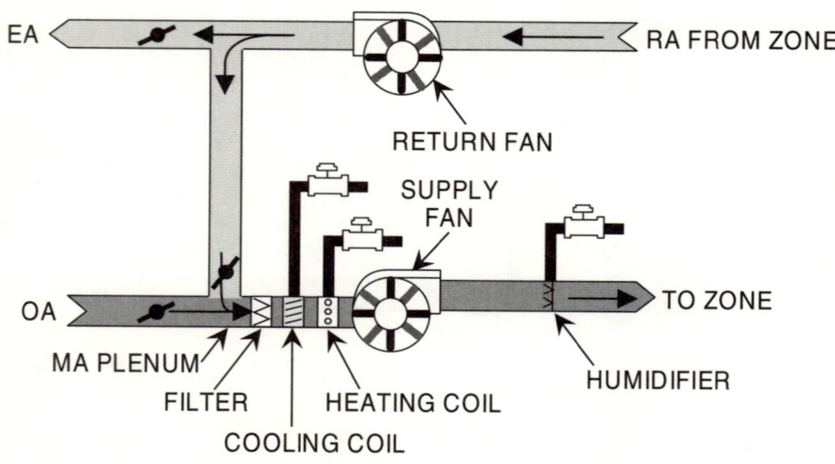

Fig. 7-2 Central System with Return Fan

tioners, through-the-wall air conditioners, rooftop systems, air conditioners commonly mounted on a concrete slab outside the building, air-to-air heat pumps and water source heat pumps. Packaged or unitary units are common, but since most have their own predesigned configurations, they will not be covered in detail.

HVAC Zones and Rooms

When analyzing any system, it is helpful to recognize whether it is conditioning zones or rooms, or both. A **zone** is an area that requires separate thermostatic control, while a **room** is a separately partitioned area that may or may not require separate thermostatic control. Large commercial and institutional buildings, for example, will have many zones, and each zone will likely have several rooms. Your house or apartment will have several rooms, but because it has only one thermostat it will have only one zone.

Because the cooling and heating needs of large buildings are often complex, they require equipment that can serve multiple zones. Satisfying these zoning needs is the force behind the creation of the wide range of HVAC system types.

The All-Air System

An **all-air system** distributes heated or chilled air through ducts to the building zones. Refer to Figure 7-1 for a diagram of the central equipment of a typical all-air system. In an all-air system:

- Each zone receives a cold air supply that contains the complete sensible and latent cooling capacity. No additional latent or sensible cooling equipment is available in the zone.
- However, some all-air systems also add heat with separate water, steam or electric heating devices in the air handling unit or zone.

All-air systems are extremely flexible and work well in buildings of many zones, such as offices, schools, hospitals, laboratories, stores and hotels. They are also ideal for closely controlling temperature and humidity in computer rooms, clean manufacturing rooms and hospital operating rooms.

All-air systems offers many advantages. Several of these are:

1. They are adaptable to multiple zoning needs and humidity control. They easily supply clean filtered air over a wide range of humidity and sensible and latent heats.
2. They are adaptable to seasonal changeover. During the cooling season they can readily use cool outdoor air, called **free cooling**.
3. They easily accept heat recovery systems.
4. They also provide ventilation requirements year round.

On the other hand, all-air systems do have some disadvantages:

1. They need much more distribution space than do water systems. They require additional space throughout the building for duct risers, ceiling ducts and fan rooms. This is due to the low specific heat and density of air, compared to water.
2. Because of air's low specific heat and density, these systems operate longer during unoccupied hours to maintain preset temperatures.

3. Their air distribution and performance may be difficult to balance.

Introduction to Single-Path and Dual-Path All-Air Systems

When describing all-air systems, ASHRAE evaluates several features. They include the air flow path, the distribution system, the number of zones served, constant or variable air volume, and reheating if used. These terms are summarized in Table 7-1.

All-air systems are either single-path or dual-path. Each type has specific advantages. In the **single-path, all-air system** the supply air flows in series directly from one conditioning device into the next. This conditioned air stream then flows through a **single duct system** that uses one duct to distribute the air to each zone. See Figure 7-3.

Like the heating/cooling system in your home, in its basic configuration the single-path system delivers conditioned air with a very limited range of sensible and latent heat. Regardless of zone needs, it delivers only one heating supply air condition and only one cooling supply air condition. Because of this limitation, designers created the dual-path, all-air system.

The **dual-path, all-air system**, splits the mixed supply air into two streams. One stream is chilled and the other is heated. After conditioning, the dual air streams may flow throughout the facility in parallel, **dual duct** distribution airways and into mixing boxes at each zone. See Figure 7-4.

A **mixing box** is a device for blending cold and hot air streams and attenuating the sound of the mixing process. See Figure 7-5. The zone controller regulates the amount of chilled and heated air entering the box. After mixing, the air then flows into the zone.

ASHRAE Term	Described System Component
Path	The route the air follows through the conditioning equipment. Single path is series flow; dual path is parallel flow.
Duct	A single duct distribution system carries air of one condition. A dual duct system carries heated air in one duct and cooled air in the other. A mixing box blends dual duct air at the zone. The number and arrangement of *return* air ducts are not part of the ASHRAE description.
Zone	Identifies the zones that are served with the conditioned air as either single or multiple.
Volume	The air flow throughout the system is of a constant volume or a variable volume.
Reheat	A system with reheating capability has additional heaters in the zones to warm the centrally chilled air.

Table 7-1 ASHRAE All-Air System Classifications

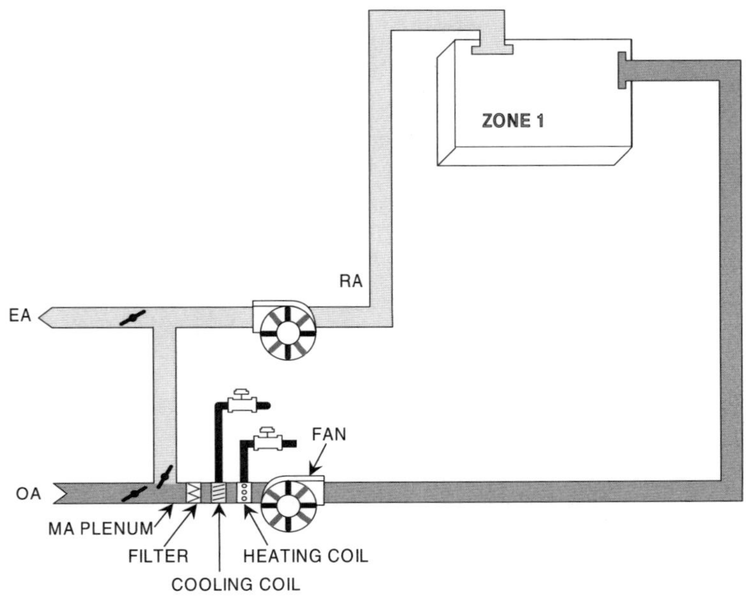

Fig. 7-3 Single-Path, Single Duct, All-Air System

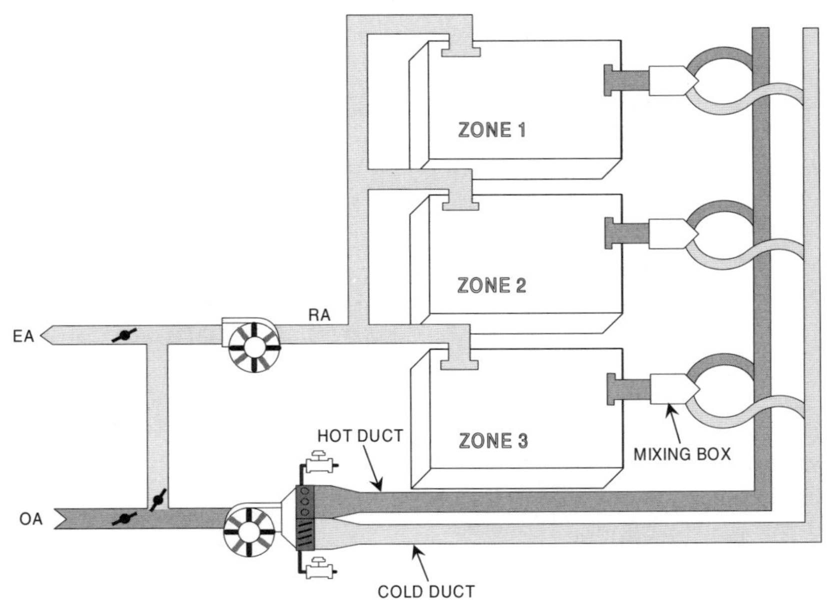

Fig. 7-4 Dual-Path, Dual Duct, All-Air System

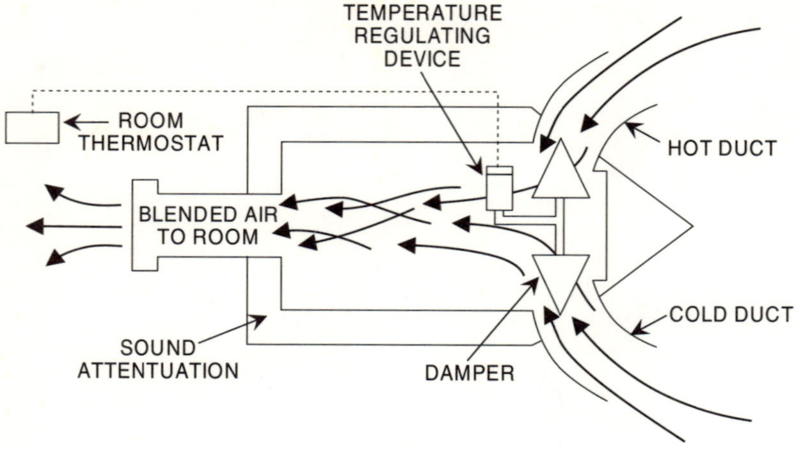

Fig. 7-5 Mixing Box, Constant Volume

As you might guess, the advantage of the dual-duct system is that it serves many different zones. In addition to having a multiple zone capacity, most of the equipment is centrally located.

Variations of the Single-Path, All-Air System

In a basic single-path, single duct, all-air system, the air flows at a constant volume and the system serves one or more zones. To expand beyond this singular zone limitation, designers have two choices:

1. They could change the temperature at the *zone by adding extra equipment. This is known as* **CVVT** or constant volume, variable temperature. This is also the basis of the zone reheat approach. In a **zone reheat** system, zone heaters temper, or reheat, air distributed from the centrally located system.
2. They could change the air flow to a variable volume. In this method, the volume of air is controlled to maintain temperature. This is known as **VVCT** or variable volume, constant temperature. This is the basis of the variable volume approach.

To summarize, the single path system has three design options:

- Single-duct, single zone, constant volume
- Single-duct, multiple zone, constant volume, zone reheat
- Single-duct, multiple zone, variable volume

Single-duct, Single Zone, Constant Volume

Figure 7-3 illustrates a typical system. A **single-duct, single zone, constant volume** system delivers a constant volume of air of the same temperature to the zones which may have multiple cubicles or rooms. When the system runs at less than full load during the cooling season, this single constant air temperature can create the problem of **sub-cooling** (overcooling) some rooms. The sub-cooling problem is solved with the next system, zone reheat.

Single-duct, Multiple Zone, Constant Volume, Zone Reheat

Imagine installing a heater in each duct as it enters a zone. Now, you can cool the supply air centrally (or use cool outside air) and then heat it to the right temperature at the zone. This heater is a **reheat coil** and this system is known as the **single-duct, multiple zone, constant volume, zone reheat** system. See Figure 7-6.

With the addition of zone reheat, the single duct system takes on a multiple-zoning capability, at a price. The reheat system wastes energy, especially during warm weather. It takes energy to cool the air and more energy to reheat it to the comfort temperature.

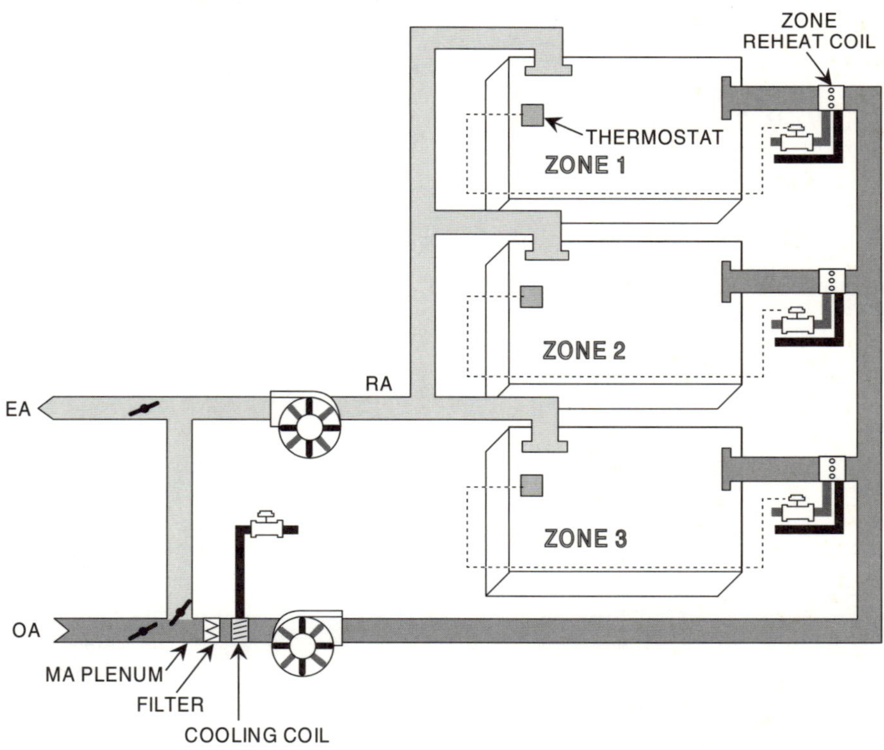

Fig. 7-6
Single-Duct, Multiple Zone, Constant Volume, Zone Reheat System

Single-duct, Multiple Zone, Variable Volume

Another way of adding multiple-zone capability to a single-duct system is to vary the *volume* of air. You'll remember from Figure 6-14 that mixing air streams is an effective way of controlling sensible heat. The **single-duct, multiple zone, variable volume** system varies the volume of the air entering and leaving the zone, effectively using the entire space as a mixing box. Responding to the heating or cooling load, the thermostat controls a damper that in turn regulates the volume of supply air entering the zone. See Figure 7-7.

Varying the air volume is known as **variable air volume** or **VAV**. The concept also saves considerable energy compared to reheat systems. With VAV, chillers and fans usually operate at less than maximum capacity and energy is not wasted to reheat conditioned air.

Variations of the Dual-Path, All-Air System

Although dual-path systems are multiple-zone by design, several enhancements add to their performance. The two major variations in design are single or dual distribution ducts and the use of constant or variable air flows. These variations reduce construction, operation and maintenance costs, reduce energy use, and provide better zone control. The four dual-path systems are as follows:

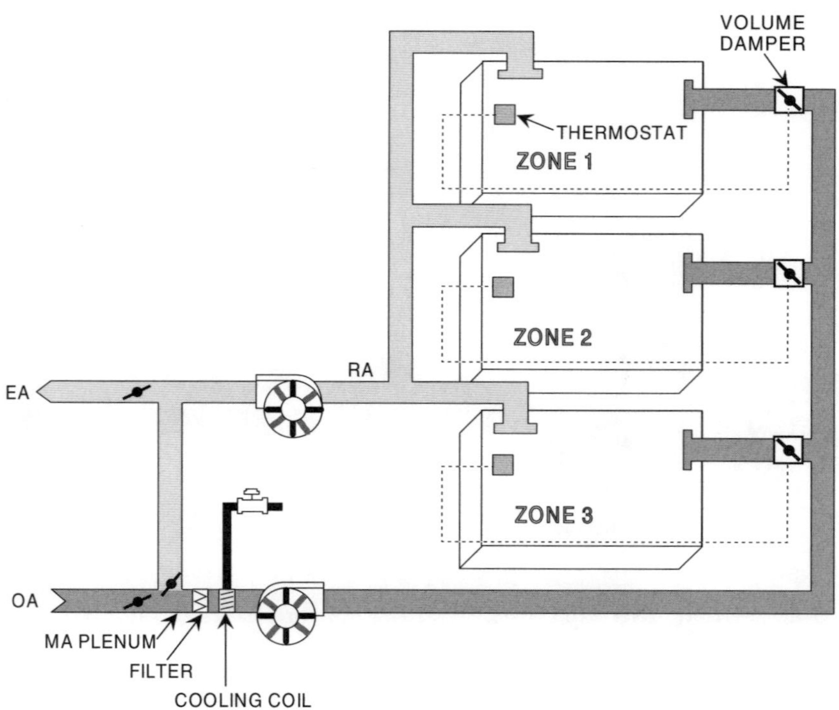

Fig. 7-7 Single-Duct, Multiple Zone, Variable Volume System

- Dual path, multi-zone, constant volume
- Dual path, multi-zone, variable volume
- Dual path, dual duct, multiple zone, constant volume
- Dual path, dual duct, multiple zone, variable air volume

Note that both the terms *multi- and multiple* are used in describing these systems. **Multi-zone** systems mix the several air streams centrally and then distribute the mixtures to each zone. **Multiple zone** systems, in contrast, distribute both hot and cold air streams throughout the facility and then mix the air at the zone.

Dual path, Multi-zone, Constant Volume

The **dual path, multi-zone, constant volume** system mixes the hot and cold air streams centrally. It then distributes each mixture through single ducts to each zone. Obviously, this system saves the cost of a second duct system. Study a typical setup in Figure 7-8.

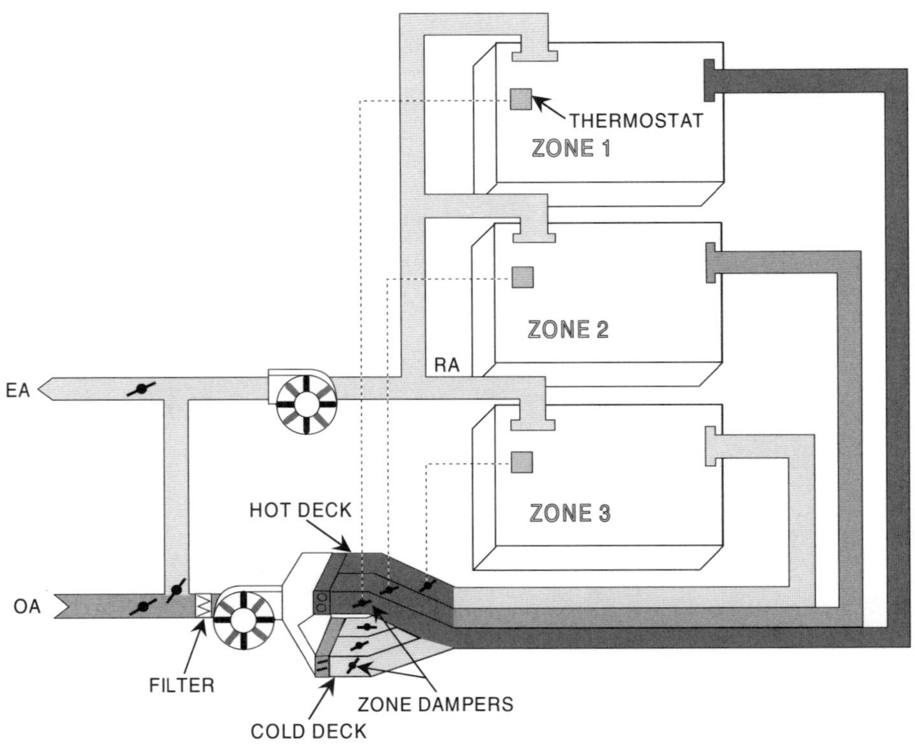

Fig. 7-8 Dual Path, Multi-Zone, Constant Volume System

Dual Path, Multi-Zone, Variable Volume

This system adds variable volume air distribution to the former system. The **dual path, multi-zone, variable volume** system mixes the heated and cooled air streams in the central system. The mixtures are distributed by the central variable volume fan to the various zones. At each zone, variable volume dampers complete the system as shown in Figure 7-9. This system is rarely used.

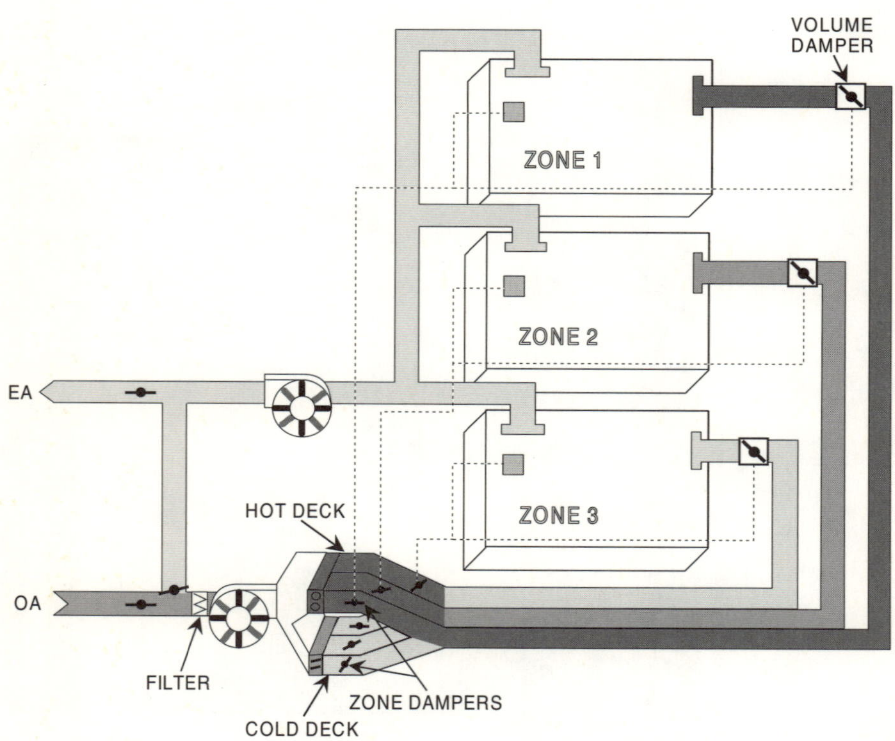

Fig. 7-9 Dual Path, Multi-Zone, Variable Volume System

Dual Path, Dual-Duct, Multiple Zone, Constant Volume

The multiple zone systems distribute both hot and cold air streams and mix them at each zone to meet the zone requirements. The **dual path, dual-duct, multiple zone, constant volume** system distributes equal volumes of hot and cold air to each zone. The zone thermostat controls the mixing of the air streams in a mixing box. After mixing, the air then flows into the zone. See Figure 7-10.

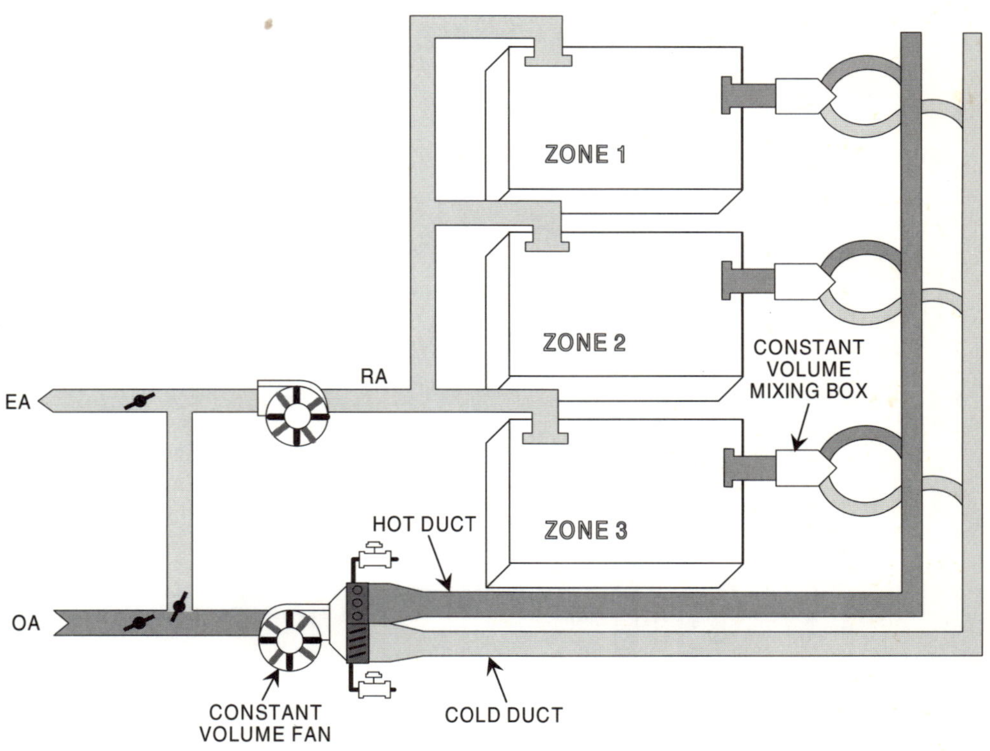

Fig. 7-10 Dual Path, Dual-Duct, Multiple Zone, Constant Volume System

Dual Path, Dual-Duct, Multiple Zone, Variable Volume

Adding variable air volume equipment expands the basic dual duct design into a **dual-duct, multiple zone, variable volume** system. See Figure 7-11. Now, the dual-duct design can meet sophisticated demands, while also realizing the savings inherent in variable air volume equipment.

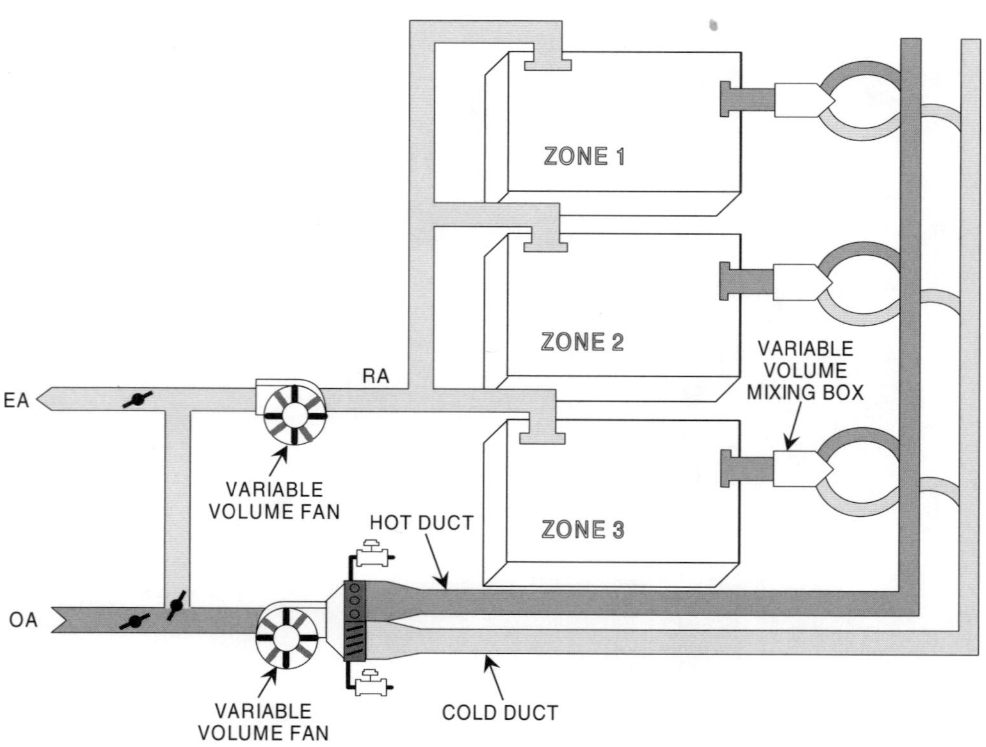

Fig. 7-11 Dual Path, Dual-Duct, Multiple Zone, Variable Volume System

The 100% Outdoor, All-Air System

For some very sensitive research, hospital and manufacturing operations, the cleanliness of the HVAC air is very important. It must be as free of air borne contamination as possible. All-air systems for these facilities exhaust 100% of their return air and draw in 100% of their supply air. These systems are called **100% outdoor air**, all-air systems and represent yet another variation of all-air systems. See Figure 7-12.

Any of the all-air systems can be operated with 100% outdoor air. However, because of the climatic extremes of outdoor air, these systems require special filtration and pre-heating equipment to work well. Installing and operating this extra equipment makes 100% outdoor systems very costly.

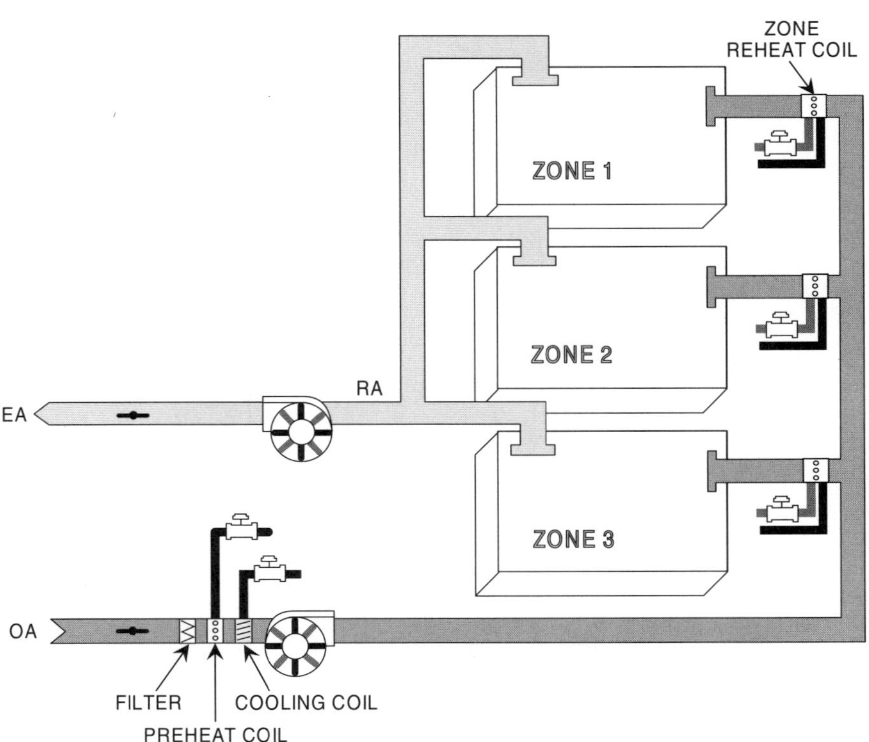

Fig. 7-12 100% Outdoor, All-Air System

The All-Water System

An **all-water** system pipes steam, hot water or chilled water from a central system to the building zones. It is also known as a **hydronic system**. Terminal units in the zones provide the heat transfer from the steam or water to the surrounding air.

Water has a much higher specific heat and density than air. Thus, an all-water system uses much less water than air by volume to produce a comparable amount of heat transfer. The principal advantage of an all-water system is this:

1. Its pipes have a much smaller cross-section than the ducts of a comparable all-air system. Because of smaller size, an all-water distribution system is cheaper to build and it saves valuable building space.

On the other hand, all-water systems also have some distinct disadvantages:

1. These systems are not well equipped to filter air or ventilate zones. Terminal units often have openings for outside air. However, wind pressure affects the free flow of air through these openings. Furthermore, the terminal equipment must be protected from the rain, snow and freezing temperatures that may enter the openings. Finally, the filters in the terminal units can clean only a modest amount of air.

2. All-water systems lack humidity control. Adding humidification equipment to each terminal unit involves considerable cost.

3. When the seasons change, they must be converted from hot to cold water, or vice versa. It is difficult to provide good temperature control on the cooler-than-normal spring days and the warmer-than-normal fall days.

4. Finally, terminal units, especially those with fans, require a high degree of maintenance. As the units age, their fans usually become noisy. Maintaining several hundred of these units can be a chore for even the best maintenance staff.

Variations of the All-Water System

As in all-air systems, there are many arrangements for delivering the conditioned water in all-water systems. Each piping design meets unique zone demands in a versatile and cost-effective way. Several of the basic piping arrangements are:

- Series loop
- One-pipe main
- Two-pipe direct return
- Two-pipe reverse return
- Three-pipe system
- Four-pipe system

Series Loop

In the **series loop** design, all the water flows through each terminal unit. Water cannot flow around any unit. The main advantage of the series loop design is that it is simple and inexpensive to construct. See Figure 7-13.

Since each terminal unit is an integral part of the flow network, this design has several disadvantages:

1. It is impossible to control the temperature of individual terminal units.
2. Since hot water gives up its heat as it flows away from its heat source, distant terminal units do not heat as well as closer units. Similarly, distant units do not cool as well.
3. The entire system must be shut down to service or repair *one* terminal unit.

Because of these piping/operational limitations, only very small commercial facilities or homes use the series loop system. However, splitting the supply loop into several smaller loops offset some of the disadvantages. Then each loop can be operated independently.

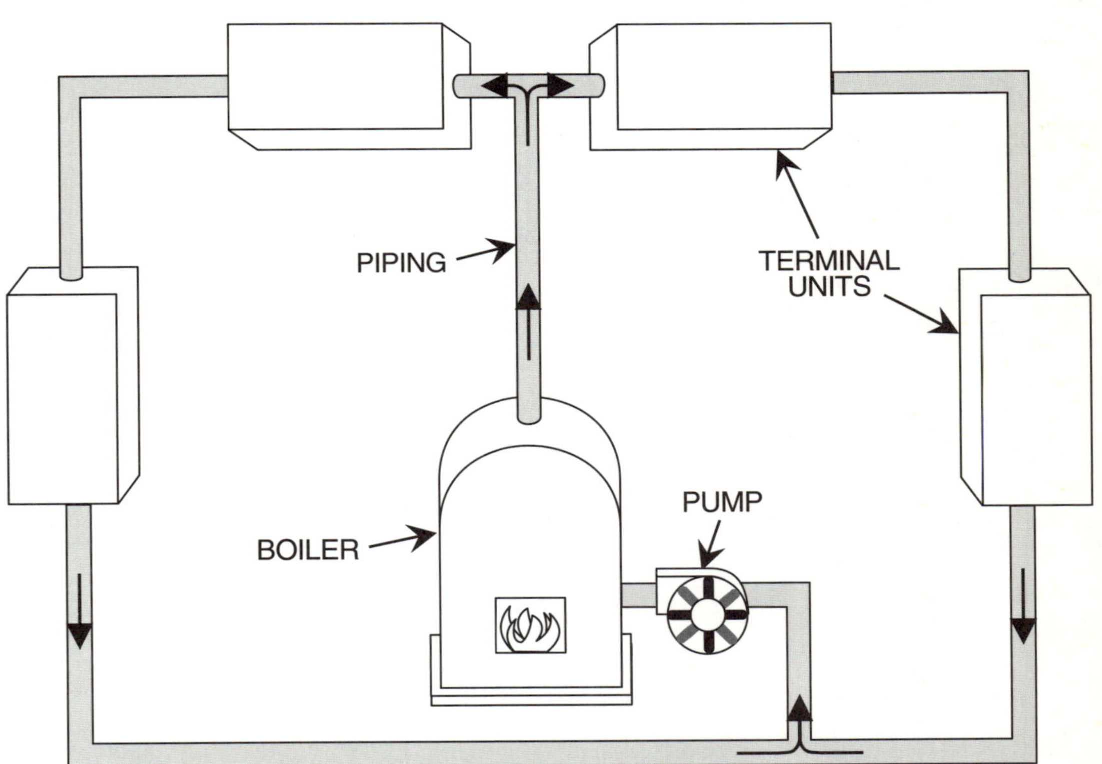

Fig. 7-13 Series Loop, All-Water System

One-Pipe

This arrangement solves the interdependence problem of the series loop system. In the **one-pipe** design, supply and branch pipes and valves separate each terminal unit from the main supply line. See Figure 7-14. This design also permits separately controlling and servicing each terminal unit.

As with the series loop system, the terminal units most distant from the conditioner often heat or cool less effectively than the closer units. Because the supply line is also the return line. For example, during the winter, warm water in a terminal unit cools as it loses heat. The cool water then flows out of the unit and into the supply line, slightly cooling the supply line water. By the time the supply line reaches the last terminal unit, it also contains the outflow from every earlier terminal unit. In Figure 7-14, water leaving the conditioner at 180°F may cool to 160°F before reaching the last terminal unit. This gradual temperature drop drastically affects the efficiency of the terminal units near and at the end of the loop.

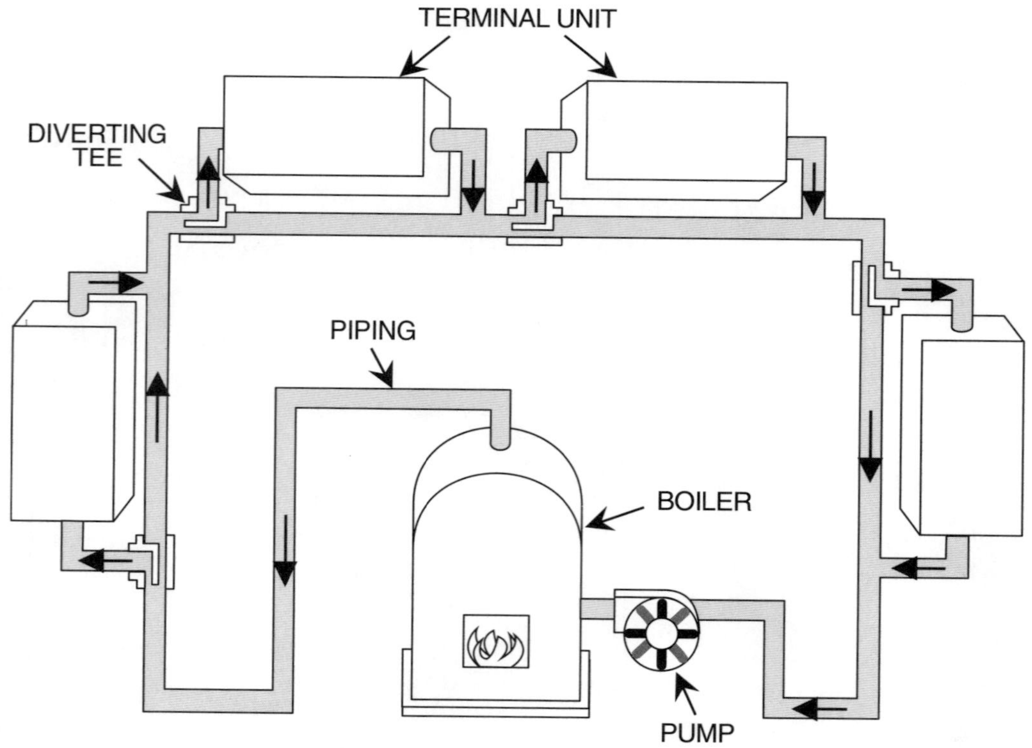

Fig. 7-14 One-Pipe, All-Water System

Two-Pipe Direct Return

The **two-pipe direct return** design uses a supply line and a return line to improve terminal unit efficiency. Now, each terminal unit receives water directly from the supply line. See Figure 7-15.

While this piping arrangement solves the problems of the one pipe system, it introduces another. Water always flows along the path of least resistance. In this design, most of the water flows into the first terminal unit and then back along the return line; it's the shortest path. To prevent this short-circuiting, balancing valves placed at every intersection along the supply path maintain proper flow into the terminal units and along the supply line.

Balancing one- and two-pipe direct return systems is difficult. Solving this difficulty led to the development of the next piping arrangement.

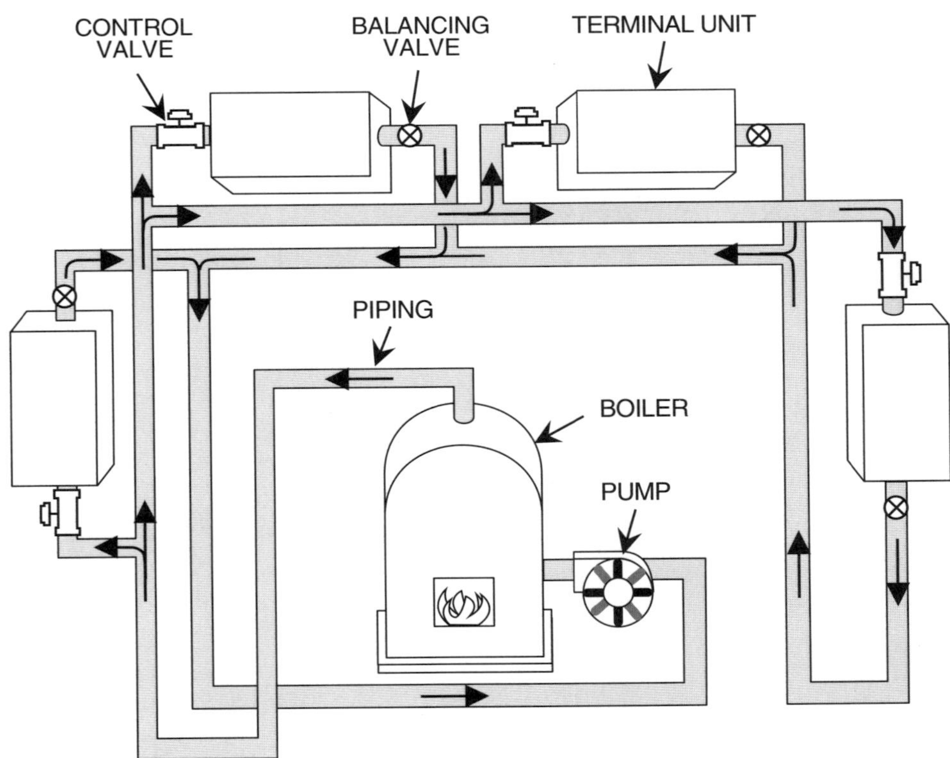

Fig. 7-15 Two-Pipe Direct Return, All-Water System

Two-Pipe Reverse Return

The key to solving the short-circuiting in the two-pipe direct return arrangement is to equalize the supply and return flow resistances. Flow resistance is directly proportional to the pipe length. So, the **two-pipe reverse return** system keeps the supply and return lines about the same length. Now, each terminal unit is an equal piping distance away from the conditioner. The terminal unit with the shortest supply line has the longest return line and the unit with the longest supply line has the shortest return line. See Figure 7-16.

This approach adds extra return piping, but the ease of balancing and operating the system makes it preferable to the two-pipe direct return system.

The one- and two-pipe systems can carry only hot or chilled water at one time. Neither system can serve diverse zoning requirements. For example, in many buildings the interior rooms require winter cooling even while the exterior rooms require heating. Since only one medium is available, the needs of one of the zones cannot be met.

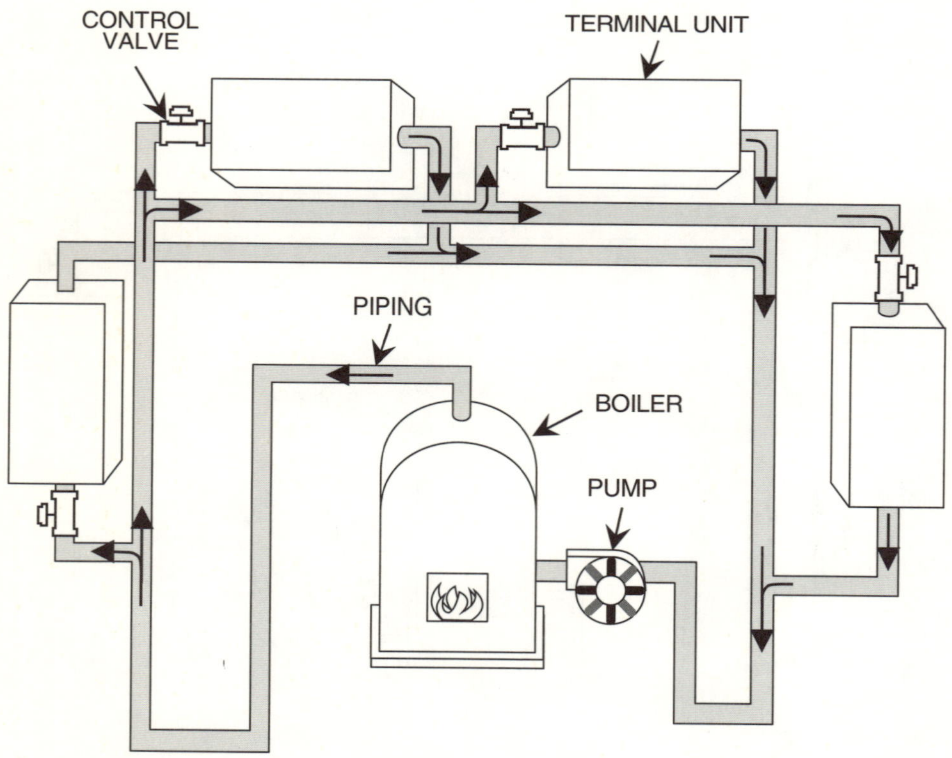

Fig. 7-16 Two-Pipe Reverse Return, All-Water System

Three-Pipe System

The solution to the problem of different zones requiring heating or cooling simultaneously is the **three-pipe system**: one pipe supplies chilled water, a second pipe supplies heated water and the third pipe carries the return flow. A three-way valve on the supply line to each terminal unit determines whether the unit receives hot or cold water. The system uses either direct or reverse returns. See Figure 7-17.

Unfortunately, the three pipe system wastes energy and causes primary equipment problems because it mixes the returning hot and cold water in the same return line. For this reason, designers seldom use three-pipe systems.

Four-Pipe System

The **four-pipe system** avoids this energy waste by using a separate return line for both hot and chilled water. Four-pipe systems are designed with either direct or reverse returns. Obviously, because of its four, parallel pipe arrangements, the four-pipe design is also the most expensive all-water system. See Figure 7-18.

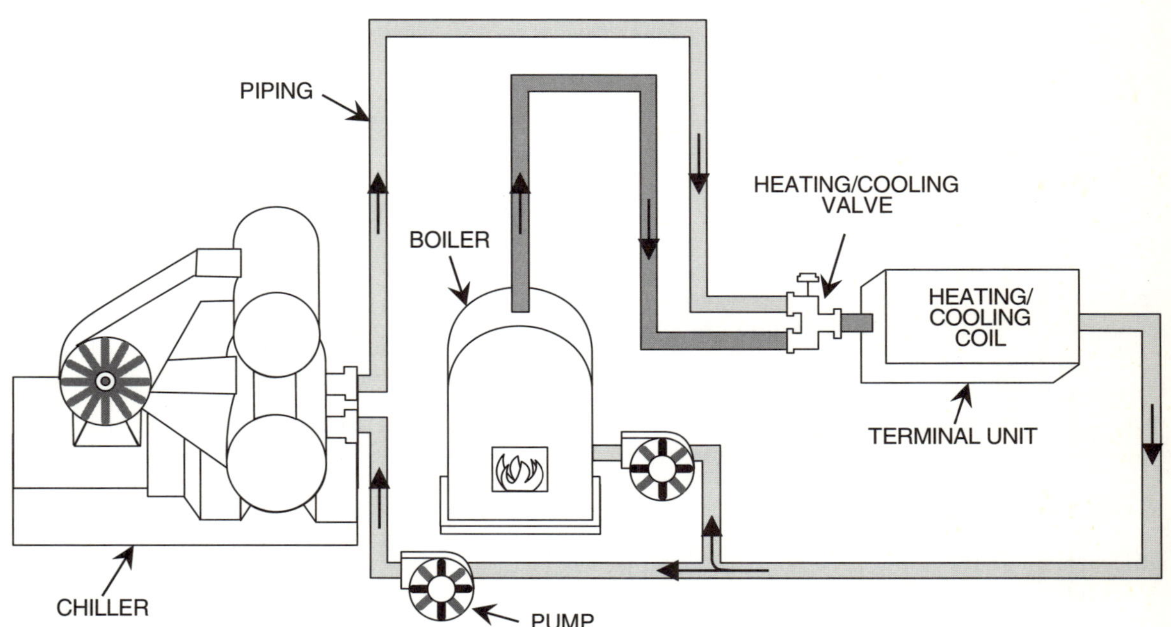

Fig. 7-17 Three-Pipe, All-Water System

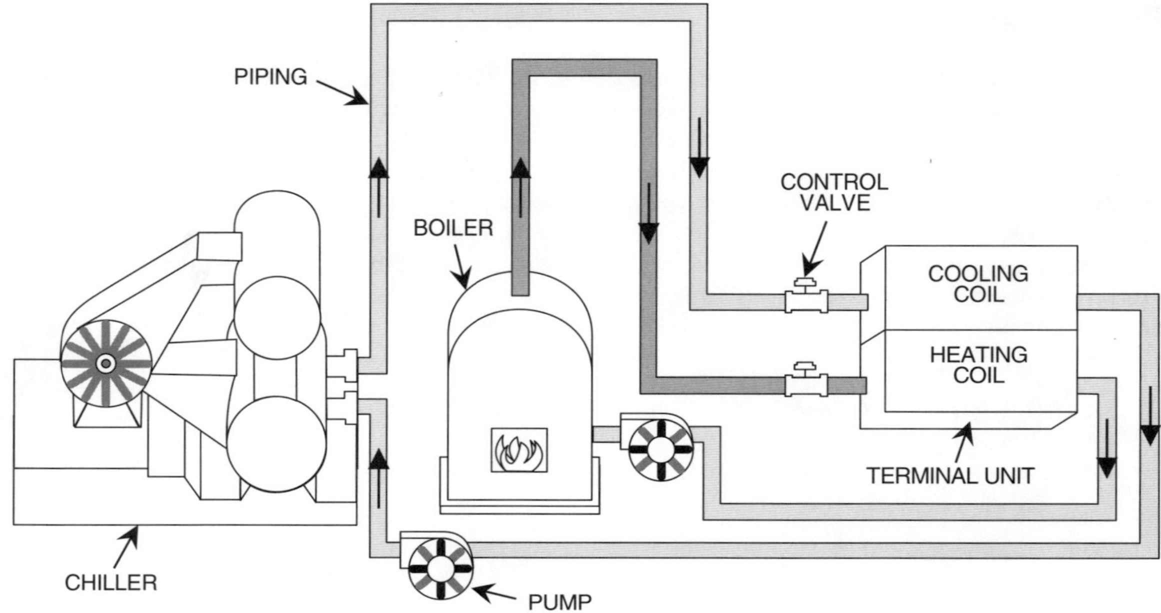

Fig. 7-18 Four-Pipe, All-Water System

Terminal Units

As described earlier, terminal units are the exchangers that transfer heat between the conditioned water and the room air. Some units can only heat and others can heat or cool. Terminal units come in many shapes, sizes and functions. Following is a list of the common terminal units, arranged by their use:

Heating Only:

Radiators, Convectors, Baseboard, Fin-tube, Unit heaters, Radiant panels

Heating and Cooling:

Unit ventilators, Induction units, Fan coil units.

Although these terminal units will be presented in the context of all-water systems, remember two points:

1. In other applications, unit heaters and unit ventilators may also use natural gas or electricity as their heating medium.
2. Induction units are only used in air-water systems since they require a primary air source.

Radiators

You are probably familiar with the large, ornate cast iron radiators used in the early 1900s. Although called radiators, these units actually heat more by convection than by radiation. Air flowing around a **radiator** warms by convection and rises in the room, creating a natural circulation. While modern radiators are slimmer, shorter and more efficient, few designers choose them today because they are still bulky and expensive.

Convectors

Except for radiators and radiant panels, the term convector more properly describes all the terminal units in this section. However, the HVAC industry also uses the term convector to describe a specific convection terminal unit. A **convector** is a finned tube or cast iron heat exchanger enclosed in a sheet metal cabinet. Conditioned water or steam flows into the exchanger. Cool, dense room air is drawn into the bottom of the convector and flows upward past the exchanger. Warmed lighter air flows out the top, creating a natural air circulation in the room. Convectors come in free-standing, flush, recessed and wall-hung models. See Figure 7-19 for a typical free-standing model and Figure 7-20 for a recessed model.

Baseboard: **Baseboard convectors** are longer, slimmer versions of the basic convector. Their low profile makes them popular for home use. Mounted slightly above the floor, air enters the bottom of the unit and flows upward. Although the baseboard unit may be quite long for appearance's sake, the heating element can be much shorter. The finned tube heat exchanger has a tube diameter of 1/2 to 3/4 in. See Figure 7-21.

Fin-tube: The **fin-tube convector** is the commercial-industrial version of the baseboard heater. The fins and housing of the exchanger

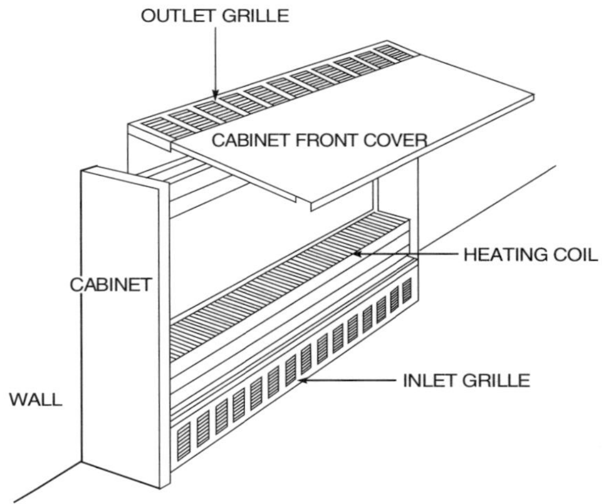

Fig. 7-19 Free-Standing Convector

AIR CONDITIONING PRINCIPLES AND SYSTEMS, 2ND ED. by Pita, Edward G., ©1989 Adapted by permission of Prentice-Hall, Inc., Upper Saddle River, NJ

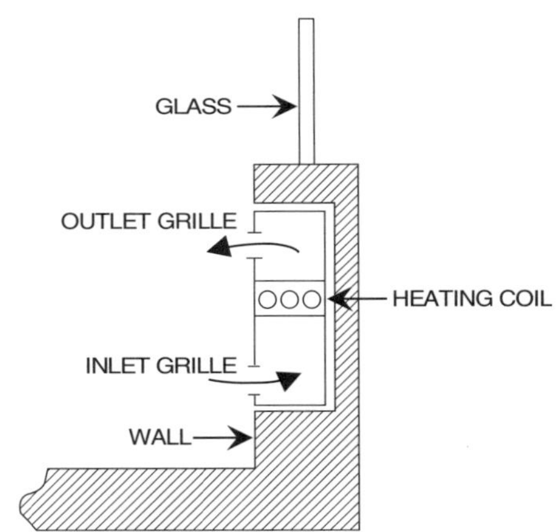

Fig. 7-20 Recessed Convector

AIR CONDITIONING PRINCIPLES AND SYSTEMS, 2ND ED. by Pita, Edward G., ©1989 Adapted by permission of Prentice-Hall, Inc., Upper Saddle River, NJ

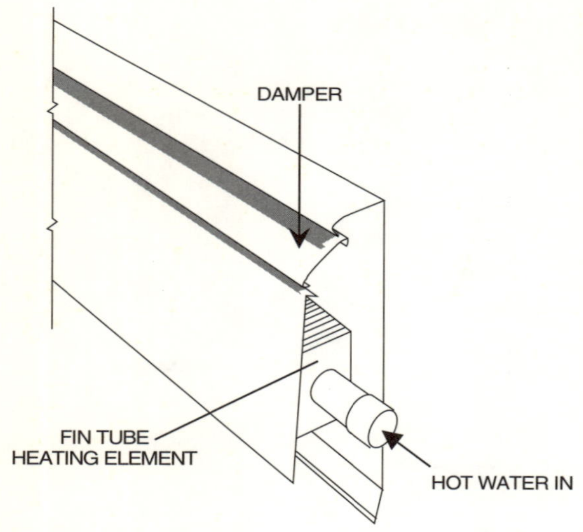

Fig. 7-21 Baseboard Convector

AIR CONDITIONING PRINCIPLES AND SYSTEMS, 2ND ED. by Pita, Edward G., ©1989 Adapted by permission of Prentice-Hall, Inc., Upper Saddle River, NJ

are of heavier gauge metal and its exchanger has a tube diameter of 3/4 to 2 in. Several exchangers may be stacked within the convector for greater output. See Figure 7-22.

Radiators and convectors (including baseboard and fin-tube designs) exchange heat primarily by convection. They also rely on convection currents to move air through the unit. Not surprisingly, radiators and convectors are also slow heat exchangers. However, with the addition of more coils and a forced air fan, these low output convectors can become high output unit heaters.

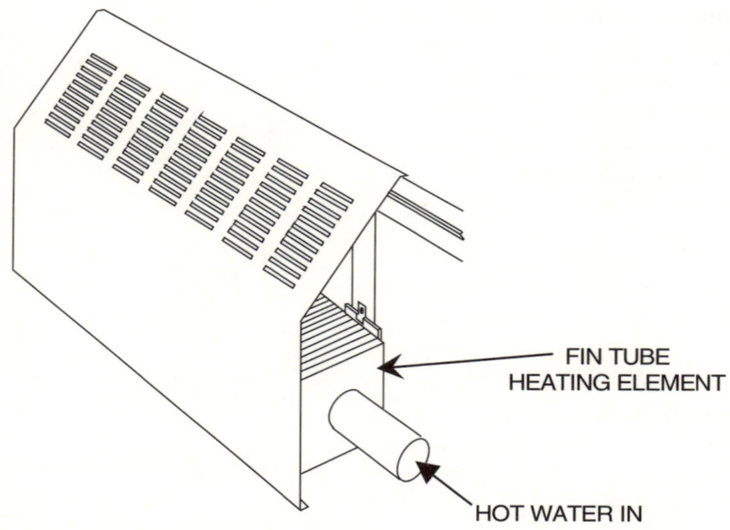

Fig. 7-22 Fin-Tube Convector

AIR CONDITIONING PRINCIPLES AND SYSTEMS, 2ND ED. by Pita, Edward G., ©1989 Adapted by permission of Prentice-Hall, Inc., Upper Saddle River, NJ

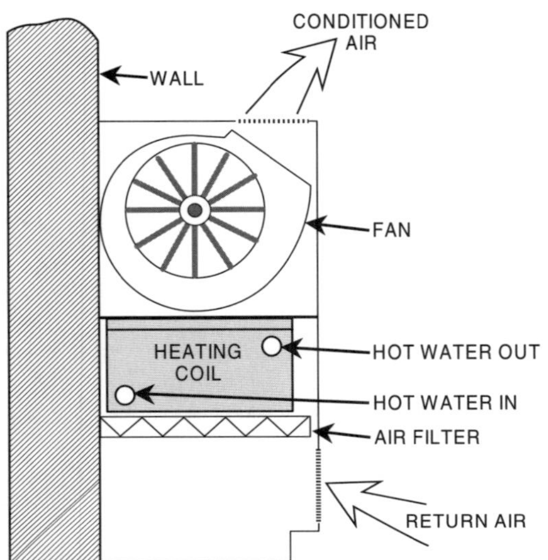

Fig. 7-23
Floor Mounted, Cabinet Unit Heater

Unit heaters

Unit heaters are terminal units with tightly spaced fin tube coils and internal fans. There are two types: cabinet unit heaters and propeller unit heaters. Cabinet unit heaters have internal centrifugal fans and have cabinets that look much like convector cabinets. See Figure 7-23 for a typical floor mounted cabinet unit heater. Notice that the addition of the fan also adds the capability for air filtration. Notice, too, that when mounted on walls or ceilings, the cabinet's adjustable louvers can direct the heat downward. Commercial users like cabinet unit heaters because they are quiet and attractively designed.

Propeller unit heaters have internal propeller shaped fans and have either a horizontal or vertical discharge design. See Figures 7-24 and 7-25. Because they are bulky and noisy, they are suitable mainly for industrial uses.

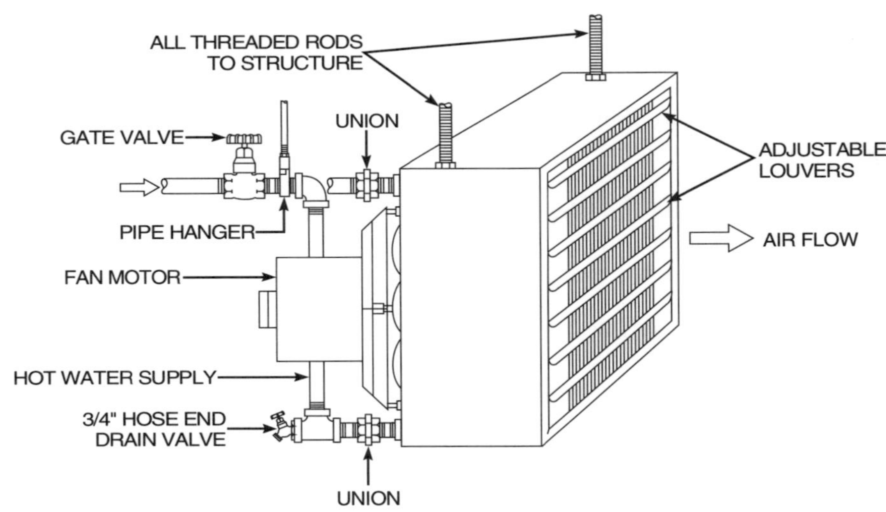

Fig. 7-24 Horizontal Propeller Unit Heater

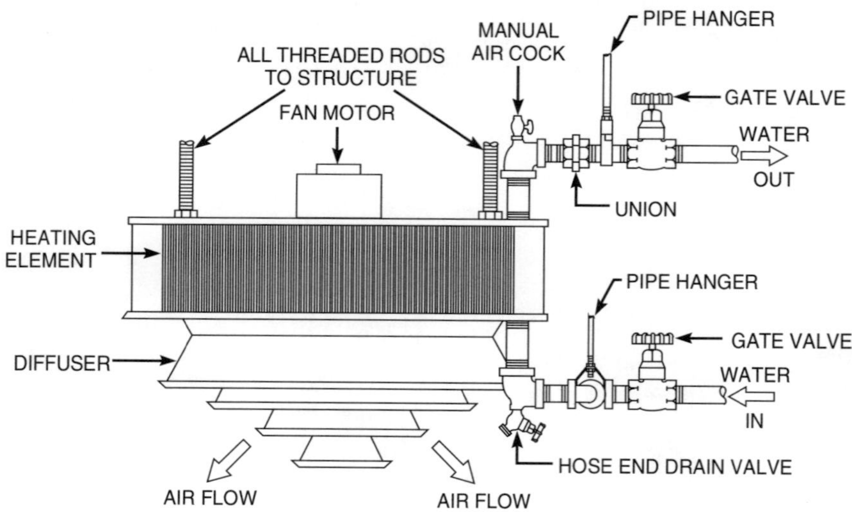

Fig. 7-25 Vertical Propeller Unit Heater

Radiant panels

Finally, these are the true radiators. **Radiant panels** are grids of tubing installed in floors, walls and ceilings. See Figure 7-26. Since air can't circulate through the panels, they exchange heat only by radiation. The systems promote a very uniform distribution of heat, but with a slow response. They are also very expensive because each installation is custom built to the specific floor, wall or ceiling. Radiant panels are not commonly used today.

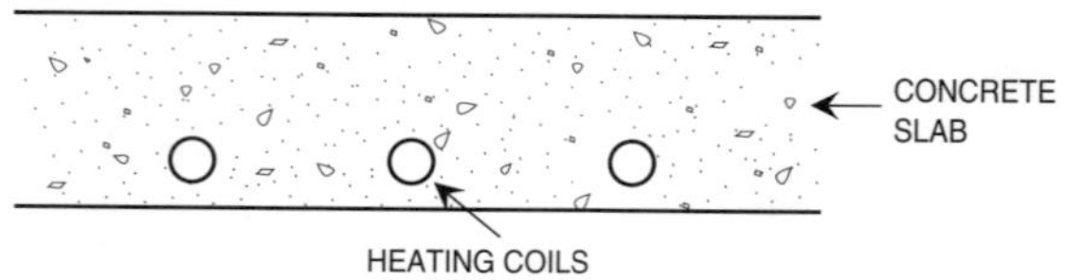

Fig. 7-26
Radiant Tubing in a Concrete Slab

The terminal units described so far are only capable of heating a room or zone. The next two terminal units, unit ventilators and induction units, can also cool. Remember that induction units are specific to air-water systems because the units need a source of primary air.

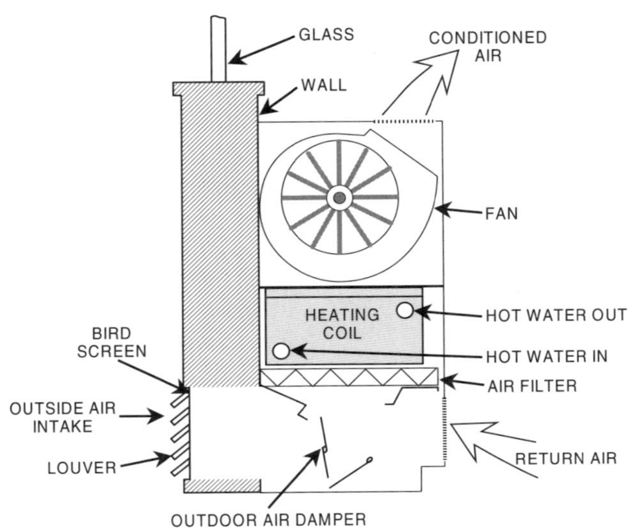

Fig. 7-27 Unit Ventilator

Unit ventilators

See Figure 7-27. Similar to the unit heater, a **unit ventilator** has a serpentine-shaped coil, a fan and an air filter. The difference is that the coil may carry hot or cold water. These units also offer a solution to the lack of fresh air ventilation inherent in the other terminal unit designs. Unit ventilators have an opening to the outside to bring in fresh air. The outside air can be heated, cooled or used directly as free cooling. This terminal unit design is very popular for school use.

Induction Units

These terminal units are used in air-water systems. They receive both primary air and secondary water from the central air-water systems. **Induction units** have a fin-tube heat exchanger capable of carrying hot or cold water, an air filter, and a high velocity air supply. See Figure 7-28.

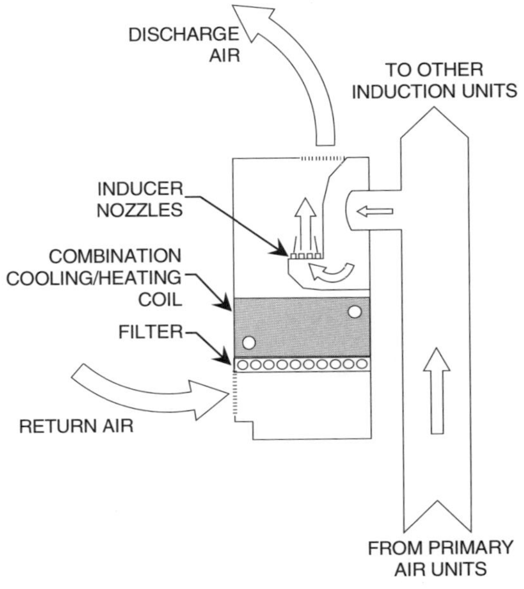

Fig. 7-28 Induction Unit

As the primary air flows upward through the small inducer nozzles, the **inducers** increase the air velocity. As this air rushes past the coil it creates a slight vacuum on the backside of the coil, thus inducing or drawing in secondary air (room air). The primary and secondary air mix before flowing out the top of the unit.

Because an induction unit requires a high velocity primary air supply, the concept does not work well with all-air systems. Designers and building owners frequently specify these units because they are quiet and require little maintenance, since they lack fans and motors.

The Air-Water System

Designers developed the air-water system, as so often happens in HVAC design, to fill a need. It takes advantage of the ventilation capability of the all-air system and the greater heat transfer efficiency of the all-water system. An **air-water system** can deliver chilled and heated air *and* chilled and heated water from central systems to each zone. Terminal units within each zone then cool or heat the space.

Heating and cooling the perimeter zones of a building are often the designer's bigger challenge. Wind-induced infiltration and exfiltra-

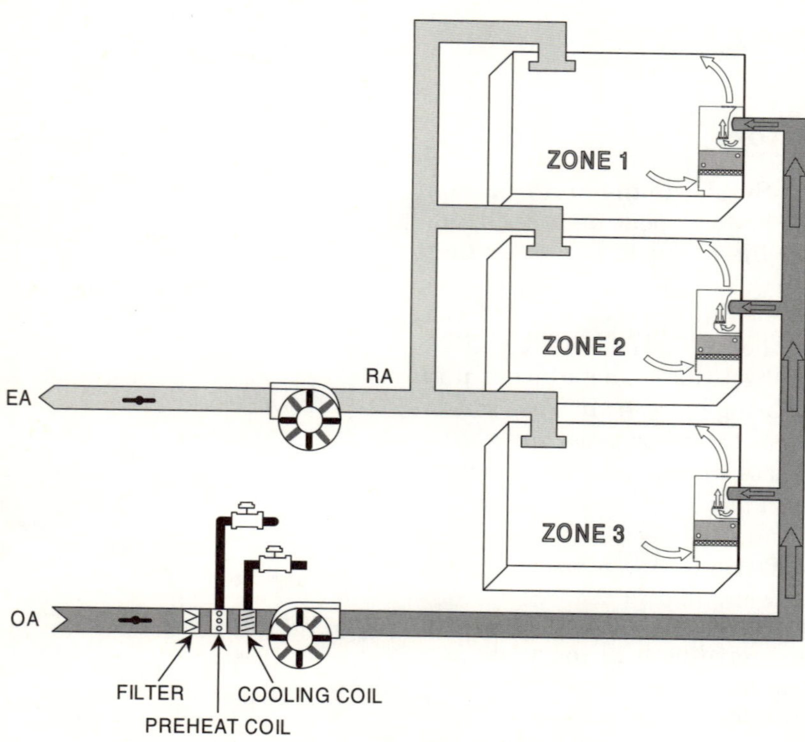

Fig. 7-29 Air-Water System

tion, conduction heat loss and solar gain can all occur simultaneously, causing wide swings in the heating/cooling load. For example, even when the outdoor temperature is 30°F, zones of southern exposure may require cooling. At the same time, zones of northern exposure may be experiencing conductive heat loss and wind-induced infiltration or exfiltration. Fortunately, the air-water system offers the designer several options.

In the air-water system, the greater part of the facility's sensible heating/cooling load is borne by the water system. The air system helps with the lesser heating/cooling load, provides positive ventilation and humidifies and dehumidifies the zones. The air-water system can simultaneously provide cooling and heating to different building zones. Figure 7-29 displays a typical air-water system using a roof-mounted unitary air unit.

There are several features unique to an air-water system you should note.

1. Secondary water carries most of the energy in this system.
2. The chilled or hot water circulating through the building piping and in the terminal units is **secondary water** and is part of the secondary system. The primary and secondary water systems have their own pumps and distribution piping.
3. The air side of the system has central air-conditioning equipment, a distribution system and room diffusers. This air is **primary air**, to distinguish it from the room air circulating over the terminal unit coils. The air system is usually a constant volume, high velocity *ventilation* system. It carries about 25% of the volume of an all-air system, just enough to ventilate the space.
4. The terminal units in the air-water system are called induction units. They are similar to fan-coil units but have no fan! They receive air from a primary air duct. They ventilate, exchange heat and create air circulation throughout the zones.

Designers usually restrict their use of air-water systems to exterior zones for two reasons. First, the exterior zones have the most wide-ranging demands. By comparison, interior zones usually have more consistent demands. Second, air-water systems are too expensive to build and operate to be used throughout a building.

The advantages of air-water systems are many:

1. Because of water's greater specific heat and density, water distribution pipes are much smaller than comparable all-air system duct work. This saves valuable building space.
2. Typically, the air system operates at high velocity. High velocities further reduce air duct cross-sections and save building space.
3. The pumping horsepower of an all-water system is usually much less than the fan horsepower of a comparable all-air system. Lower horsepower means energy savings.
4. Adding an air system provides the needed ventilation, humidification and dehumidification control that an all-water system lacks.
5. Adding an air system provides the conditioning alternatives needed during seasonal changes. In fact, with an air-water system, each room can be maintained at a different temperature.

6. Finally, the energy use of an air-water system compares very well with other system types.

An air-water system does have several notable disadvantages:

1. It is sophisticated and requires a trained operating staff.
2. The low volume air supply makes between-the-seasons operation especially tricky.
3. For most buildings, this system is limited to exterior zones.
4. Because of its low air volume, it is not ideal for spaces with high exhaust requirements, such as laboratories or operating rooms.

Summary

This chapter introduced the many HVAC systems and the ASHRAE method of classifying and describing them. In this chapter you learned that:

- HVAC systems have common elements, including a heating device, a cooling device, a distribution system, equipment for moving air, water or steam, heat transfer devices, operational items and features, and specialty systems for treating the air or water.
- The three broad ASHRAE classifications for HVAC systems are all-air, all-water and air-water. These classifications are based on the system's cooling medium.
- A zone is any space with separate thermostatic control.
- All-air systems are the most common because they are easy to design and construct. They also provide clean, filtered air over a wide range of sensible and latent heats.
- Air flows through all-air conditioning equipment in a single path (series flow) or a dual path (parallel flow). The air then flows in single or dual ducts to serve single or multiple zones. Mixing boxes, zone heaters and variable volume fans add multiple zone capability to these basic systems.
- Because of water's greater specific heat and density, all-water systems transfer heat more efficiently. However, they cannot filter, ventilate or humidify the air. Designers have developed six all-water piping arrangements to help balance supply and return flows. They are the series loop, one-pipe main, two-pipe direct return, two-pipe reverse return, three-pipe and four-pipe systems.
- Terminal units are the heat exchangers for all-water and air-water systems. Radiators, convectors, baseboard, fin-tube, unit heaters and radiant panels are heating-only terminal units. Unit ventilators and induction units can heat or cool.
- An air-water system uses the all-air system's ventilation capability and the all-water system's heat transfer efficiency. Air-water systems are restricted to building perimeters because the systems are expensive and complex.

8

Introduction to Boilers

The purpose of this chapter is to familiarize you with some of the more popular boilers in use today. Terminology common to the boiler industry is defined in this chapter along with an explanation of heat loads, boiler components, fittings and accessories. Also covered are boiler support systems, fuels and combustion, water treatment, safety, efficiency, and a brief overview of some good preventive maintenance practices. The discussion that follows concentrates on the four basic boiler types with special emphasis on fire tube boilers, which are the most popular boiler in use today for commercial heating applications. After successful completion of this chapter, you will be able to:

1. Define terminology common to boilers used in HVAC.
2. Distinguish between various boilers used in commercial heating applications.
3. Describe boiler classifications, types, common components, and accessories.
4. Identify the major systems that support boiler operation.
5. Describe principles related to combustion, fuels, combustion controls, and water treatment.
6. Explain operational considerations which refer to safety, increasing boiler efficiency, and hot water controls.

Boiler Terminology

The term **"boiler"** applies to all devices that are designed to transmit heat from an external source — usually the combustion of fuel — to a fluid contained within the boiler, typically water.

Figure 8-1 illustrates a boiler symbol commonly used in blueprints and drawings.

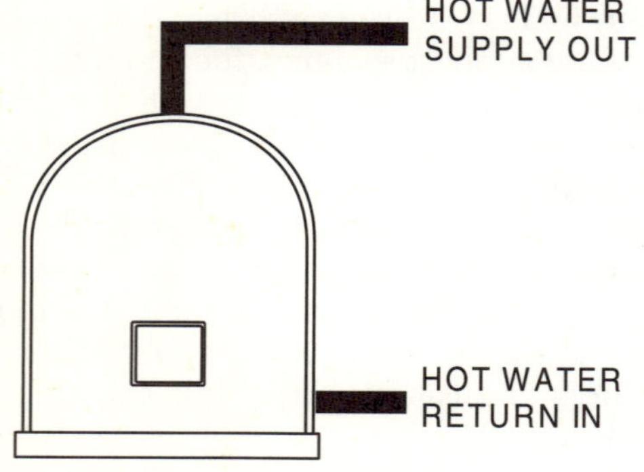

Fig. 8-1 Typical Boiler Symbol

Boilers are used for generating:

- Steam for power, processing, or industrial purposes
- The heat energy used in large absorption air conditioning units
- Steam or hot water boilers for heating purposes and hot water supply

This chapter is concerned primarily with the last purpose above because of its direct application to commercial heating.

The earliest and simplest forms of boiler produced large volumes of steam, but were quite inefficient. Only 30 - 50% of the energy available from the fuel could be extracted as usable heat in the production of steam or hot water.

This **thermal efficiency** is the ratio of heat supplied from burning the fuel to the heat absorbed by the water. Today's boilers have come a long way from their predecessors; they are now designed to be much more efficient by operating at a typical 80% efficiency. Today's state of the art condensing-type boiler can now achieve an efficiency of over 90%. It accomplishes this by extracting the nearly 1000 Btu/lb of latent heat from the vaporized water (a by-product of combustion) normally lost out of the chimney.

Boiler capacity is a measure of boiler rating based on the maximum number of Btus transferred into hot water, or the maximum pounds of steam, in pounds per hour, a steam boiler can produce. NOTE: One pound of water vaporizes into one pound of steam.

Boiler horsepower (BoHP) is an older rating term from the days of steam engines. It is still commonly used to refer to a boiler's capacity, but is rarely used in the boiler selection process. One BoHP is defined as the equivalent of evaporating 34.5 pounds of 212° feedwater to 212°F steam in one hour (at atmospheric pressure).

Heating surface is another criteria commonly used to estimate boiler rating. It is the boiler's total surface area where heat is transferred. The surface area consists of all tubes, both tube sheets, and the furnace tube of fire tube boilers. In a water tube boiler, the heating surface is only the tube exteriors.

It is common practice to use the American Society of Mechanical Engineers (ASME) method of computing area values on the boiler's *fire side* — the area exposed to flame and combustion gases. The Steel Boiler Institute Method uses the *water side* area to compute the heating surface.

Until approximately 20 years ago, 10 square feet of heating surface was considered equivalent to 1 Boiler Horsepower, but due to dramatically increasing boiler efficiencies, 5 sq ft of heating surface became the currently accepted value equivalent of 1 BoHP.

One **"MBH"** is 1000 Btus per hour - for example 9 MBH = 9,000 Btus/hr. ("M" is the Roman Numeral for 1000).

Gross output is the total amount of heat available at the boiler outlet per hour. It is commonly used when selecting boilers. A 1 BoHP boiler has a gross output of 33.475 MBH.

Net output is the recognition that the usable gross output heat from a boiler is reduced by **piping and pickup losses**. A percentage of this heat is "lost" through piping walls and insulation, while pickup is a load compensation allowed to "pick up" the temperature to the design indoor temperature from a cold start, such as reheating a cold building just before it is occupied in the morning.

Net output is the percentage of heat remaining to do the work after pickup and piping losses are considered. The commonly used allowance is 25%, set by the Mechanical Contractors Association (MCA), but the Institute of Boiler and Radiator Manufactures (IBR) recommends 13% for their pickup and piping loss allowance factor.

Examples of net output classifications include "Steam Net - MCA (or IBR)" and "Water Net -MCA (or IBR)." It is important to specify which system and which pickup/piping allowance is used to determine the required boiler gross output.

Net load is similar to gross output, but is from the viewpoint of the *needs of the system*. It is the actual heat *requirements* of a system, exclusive of piping and pickup losses. Net Load can be expressed in terms of Btu heat loss, square feet of radiation, hundreds of degree-gallons, horsepower, or pounds of steam per hour.

Net rating is an indication of the net load that may be connected satisfactorily to the boiler. It is always *less* than the gross output of the boiler minus the piping losses and

Example 8-1

A building has a net space heating load of 90,000 Btu/hr and a kitchen hot water load of 10,000 Btu/hr. The building is heated intermittently. A pick up and piping allowance of 25% is required. What is the boiler's required gross output?

Solution

Boiler Net Output	= *90,000 Btu / hr (heat) + 10,000 Btu / hr (hot water)*
	= *100,000 Btu / hr*
Pickup & Piping Allowance	= *0.25 x 100,000 Btu / hr*
at 25%	= *25,000 Btu / hr*
Required Boiler Gross Output	= *125,000 Btu / hr minimum*

pickup load, provided that they do not exceed the standard allowances (25% MCA or 13% IBR).

Heat Loads for Heating and Processes

Devices using steam or hot water produced by a boiler are called loads. Typical loads are heating and process loads, for example:

Heating Loads
Steam Coils
Hot Water Coils
Humidifiers
Shell & Tube Heat Exchangers
Radiant Heating Devices

Process Loads
Turbines
Sterilizers
Ovens
Steam Operated Equipment

A building's environmental control system is designed to provide a comfortable indoor climate year-round, through most weather extremes that the local climate can produce. *Air conditioning* is commonly thought of as a system used only for cooling air, but in reality heating is also one of its major purposes.

In the heating season, sensible heat (temperature) and latent heat (humidity) need to be replaced in a room or zone to *balance* the heat losses that occur through conduction, cold air infiltration, warm air exfiltration, and dehumidifying. See Figure 8-2.

As first mentioned in chapter 4, the materials used in construction of a building affect their "U" value and their "R" value. The four variables that influence the heat transfer rate either individually or in combination are the following:

- Temperature Difference
- Time
- Kind of Insulation
- Thickness of Insulation

The **total heating load** for a room or zone consists of the sums of the heat required to balance:

1. the room or zone temperature (add sensible heat).
2. the room or zone humidity (add latent heat).
3. to warm incoming outside air from both ventilation and infiltration sources.

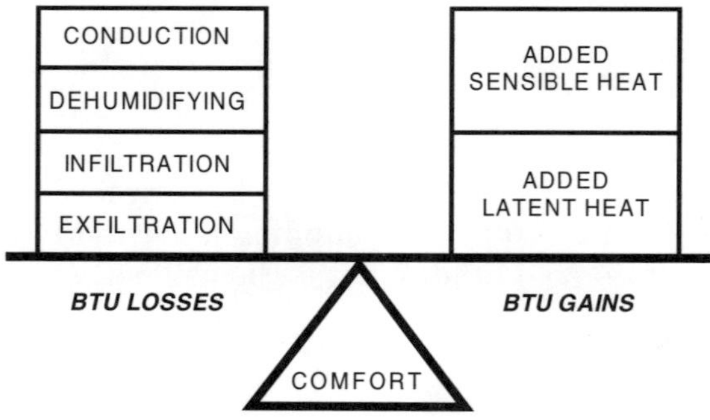

Fig. 8-2 Heat Balancing

Furnaces

A direct-fired heat exchanger, or **furnace** as in Figure 8-3, is similar to a boiler in that it is used in heating applications. It is different from a boiler because it uses its burner to heat *air*, not water, via its heat exchanger. Therefore, warm air (not water or steam) is used as the room or zone heating medium.

Furnaces are popular choices for residential and small commercial installations. The cost is generally lower because ductwork is less expensive than piping.

In a furnace, a heat exchanger is heated on the outside by a burner's radiant heat and combustion gases. A blower then passes air through this heated heat exchanger. Heat is transferred through the walls of the heat exchanger into the air stream moving through it by conduction. The warm air is then directed into the room or zone by way of ducting.

Each furnace manufacturer designs different types of furnaces to maximize the heat transfer capabilities of their furnaces. They are classified by the direction of air flow through the heat transfer surface:

- **Upflow Furnace** - Air is forced through the heat exchanger in an upward direction.
- **Counterflow Furnace** - Air direction is downward through the heat exchanger.
- **Horizontal Furnace** - Air direction flows horizontally through the heat exchanger.

A second definition of the term "furnace" refers to the area of a boiler where combustion takes place. In this sense the furnace is also known as the "firebox."

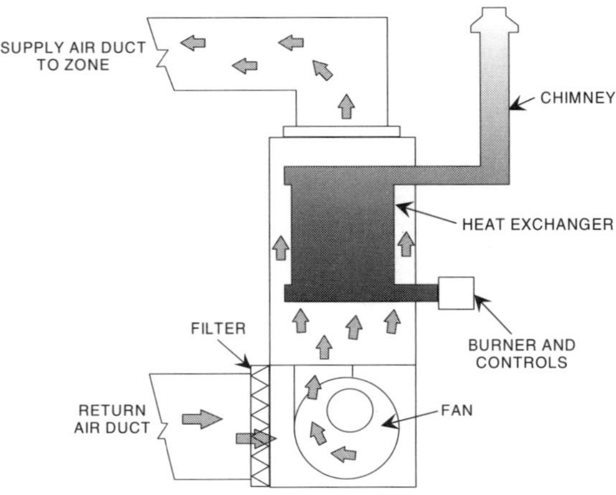

Fig. 8-3 Upflow Furnace

Boilers for Water and Steam

A boiler is a closed vessel that is essentially a large heat exchanger. It is operated to either heat water or vaporize water into steam, usually under pressure. Sometimes the water is heated with an electric heating element, but it is normally heated by the combustion of gas, oil, or coal. Since pressure and temperature are directly proportional, the higher the pressure is on the surface of the boiler water, the higher the corresponding boiling point of the water in the boiler.

For example:

Water under 15 in. Hg of vacuum will boil at 179°F

Water at atmospheric pressure boils at 212°F

Water at 15 psi will boil at 250°F and

Water at 100 psi boils at 338°F.

A hot water system can operate at temperatures of 250°F without boiling the water because at the system pressure of 30 psig, the **saturation temperature** (or boiling point) is 274°F. By placing the medium (water) under pressure, a boiler gives water the ability to absorb a greater amount of heat and release it to the system. Water at 274°F will **flash** from water into steam if exposed to atmospheric pressure, such as from a leak.

Boilers that produce steam are sometimes called **steam generators**. This steam can be used to operate machinery which supplies electricity, water, gas, or supports industrial/commercial processes. Typically, in plants where steam generators produce process steam, the plant is heated by tapping off of the process steam supply.

On a steam boiler, internal steam and water drum fittings, such as the dry pipe, scrubbers, and steam separators, are used to increase the steam quality by reducing its moisture content. **Steam quality** is a percentage which indicates the dryness of the steam. For example, 95% quality steam is 95% evaporated (dry) steam and 5% moisture. **Saturated steam** is steam at its boiling pressure and temperature.

Superheated Steam is at a temperature above its boiling point (saturation temperature). Steam is superheated by removing saturated steam from the water that it was generated from and applying more heat. The additional heat further raises its temperature.

Steam can also be superheated by using a reducing valve. As the pressure of the steam is reduced, the temperature remains nearly the same, thus leaving more heat (superheat) in the steam than its corresponding saturation temperature. Superheated steam contains more heat energy and less moisture than saturated steam. It is also called **dry steam**.

If steam were to be **subcooled** (below the saturation temperature) it would become water.

Boiler fuel contains *chemical energy* which is then converted into *thermal energy* (heat) in the combustion chamber or furnace by the process of combustion. Heat is energy that is absorbed into, and released from, hot water or steam. In steam systems, enthalpy is the total amount of heat energy contained in the steam.

Boiler Classifications

Boilers are classified in a number of different ways.

- *By Operating Pressure*: A low pressure (LP) *hot water* boiler has a maximum pressure of 30 psi.

  A low pressure steam boiler has a **MAWP** (maximum allowable working pressure) of 15 psig.

  Of all low pressure boilers, 90% are applied to the commercial comfort market.

  A high pressure (HP) steam boiler operates at pressures which exceed 15 psi and over six boiler horsepower. A high pressure water boiler operates at pressures above 30 psi.

  Of all high pressure boilers, 25% are applied to the commercial comfort market.

- *By Operating Temperature:* A low temperature boiler operates below 250°F.

  A medium temperature boiler is between 250° - 350°F.

  A high temperature boiler produces temperatures above 350°F.

- *By the Fuel Used*: Oil fired boiler propane, butane, natural gas boiler or coal fired boiler

- *By Installation*: The **package boiler**, as shown in Figure 8-4, is almost entirely manufactured at the factory. It arrives with burners, controls, safety/regulatory

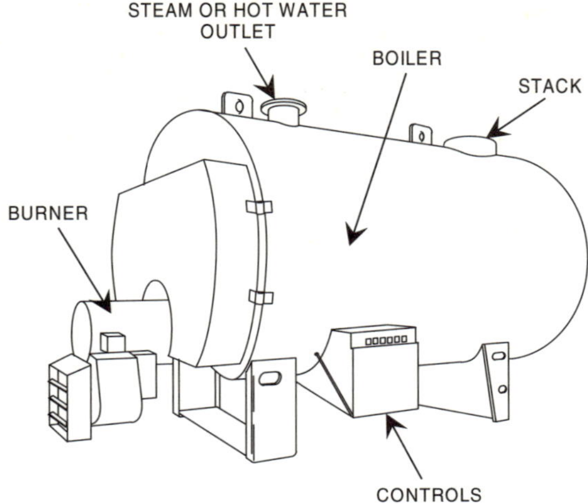

Fig. 8-4 Package Boiler

Courtesy of the Trane Company

gauges, oil pumps, fans, valves, and other associated equipment. All that is usually necessary in the field before beginning operation is a solid foundation and making the necessary fuel, steam, electrical and water piping connections.

A **field-erected boiler** must be constructed on-site because of its size or complexity.

- *By the Medium Which Flows Inside the Boiler Tubes:* Fire tube boiler and Water tube boiler
- *By Type*: Electric boilers
  Water tube boilers
  Cast iron boilers
  Fire tube boilers

Summary of Boiler Types

In an **electric boiler**, as shown in Figure 8-5, the water is heated by electric resistance heating elements, much like those in a typical electric water heater. They are available in capacities up to 150,000 pounds of steam per hour. Their principal advantages over fuel-fired boilers are the following:

- Elimination of handling, distribution & storage of fuel
- No burner cleaning or adjustments
- Use of less floor space

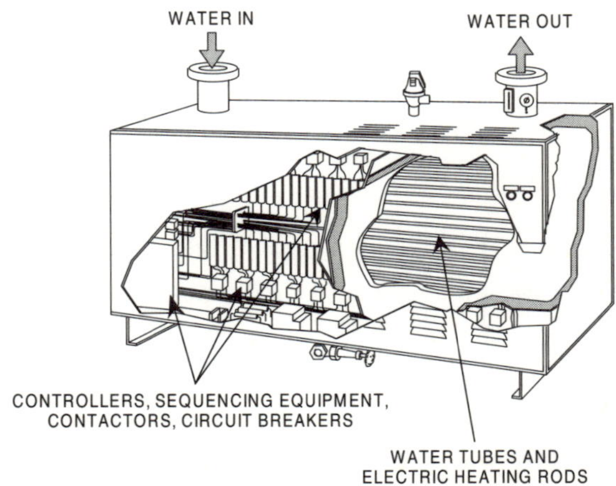

Fig. 8-5 Electric Boiler

Modern Air Conditioning Practice, 3rd Edition. by Harris, Norman C., ©1983
Adapted by permission of McGraw-Hill

- No stack heat losses
- Nearly 100% efficient

Although they are extremely efficient from an energy usage viewpoint, the electric boiler is usually more costly to operate than fired boilers because electricity is almost always more expensive than other fuels. The boiler itself is more efficient, but the "fuel" is more costly. An electric boiler may be the best choice where electric rates are extremely low.

There are two types of electric boilers: electrode and immersion. The electrode electric boiler uses two electrodes which will automatically stop the flow of current between them if the water level is low. This keeps the boiler electrodes from burning out. The immersion-type of electric boiler requires a continuous immersion of the heating element. Its element will burn out if the water level drops too low.

Water tube boilers (Figure 8-6) are usually used with systems that require high pressure, high temperature and high capacity steam volumes. Few are found in low pressure heating applications. They are used almost exclusively for pressures above 150 psi and capacities over 15,000 pounds of steam per hour. Typically, water tube boilers achieve efficiencies of 78 - 82%.

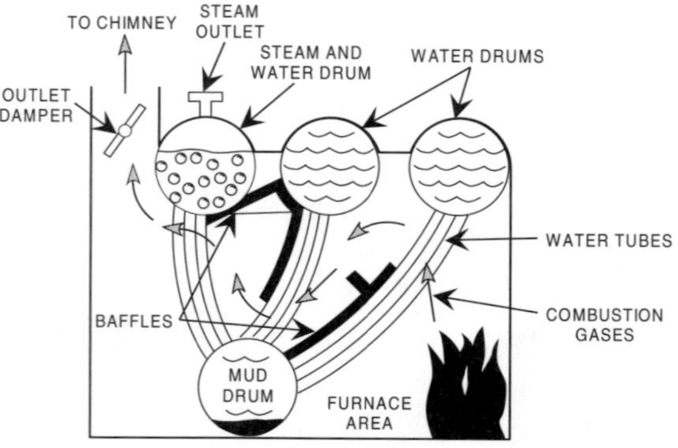

Fig. 8-6 Bent-tube Water Tube Boiler

American Technical Publishers, Inc.

Water tube boilers consist of an upper steam-water drum (and sometimes also upper water drum(s), and a lower mud drum and/or headers. The lower drum is called the mud drum because it is the collecting point for sludge, a waste by-product of the bottom rejected from the boiler. (This will be covered in the water treatment section). Tubes between the drums are either straight or bent. These tubes are the entire heating surface in water tube boilers.

The primary purposes of the drums, which provide essentially no heating surface, are to provide a reservoir for water and steam and to distribute water to and from the tubes. The design of water tube boilers allows a much greater working pressure than the working pressure of fire tube boilers.

Water circulates *inside* the tubes of a water tube boiler, while combustion gases flow around the outside of the tubes. Water is heated by the conduction of heat through the tube wall and into the water.

Combustion in the furnace warms the mud drum and tubes. This heats the water inside and causes it to become less dense and rise. The cooler, denser incoming water circulates to the lower boiler. This difference in density creates the circulation in the water sides of a boiler.

Advantages of water tube boilers include:

- Withstand extremely high pressures and temperatures.
- Respond quickly to fluctuating steam loads.
- Handle larger steam loads.

Because water tube boilers are expensive and usually require field erection, most do not serve the commercial heating market.

Cast iron sectional boilers (CI Boilers) are compact boilers that do not have any tubes. Nevertheless, they are similar to water tube boilers in many ways. Water jackets, instead of tubes, surround the furnace area. These boilers are manufactured in smaller sections which may be bolted together at the job site (field-erected). This makes them an excellent choice for retrofit applications where walls would otherwise need to be torn down for boiler replacement. Another advantage is that additional water jackets may be added at a later date to provide more heating surface, for example, after a building expansion project.

The cast-iron boiler is used primarily for low pressure heating system applications, 15 psig maximum for steam and 30 psig maximum for hot water. Cast iron boilers have excellent operating reliability, low maintenance cost and long life. They can be used for all applications up to 8,300 pounds of steam per hour. Some rated efficiencies are as high as 87%.

Two different designs are available: **vertical cast iron sectional boiler** sections stack upward, so the boiler requires little floor space. The **horizontal cast iron sectional boiler** is preferred for larger applications. See Figure 8-7.

Fire tube boilers (Figure 8-8) are often the choice for HVAC comfort applications for heating. They are also known as "immersion tube" or "immersion fired" boilers. These boilers are used mainly for steam or water heating systems. As both of their names imply, the tubes are immersed in water and the heat from the combustion gases flow through the tubes.

Fire tube boilers are generally of the low pressure type, either 30 psi hot water or 15 psi steam with a maximum working temperature of 250°F. The largest fire tube boilers can produce up to a maximum 250 psi and range up to 800 boiler horsepower. The efficiencies of modern fire tube designs have increased to about 80 - 85%.

The steel fire tube boiler is designed around its furnace and tube passages. Many arrangements have been developed. Tubes are placed in horizontal, inclined or vertical positions, with one or more passes for combustion gases.

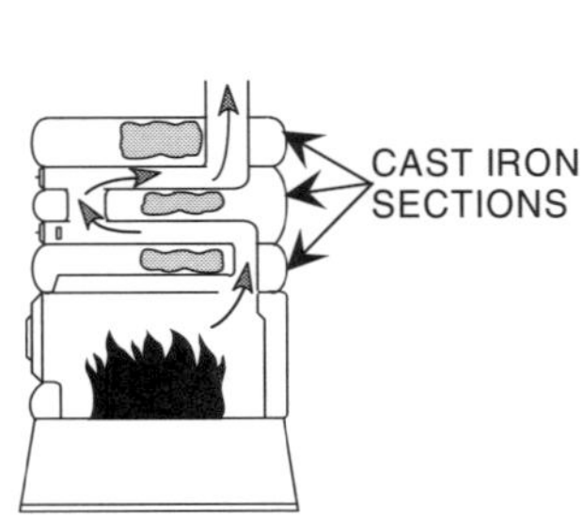

VERTICAL CAST IRON BOILER

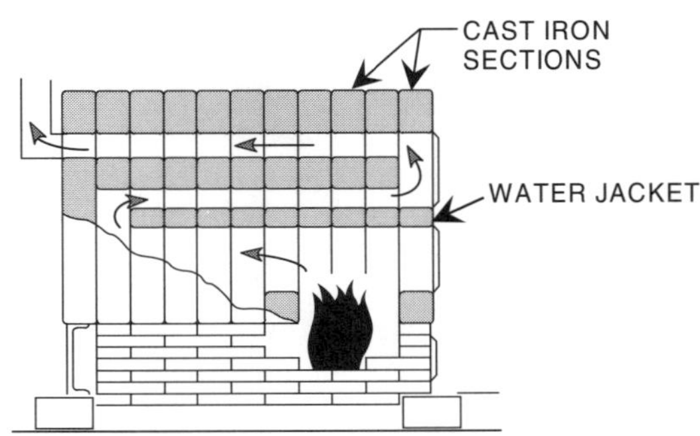

HORIZONTAL CAST IRON BOILER

Fig. 8-7 Cast Iron Sectional Boilers

Courtesy of the Trane Company

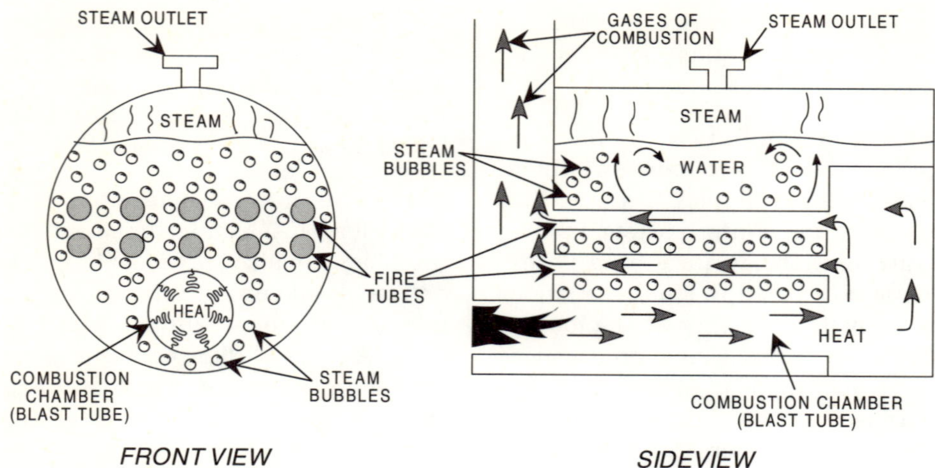

Fig. 8-8 Fire tube Boiler Operation (Steam)

American Technical Publishers, Inc.

A **pass** is the horizontal run that flue gases take through the fire tubes before exiting the flue of any fire tube boiler. By inserting baffles to direct the combustion gases, a boiler can be made into a two-, three- or four-pass boiler.

More heat can usually be extracted from combustion gases by adding more passes, but the efficiency is not always increased. A low combustion gas (also known as stack gas or flue gas) outlet temperature is one measure of boiler efficiency. Some longer two-pass boiler designs are as efficient as four-pass boilers, but the multi-pass boiler design permits a desirable smaller "footprint." A shorter compact boiler requires less floor space at installation.

The fire tube boiler is limited in physical size by the stresses that a large amount of hot pressurized water and steam exert on the tubes and welded joints.

Advantages of fire tube boilers include:

- They can be assembled at the factory as a packaged unit, resulting in better quality control.
- Initial cost is less than a water tube boiler.
- They require less floor space and headroom.
- They require little or no setting (brickwork).

Types of Fire tube Boilers

The **horizontal return tube boilers**, or "HRT Boiler," as shown in Figure 8-9, had been a very popular boiler design for heating and small industrial processes for many years. Some HRTs are still in use today, but with efficiencies of around only 70%, they are being widely replaced with more efficient designs. A variation of the HRT boiler is the two-pass boiler.

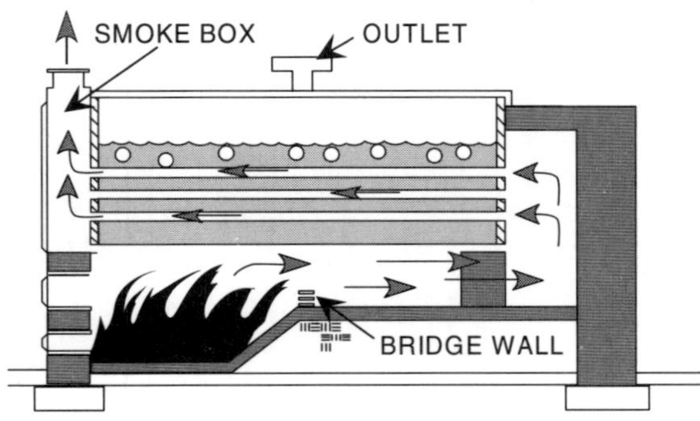

Fig. 8-9 Horizontal Return Tube Boiler

Courtesy of the Trane Company

An HRT is an externally fired boiler which has a water drum suspended over a firebox. Flames and hot gases warm the bottom of the shell, then the gases flow past a bridgewall to the rear of the boiler and through the horizontal tubes.

Vertical tube boilers are one-pass boilers which cannot use gas baffles. They require less floor space than an HRT boiler, but are not commonly used for commercial heating.

Scotch marine boilers are the most predominant style of steam and hot water fire tube boilers manufactured today. Yes, they were originally developed in Scotland to satisfy the marine industry's need for a compact, highly efficient boiler. Its compact size was accomplished by placing the combustion chamber (also called a "blast tube") within the boiler shell—completely surrounded by water. See Figure 8-10.

The unique placement of this combustion chamber not only provides the utmost in primary heat transfer surface, it eliminates the need for refractory walls! All the usable heat that used to be "lost" into the furnace brickwork is now transferred directly out of the walls of the blast tube and into the water.

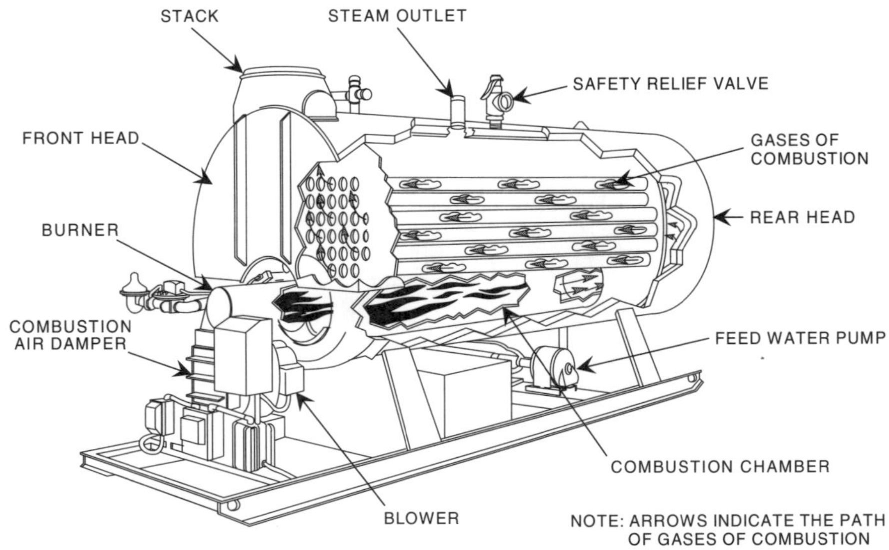

Fig. 8-10 Scotch Marine Boiler

Scotch marine boilers are used in stationary comfort applications where small size and low headroom are prime considerations. The largest hot water boiler's net output can reach nearly 23,300 MBH, and 100 psi hot water boilers are available. Steam boilers can produce a maximum of 27,600 lbs. steam per hour, with operating pressures in the ranges of 0-15 psi, 15 - 150 psi, or 150 - 250 psi.

Figure 8-11 illustrates two variations of the scotch marine boiler, wet-back and dry-back design.

- **Wet-back** refers to how the rear combustion chamber (where the combustion gases reverse direction to enter the tubes) is protected from overheating by a water wall.
- The **dry-back** design is protected by a refractory lining in the rear. Dry-backs are somewhat easier to maintain, but their refractory back is more expensive.

ASME Boiler and Pressure Vessel Code

The **American Society of Mechanical Engineers (ASME)** has developed the widely accepted "ASME Boiler and Pressure Vessel Code". ASME works closely with boiler manufacturers and The National Board of Boiler and Pressure Vessel Inspectors, who enforce the ASME codes with annual boiler inspections.

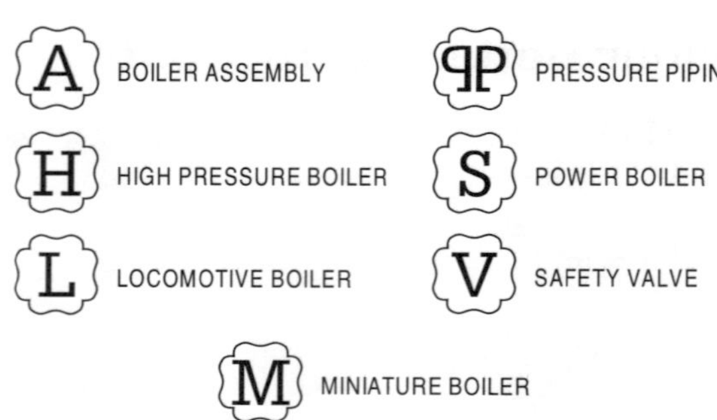

Fig. 8-12 ASME Symbol Stamps

ASME symbol stamps, such as those shown in Figure 8-12, are required to be displayed on boiler pressure vessels and vital components to certify that the boiler or component was manufactured up to ASME Code and is capable of withstanding the stress that it was designed for.

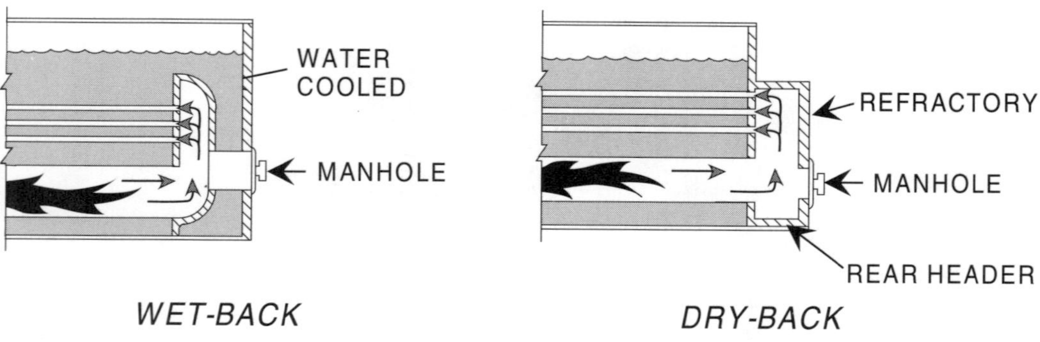

Fig. 8-11 Scotch Marine Boiler Variations

Courtesy of the Trane Company

Major Boiler Components

The **shell** of the boiler is the portion of metal welded to the tube sheets in sections to form its basic structure. It is either hot or cold rolled steel, depending on the application.

A **breeching** is a duct or passageway that directs flue gases from the boiler to the chimney. In gas-fired equipment, it is commonly called a "flue (or vent) connector."

The **furnace** is that portion of the boiler where combustion occurs. It may occur in a furnace tube, as mentioned above or inside a refractory firebox, such as on the HRT boiler.

Tube sheets shown in Figure 8-13 are the steel ends of the fire tube pressure vessel where the holes are drilled for the tubes, which are expanded into them. These holes are staggered by design to improve both the circulation of water and the heat transfer efficiency. A tube sheet's thickness is usually about 7/16 inch for low pressure boilers and 9/16 inch for high pressure boilers. Flexing of the tube sheet compensates for the lengthwise tube expansion which occurs when the tubes are heated.

A **ligament** is that portion of the tube sheet (approximately 3/4 inch in width) located between the tube holes. Ligaments are susceptible to the damaging effects of thermal shock, which is caused by uneven water circulation or adding cold return water to the boiler. If the differential temperature across the tube sheet is allowed to become greater than 50°F, the ligaments will twist. A 3° twist on a ligament will usually crack that tube sheet's ligaments See Figure 8-13. This crack will result in a leaking boiler.

An operator must ensure that the returning boiler feed water temperature and flow remain constant not only to maintain a good water circulation pattern through the boiler, but also to keep the returning water temperature from dropping too low. The correct flow and temperature are sometimes maintained by a three-way valve controlled by the return water temperature. This valve can recirculate some hot water produced by the boiler back to the return to maintain both a constant return temperature and flow, regardless of the system load. *NEVER* adjust this valve to meet lower system hot water temperature needs; it is required for proper boiler operation.

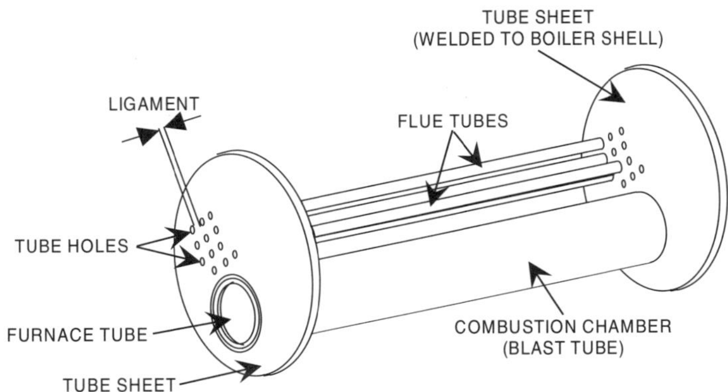

Fig. 8-13 Tube sheets

Courtesy of the Trane Company

Along the same lines, never drop the boiler return temperature below the minimum set by the manufacturer as an attempt to "save money." Repairing thermal shock damage to a tube sheet is not worth the pennies saved by reducing the boiler return water temperature.

Tubes are the primary heat transfer surface area in boilers. They are measured by their outside diameters (O.D.). Tubes are commonly made of steel, but stronger alloys of steel are used to withstand the forces on high pressure and high temperature boilers. The furnace tube (or blast tube) is the larger diameter tube where combustion actually occurs. Fire tubes may be referred to as **flue tubes**.

Drums are water tube boiler collection points for water and steam. They also distribute water to the tubes. An additional function of the lower mud drum is to serve as a collection point for sludge that has settled out of the water.

Blowdown pipe and valves are required on steam boilers to remove sludge, salts and other impurities from the water. They are connected to the bottom of the mud drum on water tube boilers and to the lowest part of the water side shell on a fire tube boiler.

Headers are the same as drums, but much smaller. A person can crawl into the **manhole** of a drum, but can only gain access a header through a **handhole**. Headers are usually located in the lower areas of the boiler.

Refractory is also known as "brickwork" or "firebrick." It retains the heat inside the boiler and protects to walls from overheating.

Stays used in fire tube boilers help support areas under stress, such as the tube sheet, to prevent bulging when the boiler is pressurized. Three types of stays as shown in Figure 8-14 are the following:

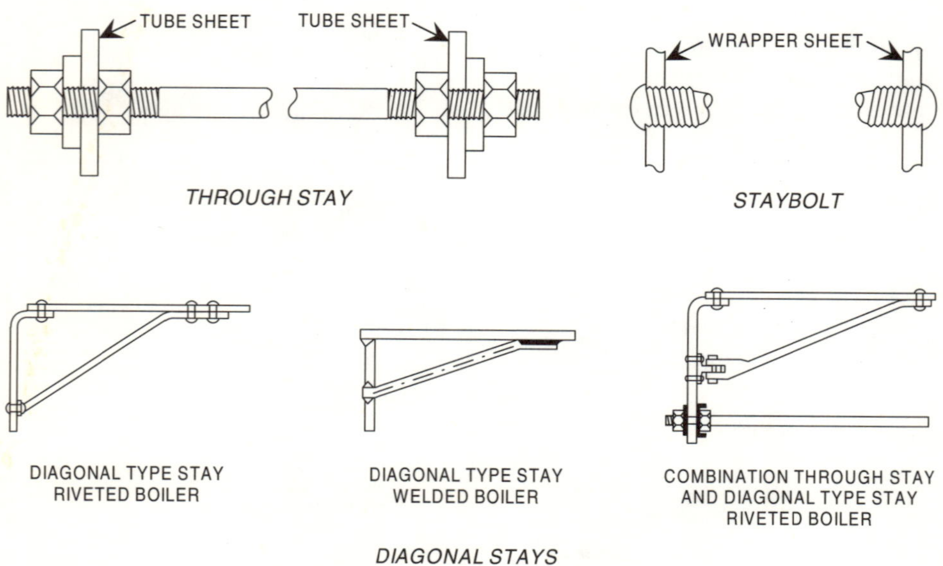

Fig. 8-14 Boiler Stays

American Technical Publishers, Inc.

- Diagonal Stays
- Staybolts
- Through Stays

Dampers control the flow of either combustion air into the boiler or combustion gases out of the boiler.

Baffles are sections of refractory (firebrick) that are strategically placed in the boiler combustion gas path to prevent a direct path to the **flue stack** (also called a **chimney**).

Baffles increase efficiency and maximize the heat transfer from gases to water. They steer hot combustion gases through the boiler passes and around the entire the heat transfer surface area of the tubes and boiler. Deteriorated baffles reduce boiler efficiency by allowing some combustion gases to bypass sections of tubes, or even an entire pass. Flue gas temperature rising above normal in the breeching is an indicator of bad baffle refractory.

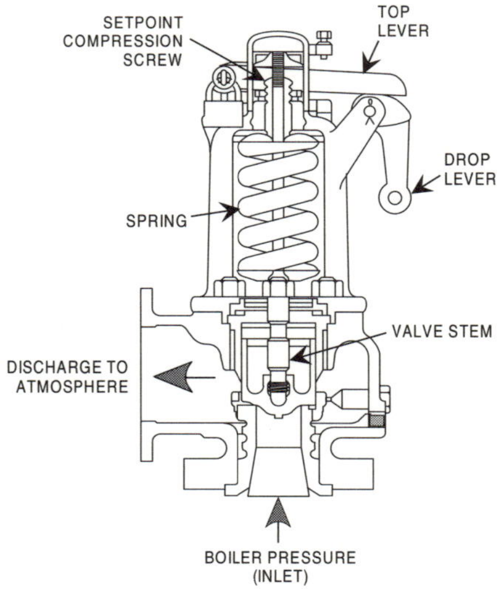

Fig. 8-15 Boiler Safety Valve

American Technical Publishers, Inc.

Boiler Fittings, Accessories and Instruments

Safety valves, also known as "pop valves" and "relief valves," relieve an overpressure in a boiler, and are considered by both operators and inspectors as *the most important piece of safety equipment on a boiler*. See Figure 8-15. These valves are designed to "pop" (open fully) at the set pressure and relieve pressure down to a lower safe pressure, where it then snaps shut. Relief of steam below the setpoint is known as **blowdown**. Their capacity must be enough to discharge all the steam that the boiler is capable of generating at maximum firing rate without allowing the boiler pressure to rise more than 6% above the maximum allowable working pressure (MAWP).

ASME code states that steam boilers must have one safety valve for each 500 sq ft of heating surface, and that low pressure boilers (below 15 psi) must have at least one safety valve.

They are set and sealed to open at a pressure not to exceed the rated pressure of the boiler.

A **water column** is a fitting connected to the drum at the **NOWL** (normal operating water level). It is designed to reduce turbulence and fluctuations of boiler water level where it is

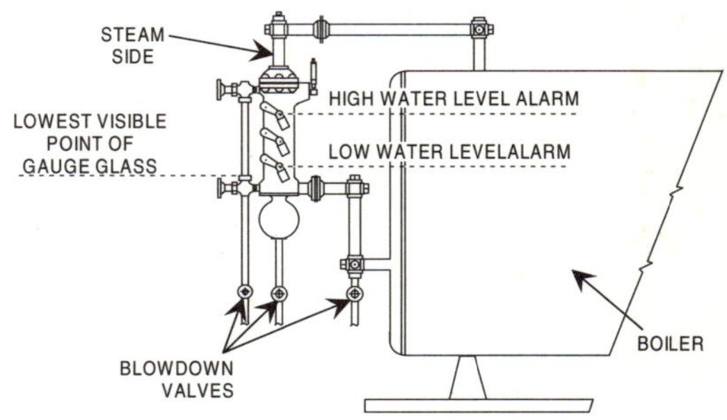

Fig. 8-16 Water Column (Steam Boiler)

American Technical Publishers, Inc.

sensed. It allows attachment of water level monitoring devices, such as a gauge glass, the try cocks, and the high and low water level alarms. Gauge glass blowdown lines are also connected to the water column. Figure 8-16 shows the relative position of a water column looking at the side of a boiler.

Boiler vents, or "air cocks," are valves located at the highest point of the boiler. They allow air to escape when the boiler is filled and when it is warming up. Vents also prevent a vacuum in the boiler when it cools or is being drained.

Pressure and **altitude (temperature) gauges** monitor operating conditions inside the boiler. The boiler steam pressure gauge must be very accurate. If this gauge is not accurate within 2% of the actual working pressure inside the boiler, it must be recalibrated.

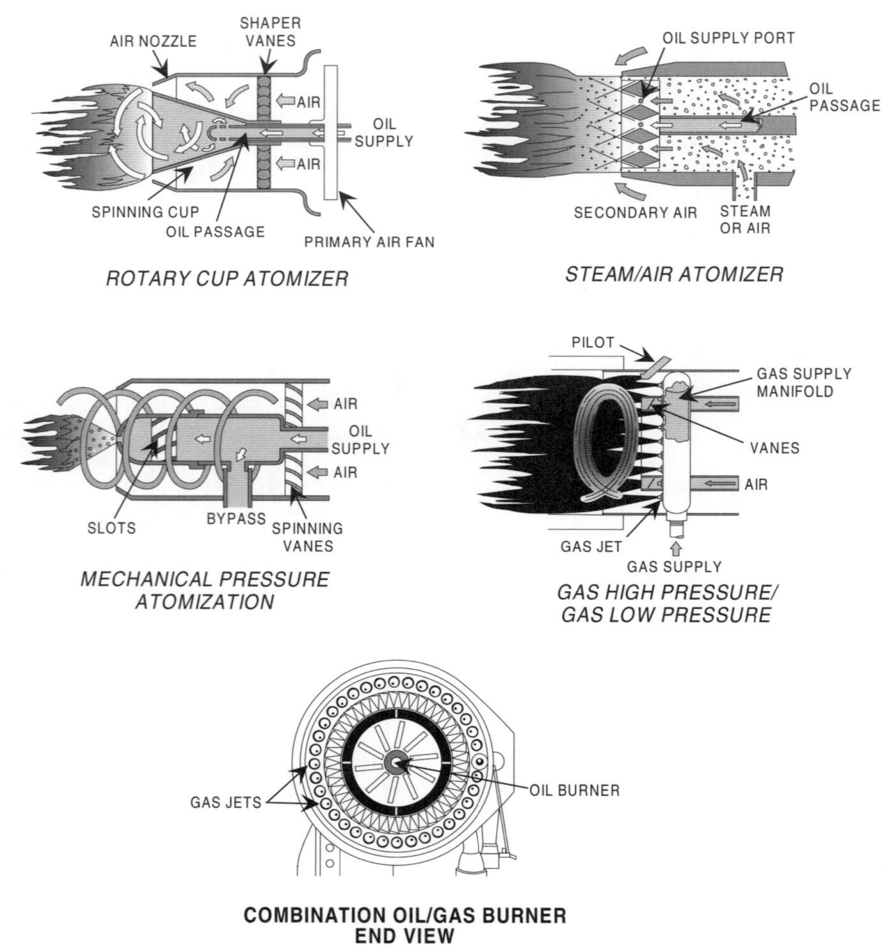

Fig. 8-17 Boiler Burner Types

Courtesy of the Trane Company

Boiler burners allow fuel to enter the boiler in a manner that will allow uniform burning and the most efficient mix with air for proper combustion. Combustion and controls will be covered later in this chapter, but here are seven common burner types. See Figure 8-17

1. Rotary Cup Atomizer
2. Mechanical Pressure Atomizer
3. Steam/Air Atomizer
4. Gas High Pressure/Gas Low Pressure
5. Combination Oil/Gas
6. Power Burner
7. Atmospheric Burner

Feedwater regulators automatically control the level of boiler water to maintain Normal Operating Water Level (NOWL). Three common types are:

- Thermoexpansion Feedwater Regulators,
- Thermohydraulic Feedwater Regulators, and
- Float Feedwater Regulators.

The float-type is commonly used on small packaged boilers. See Figure 8-18.

Basic Boiler Support Systems

The **feedwater system** shown in Figure 8-19 essentially "feeds" the boiler with water at the proper temperature and flow rate. It must maintain the proper pressure to maintain NOWL and the proper temperature to reduce the possibility of thermal shock to the boiler. You will recall, in our discussion of tube sheet ligaments that cold feedwater (50°F difference or greater) can shock the boiler, twist the tube sheets and crack their ligaments.

The system consists basically of a motor-driven centrifugal pump, a feedwater regulator, a tank, system piping, and sometimes a dearator (also known as a feedheater).

If a steam boiler is generating 20,000 lb of steam per hour, it must be supplied with 20,000 lb of water per hour. To find how many gallons the boiler will need, a conversion from pounds to gallons is necessary. Since one gallon of water weighs approximately 8.3 pounds (depending on temperature), the boiler requires approximately:

20,000 lb 8.3 lb/gallon = 2,410 gallons of water required per hour.

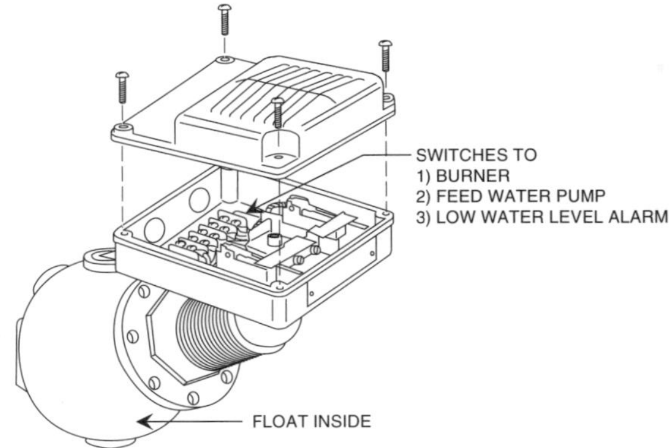

Fig. 8-18 Float-type Regulator

Courtesy of ITT McDonnell & Miller

The **fuel system** is designed to safely deliver fuel at the proper temperature and pressure from either a storage tank, or the local utility to the boiler burner(s). Components commonly found in fuel systems include pumps, piping, tanks, pressure gauges, fuel safety shut-off devices, pressure regulating valves, solenoid valves and occasionally fuel oil heaters if the boiler is required to burn the heavier grades of oil. See Figure 8-20.

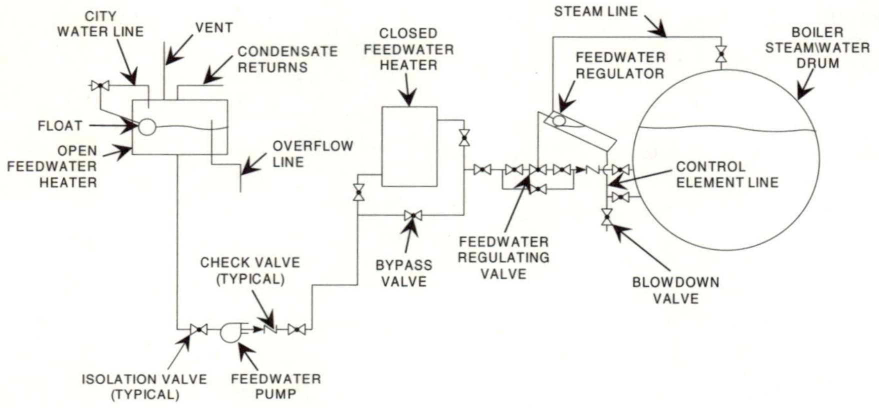

Fig. 8-19 Typical Feedwater System

American Technical Publishers, Inc.

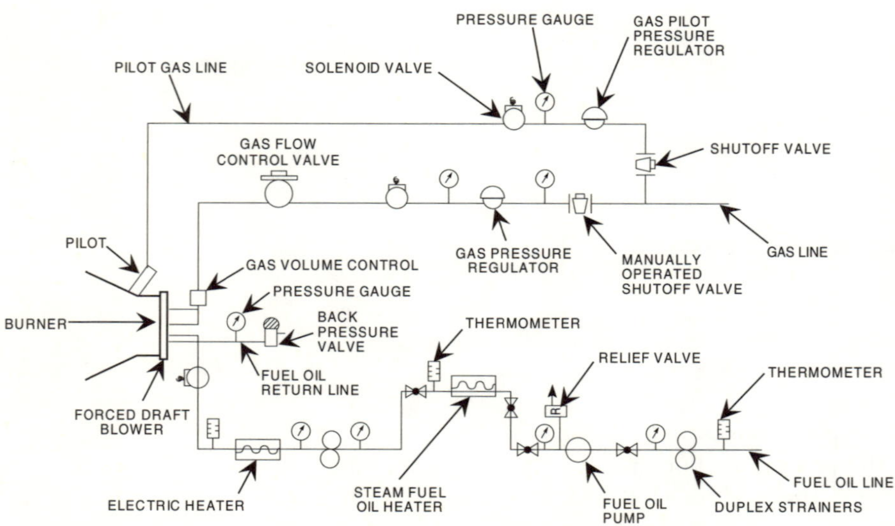

Fig. 8-20 Typical Combination Gas/Oil Fuel System

American Technical Publishers, Inc.

Common fuel systems are the low pressure gas system, the high pressure gas system, the fuel oil system and the combination oil/gas system. Coal is rarely used in today's heating boilers.

Air systems provide the **draft** (air pressure) required for proper combustion. They are designed according to the needs of the boiler. Efficient combustion requires providing sufficient oxygen to mix with the fuel and ignite in the furnace. There are four ways of providing a boiler with the draft it requires. Figure 8-21 shows each type.

1. **Natural Draft** depends on the height of the chimney (stack). The higher the stack, the more draft that can be supplied to the boiler. Natural draft also depends on the difference in temperature between the column of hot gas inside the chimney and the column of cool air outside the chimney. You may have seen many older industrial plants with tall smokestacks. They are providing their boilers with natural draft. Crude control of combustion can be achieved by adjusting the inlet air damper and/or the outlet gas damper.

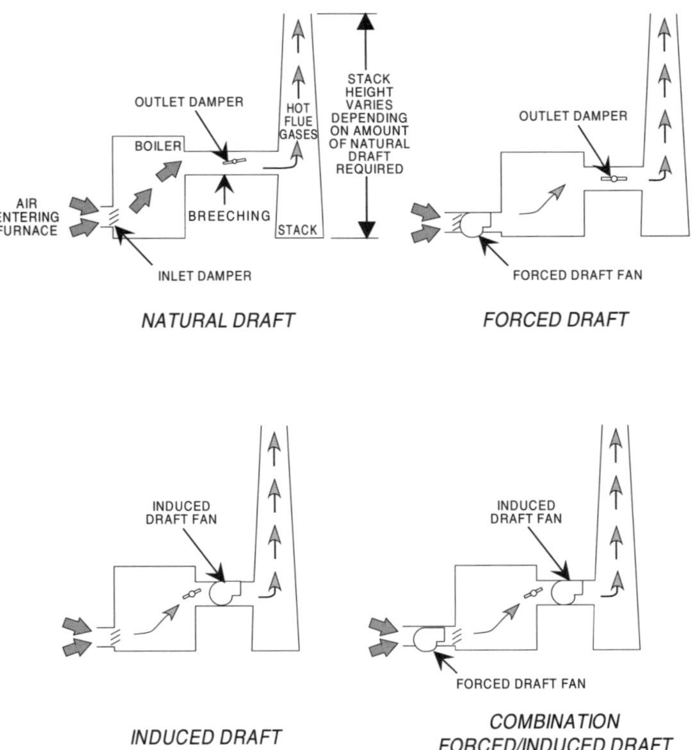

Fig. 8-21 Types of Boiler Draft

American Technical Publishers, Inc.

If a mechanical means supplies a boiler with air, you have three other options:

2. **Forced Draft** uses a fan and/or blower to force inlet air into the furnace. It provides a greater degree of control than natural draft allows. Forced draft is the most popular method of providing combustion air to the boiler.
3. **Induced Draft** places the fan inside the breeching. A slight vacuum is created in the firebox, and inlet air rushes in to fill the vacuum. This air is then mixed with fuel during the combustion process. A disadvantage is that the flue gases come in contact with the induced fan and take their toll on it.
4. **Combination Induced/Forced Draft** provides excellent control of draft conditions, but is usually only necessary for large steam plant boilers.

The **piping system** as described in Chapter 7 transfers the heating medium from the boiler throughout the building to radiation units and heating coils, which deliver the heat to rooms and zones. A steam systems delivers its medium by the pressure of the steam leaving the boiler. Steam traps are used throughout a steam heating system to prevent steam from going past its point of use and to return condensed steam back to the system. Some steam traps also remove air from the system. Poorly maintained steam traps which allow steam to pass are major energy wasters. Annual fuel budgets in some buildings have been reduced by as much as 70% by repairing or replacing defective traps.

A **hydronic** heating system uses a hot water boiler to heat water from 160°F to 200°F. The hot water heating medium may be pumped through pipes from the boiler to space radiation units, then returning to the boiler at a temperature of 160°F for reheating. When using the radiant heat of a hydronic heating system, the air that is heated is room air; it is not ducted in.

All hot water systems must have an **expansion tank** to allow for the increase in volume that results when water is heated. Most modern hot water systems use a **closed expansion tank**. When the system is filled with water, air in the tank is trapped and compressed. When water is heated, its volume increases. This extra volume further compresses the air, resulting in a slight increase in the system pressure. As the water cools the reverse occurs. Expansion tanks should always be located as a branch off of the system piping on the pump's suction side.

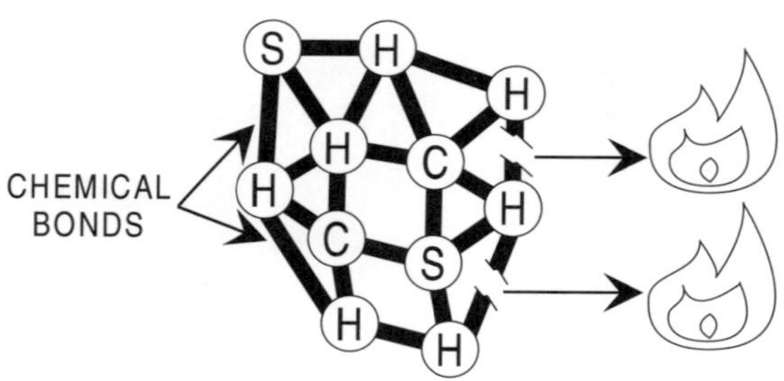

Fig. 8-22 Hydrocarbon Molecule (Fuel)

Combustion

Combustion is the rapid fracturing of chemical bonds that hold atoms together in a large hydrocarbon fuel molecule. This frees the carbon (C), hydrogen (H), and sulfur (S) fuel atoms to seek out oxygen atoms and combine with them. This splitting of chemical bonds creates an enormous release of useful heat, along with the by-products of combustion. See Figure 8-22.

Most of these by-products are the combination (or oxidation) of fuel atoms with the oxygen atoms, resulting in: carbon dioxide (CO_2), water vapor (H_20), nitrogen and sulfur dioxide (SO_2).

The **fire triangle** shown in Figure 8-23, is a model of combustion. Its three legs represent the three elements that produce combustion when combined. Any element removed would extinguish the flame. They are:

1. Fuel - Fuel is a hydrocarbon, which is many hydrogen, carbon, and sometimes sulfur atoms combined into one molecule. Examples are oil, gas, wood — anything that will burn.

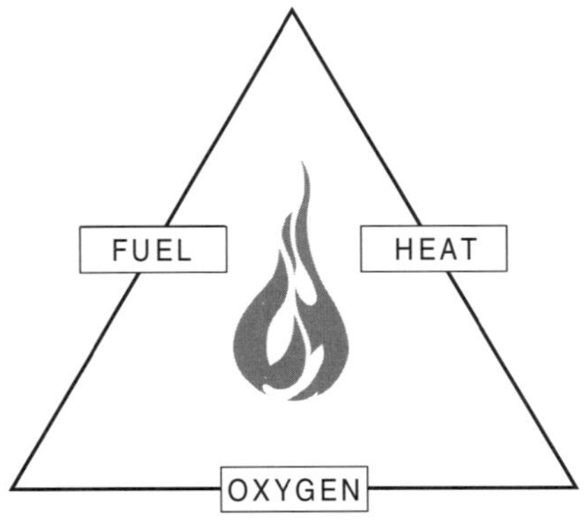

Fig. 8-23 Fire Triangle

2. Oxygen - Air is only about 21% oxygen, most of the remaining 79% of air is nitrogen which does not contribute to the process of combustion. It is only a waste by-product, which cools the flame.
3. Heat - Each fuel has its own ignition temperature, which must be maintained throughout combustion to prevent extinguishing the fires. When firemen douse a fire with water, they attempt to remove the heat leg of the fire triangle.

Depending on the efficiency of combustion, other products of combustion could include:

- Carbon Monoxide (CO)
- Soot
- Smoke
- Fly Ash
- Nitrous Oxides NO_x (also known as "brown gas")
- Trace Gases

The **3 - Ts** are factors that affect combustion efficiency include:

- *Time* allows the maximum number of fuel molecules to come in contact with oxygen molecules. This enables most of them to complete their combustion.
- *Temperature*, as mentioned earlier, is the third leg of the fire triangle. If the ignition temperature of the fuel is not maintained to enable the fuel to ignite itself throughout the process of combustion, only some of the fuel burns.
- *Turbulence* is necessary to thoroughly blend air and fuel molecules. These molecules can combine and produce combustion only if they come into contact with each other.

Without enough of the 3-Ts (time, temperature or turbulence) during combustion, some unburnt fuel could be swept up the stack before igniting.

Perfect, or theoretical, combustion cannot be achieved in boilers. It is even extremely difficult to accomplish under exact laboratory conditions, but it is a guideline for efficient combustion under ideal conditions.

Perfect combustion would occur if all the molecules of fuel would combine with only the minimum number of oxygen molecules at their ignition temperature and combust into flame with no wasted fuel or oxygen.

Complete combustion is as close to perfect combustion as can be reasonably maintained in an operating boiler. This should be an operator's target for highest combustion efficiency. Minimum pollution and minimum waste from heated air are the benefits of complete combustion.

Incomplete combustion results in smoke. It is wasteful, can reduce boiler efficiency, and can even be harmful to the boiler. With today's higher sensitivity to environmental conditions, the Environmental Protection Agency, along with other state and local agencies, monitor stack emissions. They may fine the boiler operator and/or owner for pollution resulting from incomplete combustion.

Primary air is the air provided to the boiler to control the rate of combustion. It actually determines the amount of fuel that can be burned.

Secondary air is supplied to control how completely the fuel is burned. This affects the efficiency.

Excess air is the extra amount of air supplied above theoretical that is needed to create proper turbulence.

Fuel/air ratios are designed to provide the correct mix of air and fuel for most efficient combustion. A typical fuel/air ratio is 15:1 (15 parts air to 1 part fuel), but for any given burner, ratios may vary anywhere from a 15:1 range to 30:1 range.

Fuels

Many different fuels can be used in boilers, from garbage and medical waste to coal, oil, butane, propane and natural gas. The heating value of different fuels is expressed as the quantity of Btus that a particular fuel can produce.

Natural gas, a blend of 55-90% methane combined with ethane and a trace of inert gases, is by far the most common boiler fuel in America. It is the cleanest burning of all fuels, and since it is provided by the local utility, requires no fuel tanks on the premises. One therm is a unit of natural gas equivalent to a 100,000 Btus quantity of heat, which is about 100 cubic feet of natural gas.

Fuel oil is next in popularity. It is sometimes used as a back-up fuel if the gas is shut off. It can also be used as a cost savings measure for times of the day when the gas rates are most expensive. Users who choose this *dual fuel* option have combination oil *and* gas burners. The incentive here is that these users would then qualify for a lower *interruptible* gas rate.

Crude oil is distilled into different grades ranging from #1-6, with Grade #1 being the lightest, least viscous and cleanest burning. Grade #1 is the most expensive to purchase and provides the fewest available Btus per pound. Grade #6 is the least expensive to

Grade of Fuel Oil	Heating Value - Btus/Gal.	Color
#1 Fuel Oil	137,000	Light
#2 Fuel Oil	141,000	Amber
#4 Fuel Oil	146,000	Black
#5 Fuel Oil	148,000	Black
#6 Fuel Oil	150,000	Black

Notes: Grade #1 is rarely used due to cost.
There is no #3 Grade fuel oil.
Grade #6 is commonly called "Bunker C" fuel oil.

Fig. 8-24 Fuel Grade Chart

purchase and produces the most Btus per pound, but is most viscous. It must be heated to flow to the burner(s) and pollution control equipment, such as flue gas scrubbers.

Coal is rarely used as a fuel in commercial heating applications, but it is popular in industrial boilers, such as fueling large power plant boilers.

Boilers Fuel Heating Values

- *Natural Gas* - 800 to 1400 Btu per cubic foot, depending on the mixture. It is usually considered to provide a heating value of 1000 - 1100 Btu/cubic foot.
- *Propane* - 2500 Btu/cubic foot
- *Fuel Oil Grades* (See Figure 8-24)

Boiler Controls

Most boilers come complete from the manufacturer with self-contained control packages, including temperature and safety controls.

Boiler operating controls operate in one of the four firing rate control schemes, any of which may be used for either steam or hot water boilers:

1. **On - Off** starts the burner at one pressure or temperature (for example, 5 psi) and shuts-off at some higher pressure or temperature (for example, 7 psi). An On - Off burner on a steam boiler uses one *pressure* control, while a hot water boiler requires one *temperature* control. These controls are typically used on boilers up to 50 BoHP.
2. **Low - High - Off** starts the burner, but with a low fire output until ignition is proven, then it steps up to high fire after proof of ignition. It is used in the 15 – 60 BoHP range. One temperature (hot water) or pressure (steam) control and a slow opening gas valve are used.

3. **Low - High - Low - Off** offers some degree of modulation. It starts in low fire, proves flame, then steps up to high fire until the pressure reaches the setpoint of the second operating controller (for example 106 psi). It then it drops to low fire until the next setpoint is reached (for example 110 psi), causing the burner to shut off. When the pressure falls back to setpoint (for example 100 psi), the burner is returned to high fire. Boilers from 60 - 150 BoHP commonly use this control. This type of firing rate control uses two temperature or pressure controls.
4. **Modulating** control automatically modulates the burner between low and high fire after starting on low fire. As the steam pressure approaches setpoint, the modulating controller reduces firing rate to stabilize boiler capacity at a point that will just balance the system load. It will then modulate to maintain this balance. It uses one modulating temperature or pressure control. It is typically used on boilers above 100 BoHP. The boiler manufacturer generally provides a proportional signal (0-135 ohm, for example) to a step controller or other boiler sequencing device.

The burner should always be fired for longer periods than it is not fired. This actually increases boiler efficiency by maintaining a constant furnace temperature. Another advantage of constant temperature is that it reduces refractory maintenance from destructive repeated heating/cooling cycles.

Additionally, on hot water boiler systems, the temperature control contractor will usually provide controls and a three-way valve that

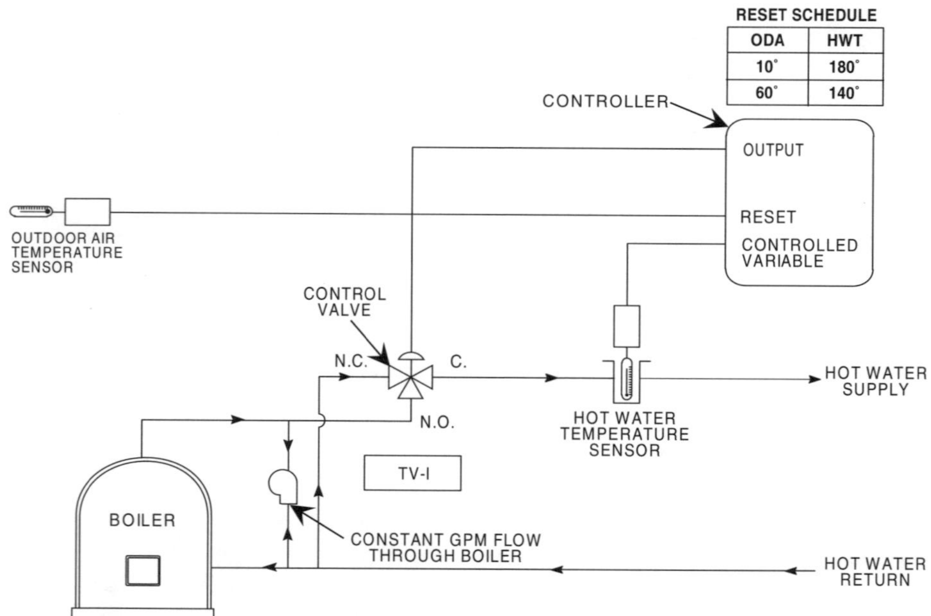

Fig 8-25 Boiler Hot Water Temperature Control

mixes boiler hot water with the returning hot water to the boiler, as covered earlier. This provides a hot water supply to the boiler which reduces the possibility of thermal shock to the boiler. Most hot water heating boilers maintain a fixed setpoint, such as 180°F. To conserve energy, this 180°F hot water is mixed with return hot water to achieve a desired hot water system temperature. A typical reset schedule based on outdoor temperature conditions is shown with the associated control scheme in Figure 8-25.

Combustion air/flue gas damper controls hold the damper closed when the burner is OFF and modulates it according to the load (steam or hot water demand) when the burner is ON. Safety controls should provide an interlock between the burner and the damper that is actuated when the damper opens and is wired into the burner circuit.

Flame safeguard control is designed to supervise and verify each step in a programmed series of control events that produce the electrical switching required for boiler start-up and safe operation.

The flame safeguard program is conducted by either digital electronics or motor-driven, cam-operated sequencing switches that are set to open and close circuits at predetermined time intervals which start:

- Tests of control circuitry
- **Pre-purge**, the 5 volumetric changes of air through the boiler furnace lasting from 30 - 60 seconds. It is used to remove any explosive gases which may have settled in the firebox before ignition. Prepurging reduces the chance of furnace explosion on boiler lightoff.
- **Pilot ignition** and trial for ignition
- **Main flame and trial** for main flame ignition.
- Transfer of control from programmed to automatic
- **Burner shutdown and postpurge**, or running the primary air fan after the burner fuel is shut off. Postpurging requires the same amount of time as the pre-purge.

Burner management controls provide for ignition, main flame failure protection, control of the blower motor, main fuel valves and modulating motor. A newer design is called the **microprocessor-based burner management control system** which also incorporates direct

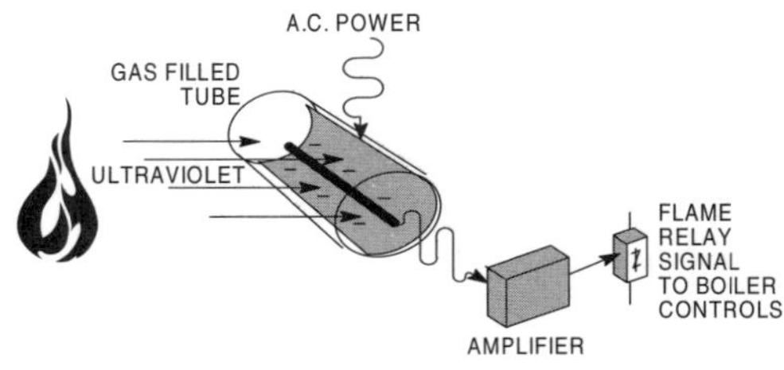

ULTRAVIOLET DETECTION

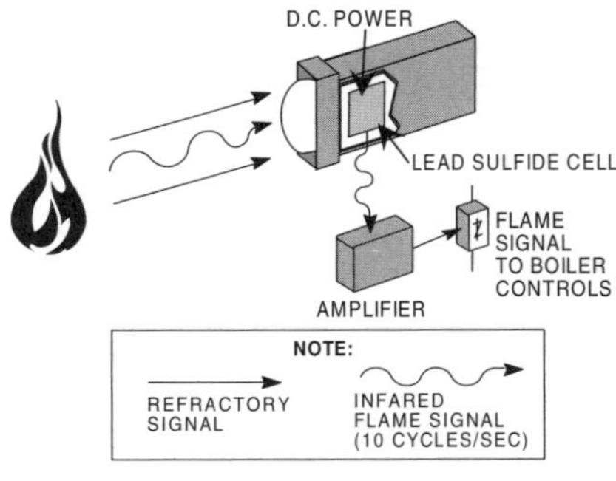

INFRARED DETECTION

Fig. 8-26 Flame Detection Methods

Courtesy of the Trane Company

connection of limit and operating controls, damper position interlocks (such as air flow, fuel pressure, and temperature) burner motor, ignition, pilot valves, and alarms.

Positioning control systems control furnace combustion by sensing changes in steam header pressure. These control systems control the air and fuel supplied to the furnace. Older pneumatic systems have been replaced by electric or electronic control.

If the flame is not lighted or suddenly goes out, **flame failure controls** guard against a furnace explosion by scanning the furnace to prove that the pilot or main flame is lit. The scanner will then either not allow the fuel valve to open or will close it. A closed valve prevents fuel from entering the furnace. Two predominant models are shown in Figure 8-26.

1. **Infrared detectors** sense the fluctuating visible and infrared radiation (IR) at a frequency of 10 cycles per second, a frequency peculiar to all flames. This "flame signal" is then amplified and goes on to close certain circuits, which indicate the presence of a flame.
2. **Ultraviolet detectors** sense the ultraviolet radiation (UR) of a flame on a gas-filled glass tube containing two electrodes connected to an AC power source. Sensed UR falling upon the electrodes causes the gas to allow current to flow from one electrode to another. This current energizes the flame relay and the system has verified the existence of a flame.

Boiler Water Treatment

If a boiler was operated with untreated or improperly treated water, it would live a short life. Untreated water scale forming salts, such as calcium and magnesium, not only damage the boiler, but also have a staggering insulating value. One-half inch of scale drops efficiency by 60%, wasting 600 of every 1000 gallons of fuel!

Proper boiler water conditions will either keep these solids in suspension in the water (instead of allowing them to adhere to the tubes) or make them settle into the mud drum as sludge. Proper conditions will also neutralize pH preventing acid attack, and remove free oxygen (prevents oxygen pitting).

Bottom blowdown removes sludge build-up from steam boilers periodically by opening a valve and removing some of the boiler water to a blowdown tank. **Continuous blowdown** does the same job as bottom blowdown on steam boilers, but accomplishes it by allowing a continuous trickle of blowndown water to a **flash tank**. A heat recovery unit can recycle the waste heat generated by blowdown and use it to heat boiler feedwater.

Surface Blowdown removes floating scum from the boiler water. It also lowers the boiler water **conductivity** (a measure of the current carrying capacity of water a reference to corrosive potential.) Water with excessively high conductivity can cause foaming.

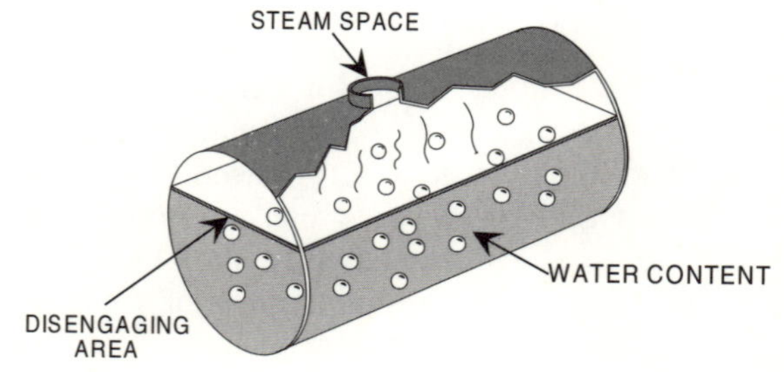

Fig. 8-27 Disengaging

Courtesy of the Trane Company

Foaming is a water problem that prevents steam bubbles from breaking in the disengaging area. It can cause carryover of water with the steam into the system.

The **disengaging area** in Figure 8-27, is the surface area of water at the waterline of a steam boiler. It is where steam bubbles "disengage" from the boiler waterline as the water is heated. Fire tube boilers typically have enough disengaging area to produce 98% quality dry steam; wet steam beyond 5% is rarely found on fire tube boilers.

Priming is a similar problem to foaming with the steam bubbles leaving the disengaging area. It is caused by steaming at too high of a firing rate or with high water, resulting in boiling that is too violent.

Zeolites and lime-soda softeners purify and soften the water used as boiler supply feed water by removing salts such as magnesium and calcium from the water.

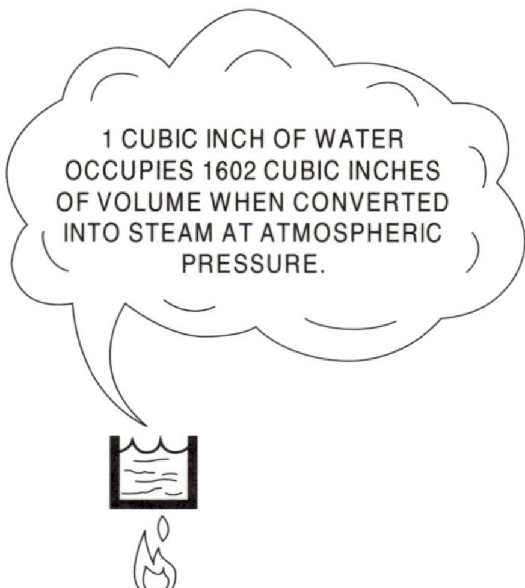

Fig. 8-28 Water to Steam Volume Change

Boiler Casualties and Safety

A furnace explosion or flareback is a boiler explosion caused by excess fuel or fuel air vapors in the firebox that result in an explosive mixture. Modern boiler controls guard against flarebacks by not allowing fuel into the boiler unless the pilot is on, as viewed by their flame failure controls. *Flame failure is the most common cause of furnace explosions.*

Fuel casualties can cause a fire if there is leakage or breaks in the fuel system piping or controls.

A **boiler explosion** will result, not from igniting fuel, but from a sudden large demand for steam, a sudden pressure drop or a ruptured pipe on a fire tube boiler without a corresponding drop in temperature.

Under these conditions, an explosion can occur because the high temperature water inside the shell immediately flashes into steam at the new lower pressure caused by the sudden large demand for steam. Since steam has a volume of about 1600 times more than that of water, the immense volume of flashing hot water will overpressurize the steel boiler shell until it eventually bursts. See Figure 8-28.

High water will damage a steam system more than the boiler itself, especially if the steam is sent to a turbine. In a steam heating system, high water will cause water hammer in the pipes. Water hammer is a problem created whenever a slug wall of water slams up against a valve or piping elbow. Steam pressure can cause this "wall of water" to develop and travel inside the pipe until it hits a valve or elbow causing an objectionable hammering sound and possibly mechanical damage. Of course, in a hot water boiler, high water is not a concern because the boiler, all piping, and heat exchange devices (i.e. coils, fintube radiation, etc.) are completely filled with water.

Low water is an extremely dangerous condition for a boiler. Since water cools the tubes, its cooling effect on the tubes is taken away if its level drops. The tubes will overheat and either warp, marry (join together) or rupture if the overheating is serious enough.

Ruptured tubes occur because of either low water overheating or milder overheating caused by excessive water side scale. This is a reason why water chemistry control is so vital.

Thermal shock, as mentioned earlier, will unevenly stress the boiler pressure parts. This can twist tube sheet ligaments and can crack them. This occurs if the temperature difference between boiler temperature and incoming water temperature is more than 50°F, or if boiler water circulation is not adequate.

Efficient Boiler Operation

Proper operation of a boiler is not only safe, but it increases the boiler's useful life and saves fuel; all of which save money!

Combustion gas analysis is a test used to measure the efficiency of combustion and its results give an indication of how to optimize the fuel/air ratio.

- **Carbon Dioxide** (CO_2) percentage is sometimes used as an indicator of complete combustion. Flue gas analysis while using natural gas fuel should yield CO_2 measurements between 8.6 - 11.5% For fuel oil, the indicator should read 11.9% CO_2.
- **Carbon Monoxide** (CO) in the flue gas indicates incomplete combustion because not enough air is being supplied to the furnace. CO is best at 0 ppm (parts per million) because zero CO indicates complete combustion, but manufacturers and local codes usually allow a maximum of 400 ppm of carbon monoxide in the combustion gases.
- **Oxygen** (O_2) is a requirement of the combustion process. Its presence in the correct percentage in the combustion gases is a good indicator of the quantity of air supplied to the burner.

If analysis reveals no oxygen after combustion, it is likely that the combustion process is being starved of air (usually resulting in smoke). Too much oxygen indicates that there is too much air entering the boiler, resulting in wasted heat. Whenever the oxygen level in the combustion gases exceeds 3 - 5% O_2 excess air, it is an indication of an overabundance of air supplied to the burner. This inefficiency stems from heating both the 77% nitrogen component of air *and* the excess unused oxygen after combustion, both of which escape as heat wasted out of the stack. By adjusting the fuel-air linkage on the burner so that 3-5% O_2 is present in the combustion gas, the boiler burner then has enough oxygen to support combustion, but not so much that heat is wasted. In general, equipment usually requires between 16-35% excess air before firing, correspondingly combustion gas analysis should reveal the 3 - 5% excess O_2.

As important as the percentage of combustion gas by-products, **flue gas temperature** is a strong indicator of the efficiency of the boiler. This temperature shows how much heat is extracted from the combustion gases before they leave the boiler. A rising flue gas temperature is evidence of dirty heat exchanger surfaces or broken/missing baffles in a multipass boiler. *For every 40°F rise in stack gas temperature, boiler efficiency drops 1 percent.*

Air preheating with economizers and air heaters, located between the boiler and the chimney on some boilers, increases efficiency by warming combustion air with the heat from flue gas. Capturing this heat not only reduces the chilling effect of cold combustion air on the flame, but also reduces the stack gas temperature by extracting the last few usable Btus before the gas escapes out the stack.

Feedwater preheating uses the waste heat steam from the flash tank to heat water in the feedwater heater. *Every 10°F rise in feedwater temperature gives about a 1% savings in fuel.*

Cycling burners continually between on and off reduces boiler efficiency. By widening the temperature differential of the operating and limit controls that turn the burner on and off will reduce the number of firing cycles — the fewer the cycles, the higher the efficiency.

The most efficient modes of burner operation are low-high-low and modulating, the modes which reduce on-off cycling.

Effects of scale on water side heat transfer can have a negative impact on efficiency. As mentioned earlier, scale has an enormous insulating effect. It reduces the number of Btus of heat that can be transferred to the water. Eventually boiler tubes will overheat. Because circulating water cannot carry away the heat on the tubes, they overheat and sometimes rupture. Scale must be removed promptly and prevented from reforming by ensuring proper boiler water chemistry is maintained. *one-sixteenth inch of scale on the tube waterside surface can reduce efficiency by 11%, wasting 110 gallons of fuel for every 1000 gallons consumed by the boiler.*

Effects of soot on fire side heat transfer can have a negative impact on efficiency. Soot that accumulates on the tubes is an excellent insulator. Poor burner performance that results in incomplete combustion will coat the tubes with a fuel wasting soot residue. Since soot is actually unburned fuel, it is also a fire hazard inside the boiler. One-sixteenth inch of soot on the tube's fire sides can increase fuel consumption by 4.5 percent. See Figure 8-29. This reduction in efficiency amounts to wasting 45 gallons of fuel for every 1000 gallons burned.

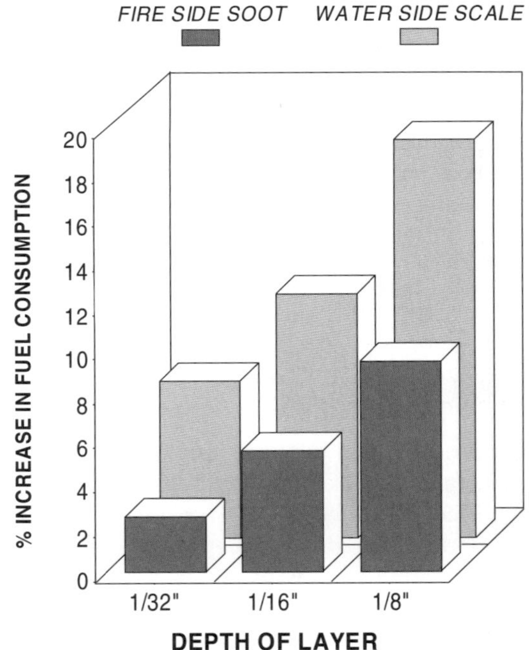

Fig. 8-29
Boiler Heating Surface Insulators

Summary

This chapter introduced common boiler terminology, components and operation. The discussion concentrated on the four basic boiler types with special emphasis on fire tube boilers which are the most popular for commercial heating applications. In this chapter you learned:

- Common HVAC boiler terminology such as boiler capacity, boiler horsepower, heating surface, gross output and net rating.
- Furnaces and electric, water tube, and cast iron boilers all can be used to heat buildings, but fire tube boilers are often the choice for HVAC comfort applications.
- There are different types of fire tube boilers, such as horizontal return tube and vertical tube boiler, but scotch marine boilers are the most dominant style of steam and hot water fire tube boiler manufactured today.
- The function of the major components of a boiler including shell, breeching, furnace, tube sheets, ligaments, tubes, drums, blowdown pipe and valves, headers, refractory, stays, dampers and baffles. Ligaments are susceptible to cracking from stress caused by temperature differentials across the tube sheet of greater than 50°. Boiler fittings, accessories and instruments were also discussed.
- Basic boiler support systems include the feedwater system to supply the boiler with water at the proper temperature and flow rate, the fuel system designed to safely deliver fuel at the proper temperature and pressure to the burner(s), the air system to provide the draft required for combustion, and the piping system to transfer the heating medium throughout the building.
- Combustion is the rapid fracturing of chemical bonds to release large amounts of useful heat. The factors affecting combustion include time, temperature and turbulence (3-Ts). Complete combustion, which is as close as possible to perfect combustion, creates maximum efficiency and minimum pollution and waste.
- Natural gas is the most common boiler fuel followed by fuel oil in popularity. Crude oil is occasionally used, but coal is rarely used as a boiler fuel in commercial heating applications.
- Boiler controls, including combustion air/flue gas damper controls, flame safeguard controls, burner management controls, flame failure controls, and positioning control systems, are normally installed on the boiler by the manufacturer. Boiler operating controls for either steam or hot water boilers operate in an On-Off, Low-High-Off, Low-High-Low Off, or Modulating firing rate scheme.
- Proper water treatment prevents scale formation, acid attack, and oxygen pitting which can reduce the efficiency of the boiler and shorten its life.
- The causes of furnace explosion, fuel casualties, boiler explosion, ruptured tubes, overheating, and thermal shock and the operational considerations relating to boiler safety, increased efficiency and hot water controls.
- Combustion gas analysis and monitoring flue gas temperature can indicate boiler efficiency. The efficiency can be improved by air and feedwater preheating and by reducing cycling. Scale on the water side and soot on the fire side also reduce the efficiency of the boiler.

9

Heat Exchange and Heat Recovery Equipment

Transferring heat from one location to another is the fundamental process of many air conditioning operations. During winter, heat energy from coal, oil or natural gas is transferred to the air or water used to heat a facility. During summer, heat must be removed from air or water to cool a facility. The broad categories of equipment used to facilitate these transfers are called heat exchangers. This equipment, as well as heating/cooling coils and various heat recovery devices, are the subjects of this chapter. Upon completing this chapter you will be able to:

1. Describe the design and operation of water and steam heat exchangers and convertors.
2. Describe the design considerations and operation of air heating and cooling coils.
3. Explain the operation of air-to-air, air to water and water-to-water heat recovery devices, including heat wheels, counter-flow units, heat pipes, coil loops and flat plate exchangers.

Heat Exchangers

A **heat exchanger**, sometimes called a **convertor**, heats or cools water, glycol, air, or other fluids flowing through it with steam or water of a higher temperature. Although several different fluids may be heated or cooled in an exchanger, only water will be considered for this initial discussion.

Water Type Shell and Tube Heat Exchanger

A heat exchanger is a steel shell that contains many tubes arranged in long rows. See Figure 9-1. There are separate inlet and outlet pipes to both the shell and tubes. In general, the water being heated or cooled circulates through the tubes while the medium that is cooling or heating flows between the shell and the tubes.

The tubes in a heat exchanger are typically copper. Since the tubes of a heat exchanger are submerged in the heating or cooling medium, the heat is transferred directly by conduction directly between the medium and the tubes.

Boiler water is pumped through the shells of water type heat exchangers. See Figure 9-1 again. Baffles mounted on the tube bundle direct the water flow across the tubes through the length of the exchanger. Baffles create an "S" flow pattern which encourages heat transfer. The hot water return enters at the bottom of the shell and leaves through the top. This permits dissolved air to follow the water flow out of the heat exchanger. In this way, air pockets are eliminated.

In most exchangers the tube bundles are bent in 'U' shapes. The medium being heated passes through the tubes as directed by the front head construction. Depending on this construction, flow may pass through the exchanger two, four or six times.

For the two-pass flow shown in Figure 9-1, water enters the bottom head opening, flows through the bottom section of the tube bundle and then through the top half, leaving by way of the top opening. Notice that the hottest water also enters the bottom of the shell. The internal baffles keep this hotter water in the coldest region of the exchanger, promoting the best heat exchange. For a four-pass operation, the head openings are side by side in the top

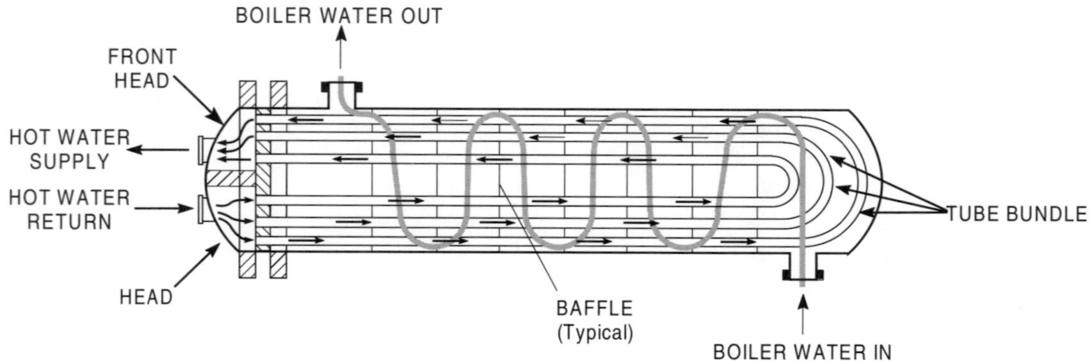

Fig. 9-1 Water Shell Heat Exchanger Cross-Section

half of the head with a vertical partition separating them. The water makes two passes through each half of the tube bundle for a total of four passes. Additional head partitions are used to provide for six-pass operation.

Water shell heat exchangers are usually installed next to the boiler to shorten the connecting piping. See Figure 9-2. The exchanger is pitched slightly upward toward the top opening for proper air venting. A circulating pump promotes uniform flow between the boiler and the exchanger. Depending upon the elevation of the exchanger above the boiler, water may flow by convection when the pump is not running. In many cases a flow control valve prevents gravity circulation when the pump is not running.

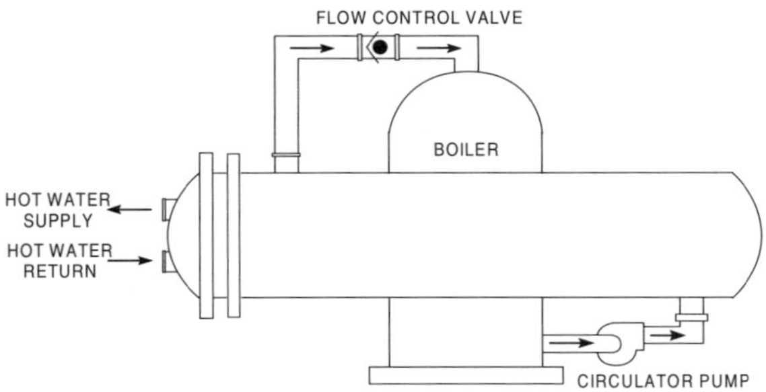

Fig. 9-2 Boiler Connections to Water Shell Heat Exchanger

Steam Convertors

Steam convertors use steam as the heating medium in the shell. The water flow in the tubes is the same in steam units as it is in the water units previously described. As the steam gives up its latent heat, it is converted to condensate and collects at the bottom of the shell. Convertors are pitched downward to direct this condensate toward the drain shown in Figure 9-3.

When the hot water supply temperature rises, the controller closes the control valve. Residue steam left in the heat exchanger gives off its latent heat, condensing and lowering the vapor pressure below atmosphere. The vacuum breaker allows air to enter the convertor allowing the condensate to escape through the steam trap. This relief is provided by air vent valves, the condensate drain apparatus and the vacuum breaker shown in Figure 9-3.

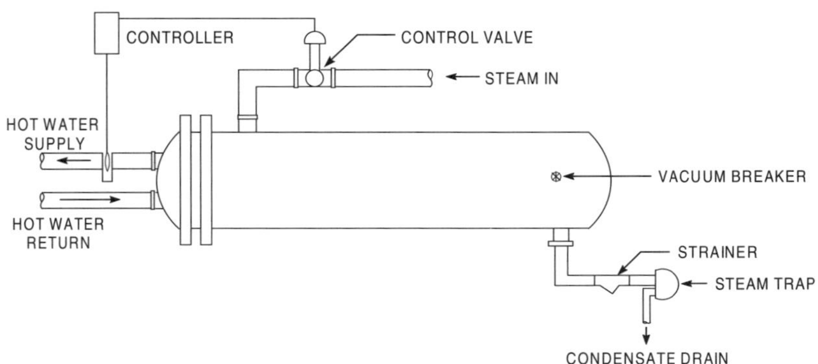

Fig. 9-3 Steam Shell Heat Exchanger

Air Heating and Cooling Coil Designs

A **coil** is simply a tube or pipe formed into a serpentine or helical shape. Unlike the coils or tubes used in heat exchangers, the coils used for heating or cooling air generally have fins bonded to the tube. These fins greatly increase the surface area for heat transfer. See Figure 9-4.

The transfer of heat from the air flowing over a pipe to the fluid flowing in the pipe is impeded by three resistances:

- The inability of the air to contact the metal coil. This is known as outside surface (or air film) resistance.
- The resistance to the conduction of heat through the pipe wall.
- The inability of all the fluid to contact the inner pipe wall.

In air conditioning applications the last two resistances are low when compared with the air film resistance. It is easy to decrease the air film resistance by adding fins to the coils. The fin may be mechanically bonded onto the tube, soldered at the fin root, or the fin may be formed out of the material of the tube itself. The bond between the fins and the tube is important in providing the rated heat transfer.

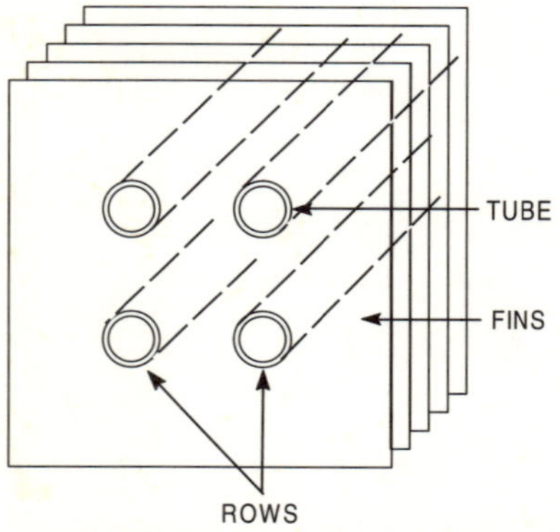

TUBE
FINS
ROWS

Fig. 9-4 Fin and Tube Details of a Two Row Air Heating/Cooling Coils

Heating coils for air are usually made of copper tubes and aluminum fins. Cooling coils containing water or direct expansion refrigerants are also made this way. If dehumidification is part of the cooling process, or if the coil will be sprayed with water for increased heat transfer, copper tubes with copper fins minimize corrosion. Coils for air-cooled refrigerant condensers are usually made of copper tubes and aluminum fins, although all aluminum coils are also common.

Heating and cooling coils have tubes with outside diameters of ⅜ to 1 inch and fin spacings of two to 14 fins per inch. The number of tubes in-line make up the number of rows. Rows may vary from 1 (as in a one-row steam coil) to as many as 12 (as in a 12-row cooling coil). The spacing between the rows is determined by anticipated dirt buildup, frost buildup and air bypassing. The heat transfer surface is made up of rows of finned piping located at right angles to the air flow. The headers and pipe connections are arranged so that heat transfer occurs as air passes from row to row. See Figure 9-5.

The finned tubes are usually mounted horizontally to promote the drainage of externally condensed moisture resulting from dehumidification. Some steam coils, however, have vertical or pitched tubes to permit rapid drainage of internal condensate. See Figure 9-6. The coil rows may be arranged in line or staggered with respect to the air flow. The staggered arrangement is usually preferred to create maximum contact between the fins and the air.

Numerous types of fin arrangements are also used. The more common are the spiral, flat, or corrugated types. The spiral fin surrounds the entire tube. Flat fins may be round or square.

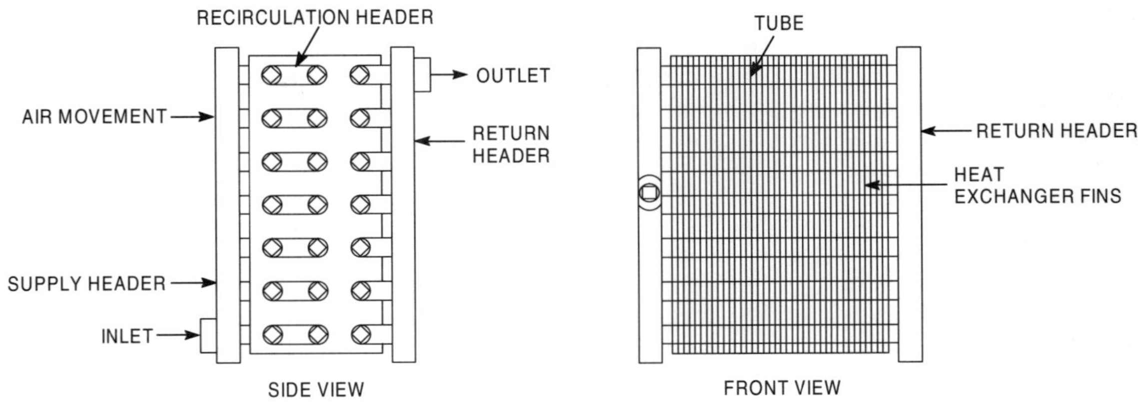

Fig. 9-5 Tube Arrangements of Air Heating/Cooling Coils

Flat and corrugated fins may be attached to individual tubes or to several rows of tubes.

To optimize heat transfer in water-filled coils, the entrapped air must be eliminated and efficient internal flow patterns must be created. To eliminate air, water is piped into the bottom and out of the top of the coil. In this way, air which naturally floats to the top is carried out of the coil. Water coils are sometimes fitted with turbulators, devices inside the tubes that swirl the water and increase heat transfer. These coils are also often protected with strainers that collect dirt. The strainers reduce the number and frequency of tube blockages.

In steam coils, both the condensate and entrapped air must be continuously eliminated to ensure an even temperature distribution within the coil. Eliminating condensate also minimizes the risk of condensate freezing at the bottom of the coil during light heating loads.

Heating Coil Operation

Hot water coils are designed for a 20 to 30°F temperature drop in the water, while maintaining the required heat transfer. Internal water temperatures of 170°F to 190°F are common in low temperature systems and

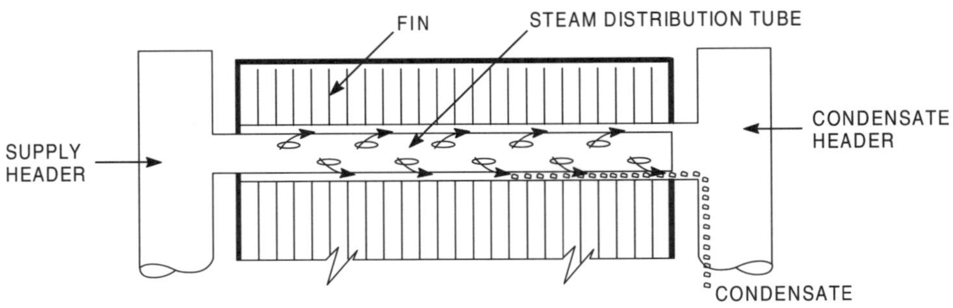

Fig. 9-6 Steam Coil

300°F is typical in high temperature systems. The water flow loss in one of these units is about 4 psig. The air flow loss is from 0.2 to 0.5 inches WG. Hot water and steam heating coils are usually rated for air flow face velocities of 500 to 750 ft/min.

Coil selection is based on manufacturer data. When selecting coils, designers consider entering and leaving air and water temperatures, coil face velocities and pressure drop through the coil.

Water reheat coils can raise the temperature of the air passing through the coil from 70°F to 120°F. They are usually oversized by 10-25% to warm the system quickly during the morning warmup period.

Steam preheat coils can raise the temperature of the air leaving the coil up to 60°F, regardless of the air's entering temperature. The air temperature rise across preheat coils is usually limited to about 30°F to ensure stable control and to minimize the risk of freezing at partial load conditions. Two preheat coils may be used in series if a greater than 30°F rise is required. Large preheat coils with a 40°F rise or larger, may require face and bypass dampers to mix heated air with bypassed air to create the right discharge temperature.

Cooling Coil Operation

Chilled water and refrigerant are the most common cooling mediums used in cooling coils. Coils are generally 4 to 8 rows deep and if cooling is accompanied by dehumidification they may be 12 rows deep. Entering air temperatures vary from 60° to 110°F dry bulb, and 50°F to 80°F wet bulb. In most applications, the design creates a leaving air temperature around 55°F. Well water designs may go as high at 60°F. Since cooling coils dehumidify the air as well as cool it, lower face velocities between 300 to 700 FPM are recommended to minimize re-entrainment moisture droplets in the leaving air.

The rate of flow through the cooling coil depends upon the required heat exchange. A flow of 2 GPM/ton of refrigeration provides a typical temperature rise in the cooling coil of about 10°F. Water flow head loss in the coil is usually about 4 psi. Air flow resistance through the coils typically varies from 0.25 to 0.75 inches WG.

Direct Expansion Coil Operation

Beginning in the upper left of Figure 9-7, refrigerant under pressure enters the thermostatic expansion valve. The thermostatic expansion valve senses the superheat of the refrigerant leaving the coil and meters the required refrigerant. As the refrigerant flows through the cooling coil (also called an evaporator), it removes heat from the passing air and the refrigerant expands. The liquid refrigerant first absorbs heat much like water, but unlike water, it vaporizes, expanding in the

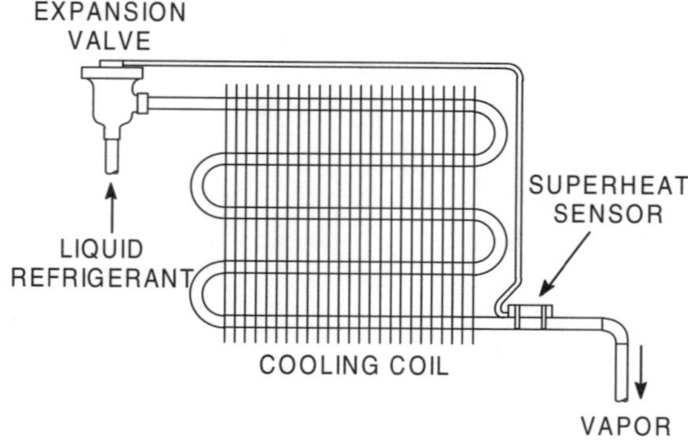

Fig. 9-7 Direct Expansion Coil (distributor tubes not shown)

Courtesy of the Trane Company

process. The cooling effect on the air is equal to the latent heat of vaporization. This process of evaporating, using the air as the source of heat is an effective way of cooling. Vaporized refrigerant leaving the evaporator flows to the compressor.

To ensure uniform refrigerant distribution in the evaporator, a distributor equally divides the mixture of liquid and vaporizes refrigerant flowing from the expansion valve among the various coil circuits. The individual liquid connections from the distributor to the coil inlets are commonly made of small tubing, all being same length and diameter in order to obtain equal friction losses between the distributor and the coil.

In direct expansion systems, the cooling coil (evaporator) capacity is balanced with the requirements of the compressor. Compressors are designed to compress *vaporized* fluids. As the temperature rises to a superheated value on the outlet pipe, the thermostatic expansion valve opens, metering more refrigerant to the cooling coil. This temperature, when superheated 10 - 20°F, will signal the expansion valve to allow more liquid refrigerant into the coil. The sensing of superheated vapor protects the compressor and provides an effective way of metering refrigerant into the coil. Liquid refrigerant cannot be compressed. If liquid refrigerant reaches the compressor, it will severely damage the compressor.

Direct expansion coils usually operate with a refrigerant temperature of 20°F to 55°F at their outlets. These temperatures may also include up to 12° of *superheat* to prevent liquid refrigerant from passing to the compressor.

Heat Recovery Methods

Concern for reducing energy costs and conserving fossil fuels has created much interest in heat recovery equipment. **Heat recovery** equipment captures energy that would often be wasted. For example, most air heating and cooling systems vent some return air outside. During the summer, this exhaust air is likely cooler than the outside air and during the winter the exhaust air is likely warmer than the outside air. In both cases, some of the energy of the exhausted air is recoverable for use in cooling or warming the incoming outside air.

There are three general categories of heat recovery methods:

- Air-to-air
- Air to water
- Water to water

Air to Air Recovery

Air-to-air heat exchangers help recover wasted HVAC energy. Four different types of heat exchangers can be used for air-to-air heat recovery. They are heat wheels, counterflow or plate-type exchangers, heat pipes, and coil loops or run-around coils.

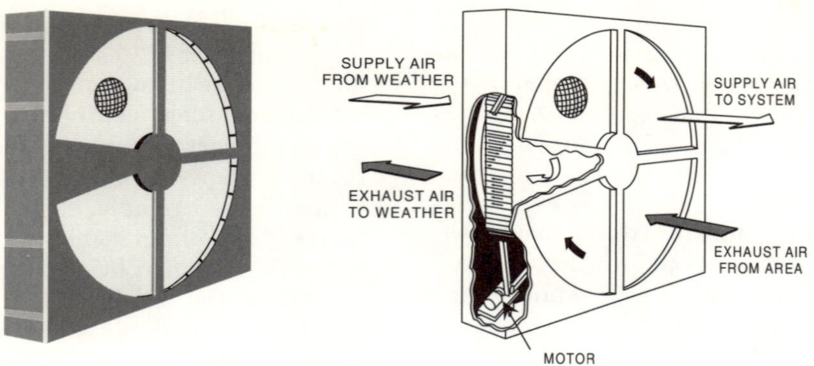

Fig. 9-8 Heat Wheel

Permission of Munters Corporation — Cargocaire Division, All Rights Reserved

Heat Wheels

Heat wheels, like the one shown in Figure 9-8, are popular heat recovery devices because their design and operation is simple and they can recover both latent and sensible heat. For latent heat transfer to occur, a desiccant material is applied to the wheel. Wheels commonly use fiberglass, aluminum foil or plastic film as the substrate to carry the desiccant material. The media is generally formed into a honeycomb shape that allows lower pressure drop and ease of cleaning. One side of the wheel is ducted to the exhaust air, and the other side is ducted to the outdoor air.

The **heat wheel** rotates slowly, absorbing the total heat energy from the exhaust air. The incoming outdoor air is pre-conditioned by the wheel as it then absorbs this energy or coolness. A small amount of exhaust air always carries over to the outside air. The wheel revolves between one and twenty revolutions per minute, controlled by the demand for heat transfer.

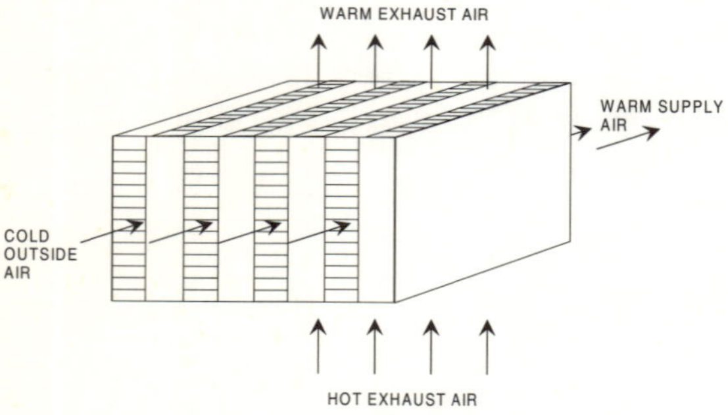

Fig. 9-9
Counterflow or Plate Type Recovery Unit

AIR CONDITIONING PRINCIPLES AND SYSTEMS, 2nd ed. by Pita, Edward G., ©1989

Counterflow or Plate-Type Units

Counterflow or plate-type recovery units use metal plates to conduct heat from one airstream to another. The exhaust and outside air streams flow in opposite directions through the layered ducts that make up the unit. See Figure 9-9. Counterflow units recover sensible heat and are between 50%-60% efficient. These units are easily cleaned and create no cross contamination between air streams.

Heat Pipes

The **heat pipe, air-to-air recovery unit** is a sensible heat recovering device with no moving parts. It looks like a dehumidification coil with a partition separating the face into equal sections. The unit is ducted to exhaust air flowing through one side and outside air flowing through the other side. Sensible heat is transferred from side to side in the individual heat pipes. See Figure 9-10 (fins not showing).

The unit is made up of an array of finned tubes and each tube is a sealed piece. The tubes are the actual heat pipes. Each pipe contains an outer pipe, an inner wick and heat transfer fluid. The fluid may be water or a refrigerant. See Figure 9-11. Heat applied to one end of the pipe evaporates the fluid from the wick. The vapor flows to the cold end of the tube where it condenses and returns along the wick to the warm end for re-evaporation. The amount of heat added or removed is equal to the latent heat of vaporization.

The heat pipe is a completely reversible device that works for both summer and winter cycles. When either end of the pipe is exposed to a different temperature, the cycle begins again as the heat flows from the warmer region to the cooler region. Low flow rates can cause the unit to vapor lock and not transfer heat. In vapor lock, both evaporation and condensation are occurring on the same side of the partition.

Heat pipe units are available with face areas of 1-12 sq ft on each side of the partition. The units are small enough that they fit existing ducts well. The exhaust and outside airstreams are totally separated so there is no cross contamination. Face velocities range between 300-600 FPM and a unit can transfer heat at a rate 1,000 times that of pure copper. Heat pipes are between 60 and 65% efficient.

Air to Water Recovery

Many heat recovery opportunities exist in air. Exhaust air or boiler stack gas are prime examples.

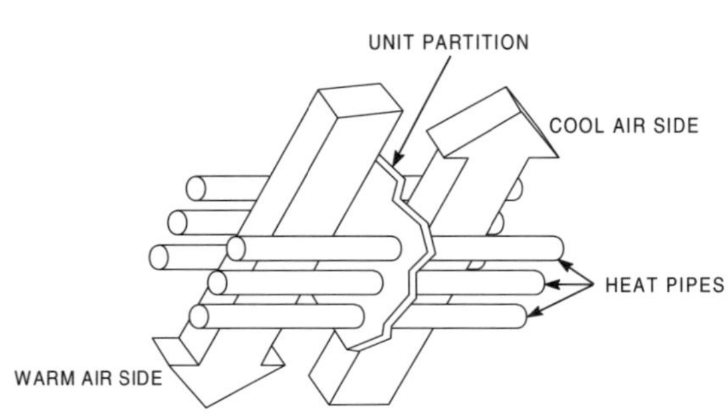

Fig. 9-10 Heat Pipe Unit

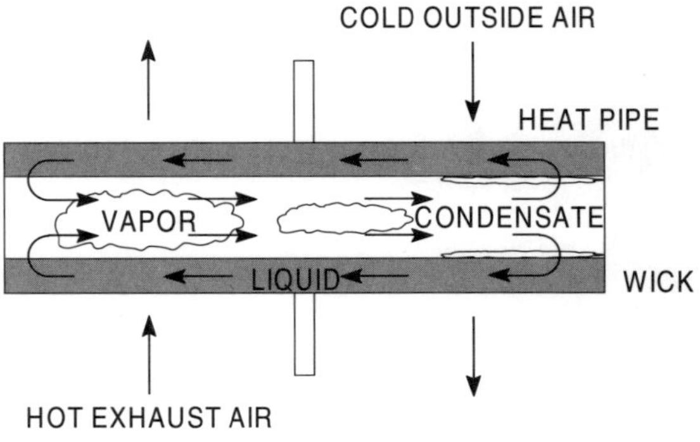

Fig. 9-11 Heat Pipe Cross Section

AIR CONDITIONING PRINCIPLES AND SYSTEMS, 2nd ed. by Pita, Edward G., ©1989

Coil Loops or Run-Around Coils

Coil loops operate similarly to the heat pipe except that the fluid is usually ethylene glycol and water. **Coil loops** have two coil sets and the fluid is pumped between them, transferring the heat from the warmer side to the cooler side. See Figure 9-12.

One advantage of the run-around coil is that the two sides of the system can be located away from each other. An external heat source is required to add just enough heat to the cool fluid entering the warm side of the unit to prevent the cooler fluid from frosting the coil. Run-around coils absorb 40-60% of the recoverable heat.

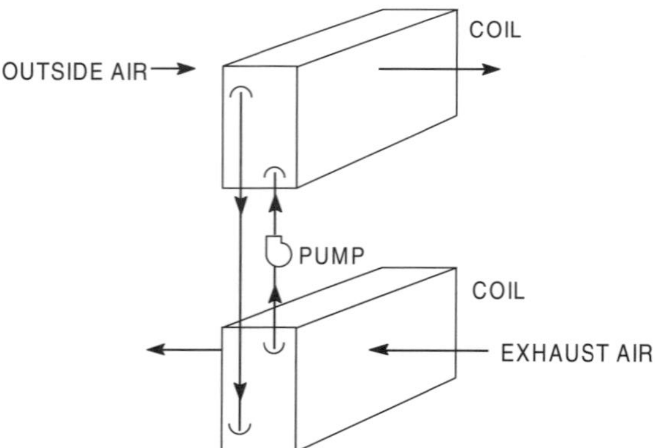

Fig. 9-12 Coil Loops or Run-Around Coils

AIR CONDITIONING PRINCIPLES AND SYSTEMS, 2nd ed. by Pita, Edward G., ©1989

Water to Water Recovery

Flat plate heat exchangers have become increasingly popular due to their high efficiency. Modular in construction, multiple plates are sandwiched together as required. See Figure 9-13. These multiple plates have a tremendous surface area, which is the main reason for their high efficiency. The plates, placed back to back, exchange heat between two fluids, typically water, for use in cooling applications. An application known as hydronic economizer, exchanges heat between cool cooling tower water, readily available during spring, fall and winter, and system chilled water. The flat plate exchanger isolates the dirty mineralized tower water from the cooling coils in the air handling equipment. In this way cooling can be created economically without contaminating the cooling coils.

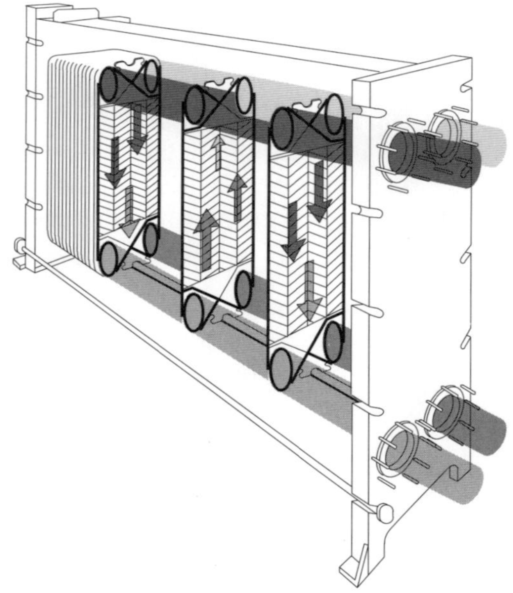

Fig. 9-13 Flat Plate Heat Exchanger

Used by permission; Paul Mueller Company Bul. AT-160104

Summary

Heat exchange and heat recovery devices transfer heat. They add or remove heat to air or water.

- Heat exchangers transfer heat between two mediums within the same vessel.
- Coils are designed and constructed to heat or cool air. They may be applied to hot water, chilled water or refrigerant direct expansion (DX) systems.
- Coils are matched to specific applications by
  - the temperature rise or drop required
  - air velocities through the coil (face velocity)
  - rows and fins per inch
  - entering and leaving water temperatures
- Heat recovery systems are classified as Air to Air, Air to Water, Water to Water.

10

Introduction to the Refrigeration Cycle and Equipment

Buildings as small as your house, up to large high-rise office buildings and apartments, hospitals and schools all have one thing in common; heat gains due to people, lights, equipment and solar radiation. These heat gains could require year-round mechanical cooling capabilities, even in cold winter areas. The components and operation of the mechanical refrigeration cycle play an important role in the operation of the HVAC industry. Upon completing this chapter you will be able to:

1. Apply the general principles of mechanical and absorption refrigeration.
2. Explain how a refrigerant absorbs and rejects heat.
3. Arrange the components of the basic mechanical refrigeration cycle, in their proper order.
4. Describe the operating characteristics of reciprocating, centrifugal, rotary and scroll refrigerant compressors.
5. Describe the operating characteristics of air- and water- cooled condensers.
6. Describe the operating characteristics of metering devices, thermostatic expansion and constant pressure expansion valves.
7. Describe the operating characteristics of direct expansion coils and other evaporators for liquid chillers.
8. Describe the operating characteristics of the absorption refrigeration cycle.

General Principles of Refrigeration

Refrigeration is "the process of extracting heat from a substance or space by any means."

A good example would be the window or central air conditioning unit in your home. During the summer, when the heat gain indoors is high, the refrigerant absorbs the heat and rejects the absorbed heat to the outdoors. In commercial systems this transfer of heat, regardless of the amount, is accomplished by the same refrigeration process.

Refrigeration Capacity

The capacity of air conditioning machines is classified by how many tons of refrigeration effect it can produce.

A **ton** of refrigeration is based on the amount of cooling 1 ton (2000 pounds) of ice can produce melting over a 24-hour period. One ton of refrigeration has the ability to remove 288,000 Btu in a 24-hour period. This number is calculated by multiplying the weight of the ice (2000 lb) by the latent heat of fusion (melting) for water, which is 144 Btu/lb, which will change 1 pound of ice at 32°F to 1 pound of water at 32°F.

1 Ton = 2000 lb x 144 Btu/lb

1 Ton = 288,000 Btus

The total amount of Btus required (288,000 Btu) to melt 1 ton of ice is then based on a 24-hour period.

1 Ton = 288,000 Btu / 24 Hours

1 Ton = 12,000 Btu/hr

One ton of refrigeration can remove 12,000 Btu/hr. If a chiller has the ability to remove 1,200,000 Btu/hr, what would be the capacity of this machine.

Tons = 1,200,000 Btu/hr ÷ 12,000 Btu/hr/ton

Tons = 100 tons

Refrigerants

A **refrigerant** is "a fluid that picks up heat by evaporating at a low temperature and pressure and gives up heat by condensing at a higher temperature and pressure."

Fluids used as refrigerants should have a high latent heat per pound to produce a significant cooling effect per pound of vapor produced. Water can be considered a good refrigerant, because if its ability to absorb 970 Btu/lb to evaporate. Refrigerants from a family of chemical compounds known as halogens are typically used in the mechanical refrigeration cycle. The standard for comparing refrigerants is based on an evaporating temperature of 5°F and a condensing temperature of 86°F. The net refrigeration effect of some of the more common refrigerants used in the HVAC industry is listed below.

R-11 66.8 Btu/lb

R-12 50.0 Btu/lb

R-22 70.0 Btu/lb

The following lists commonly used refrigerants and their saturation temperatures (boiling point) at atmospheric pressure:

1. R-11 Trichloromonofluoromethane 74.9°F
2. R-12 Dichlorodifluoromethane -21.6°F
3. R-22 Monochlorodifluoromethane -41.4°F

There are other refrigerants. R-113, R-125, R-134a, R-500 and R-502 are refrigerants that may be used in HVAC refrigeration applications.

Temperature-Pressure Relationships of Refrigerants

Understanding the effects of pressure on the boiling point temperature of refrigerants is the cornerstone in understanding how the refrigeration cycle works. As covered previously in the boiler chapter, as the pressure in the boiler increased, so did the saturation temperature. The higher the pressure, the higher the saturation temperature. Conversely, if the pressure in the boiler dropped, the saturation temperature of the water also dropped. In the refrigeration cycle, the pressures applied to the refrigerant in the different portions of the system will be varied to get the different saturation temperatures required for proper operation.

Temperature-pressure charts (Figure 10-1) can help with the understanding of the relationship between the pressure applied to the refrigerant to the corresponding boiling point temperature. The chart shows the temperature-pressure relationships of refrigerants R-11, R-12 and R-22.

R-11 is considered to be a low pressure refrigerant. This means that a small increase or decrease in pressure would result in a large change in the saturation temperature. For example, at 0 psig the saturation temperature of R-11 is 74.7°F, but if the pressure is increased to 10.5 psig the saturation temperature would increase to 104°F.

	Pressure (psig)		
Temp. °F	R-11	R-12	R-22
30	18.6*	28.5	54.9
32	18.1*	30.1	57.5
34	17.5*	31.7	60.1
36	16.9*	33.4	62.8
38	16.3*	35.2	65.6
40	15.6*	37.0	68.5
42	15.0*	38.8	71.5
44	14.3*	40.7	74.5
46	13.6*	42.7	77.6
48	12.8*	44.7	80.8
50	12.0*	46.7	84.0
52	11.2*	48.8	87.4
54	10.4*	51.0	90.8
56	9.6*	53.2	94.3
58	8.7*	55.4	97.9
60	7.8*	57.7	101.6
62	6.8*	60.1	105.4
64	5.9*	62.5	109.3
66	4.9*	65.0	113.2
68	3.8*	67.6	117.3
70	2.8*	70.2	121.4
72	1.6*	72.9	125.7
74	0.5*	75.6	130.0
76	0.3	78.4	134.5
78	0.9	81.3	139.0
80	1.5	84.2	143.6
82	2.2	87.2	148.4
84	2.8	90.2	153.2
86	3.5	93.3	158.3
88	4.2	96.5	163.2
90	4.9	99.8	168.4
92	5.6	103.1	173.7
94	6.4	106.5	179.1
96	7.1	110.0	184.6
98	7.9	113.5	190.2
100	8.8	117.2	195.9
102	9.6	120.9	201.8
104	10.5	124.6	207.7
106	11.3	128.5	213.8
108	12.3	132.4	220.0
110	13.1	136.4	226.4
112	14.2	140.5	232.8
114	15.1	144.7	239.4
116	16.1	148.9	246.1
118	17.2	153.2	252.9
120	18.2	157.7	259.9
122	19.3	162.2	267.0
124	20.5	166.7	274.3
126	21.6	171.4	281.6
128	22.8	176.2	289.1

*Indicates pressures measured in inches of mercury below one atmosphere.

Fig. 10-1
Pressure Temperature Relationship of Refrigerants

R-12 and R-22 are considered high pressure refrigerants. Unlike R-11, R-12 and R-22 need a larger pressure change (higher differential pressure) in order to get a significant temperature change. These refrigerants are less sensitive to regulating and operating mechanisms. This is an advantage in certain types of equipment. For example, at 0 psig the saturation temperature of R-22 is -41°F, but with 180 psig applied to it, its saturation temperature is 94°F. To reach the same saturation temperature as the R-11, 104°F example, the pressure applied would have to be increased to 207 psig.

Basic Mechanical Refrigeration Cycle

The basic **mechanical refrigeration cycle** consists of four major components which perform four major processes. (Figure 10-2)

1. Compressor, compressing
2. Condenser, condensing
3. Refrigerant metering device, metering
4. Evaporator, evaporating

The compressor (1) takes the low pressure vapor from the evaporator and compresses it into a high pressure vapor. The pressurization of the vapor raises the saturation temperature, allowing the vapor to be condensed at a higher temperature.

The condenser (2), a heat transfer device, removes the heat added to the refrigerant in the evaporator and the heat of compression, by transferring the heat to a medium which is lower in temperature, such as water or air. In this process the refrigerant vapor condenses. The refrigerant leaves the condenser as a liquid.

The refrigerant metering device (3) performs two functions. It controls the flow of liquid refrigerant (metering) between the condenser and the evaporator while, at the same time, dropping the high pressure refrigerant to the lower pressure used in the evaporator.

The evaporator (4), a heat transfer device, allows the low pressure liquid refrigerant to absorb heat from the indoor air, changing the refrigerant into a low pressure vapor (evaporating).

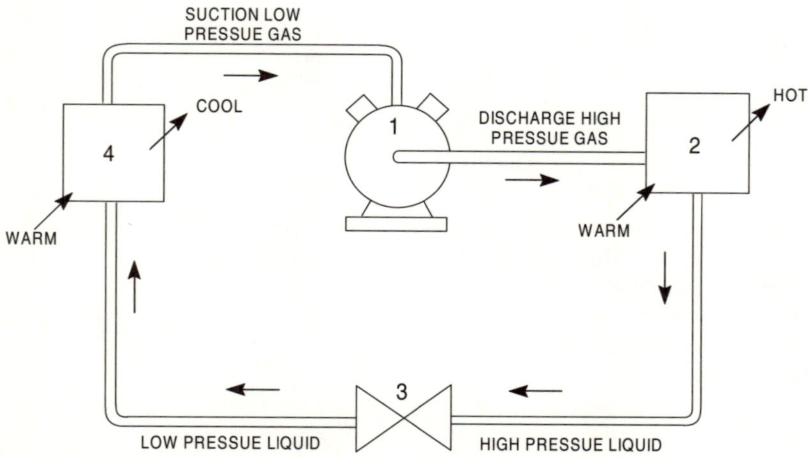

Fig. 10-2 Mechanical Refrigeration Cycle

Compressors

Reciprocating Compressors

Reciprocating compressors are positive displacement compressors which increase the pressure of the refrigerant vapor from the evaporator to the high pressure vapor needed to condense in the condenser. A reciprocating compressor can have from one to twelve or more cylinders.

The reciprocating compressor consists of a piston moving up and down in a cylinder, driven by a crankshaft connected to an electric motor, along with a set of valves that open and close. These valves are controlled by capacity control devices operated by suction pressures or devices monitoring leaving air or water temperatures.

When the piston starts its down stroke, Figure 10-3A, the pressure in the cylinder drops. When the pressure in the cylinder drops below the pressure of the refrigerant vapor in the suction line, the suction intake valve unseats, allowing the cylinder to fill up with refrigerant vapor from the evaporator.

As the piston reaches the bottom of the down stroke, the cylinder is now filled with low pressure refrigerant vapor. Next, the piston starts its up stroke, Figure 10-3B, The up stroke decreases the area of the volume of gas drawn into the cylinder, increasing the pressure by compressing the refrigerant vapor. When the piston reaches a certain point on the up stroke, the pressure in the cylinder is high enough to unseat the discharge valve, allowing the high pressure refrigerant to be discharged out to the cylinder head.

There are three types of reciprocating compressors:

1. Open
2. Semi-Hermetic
3. Full-Hermetic

This classification denotes merely the type of construction used by the manufacturer.

In an open type, the compressor and the motor are two separate units. A drive shaft extends beyond the compressor housing, Figure 10-4, to allow the motor to be connected to the

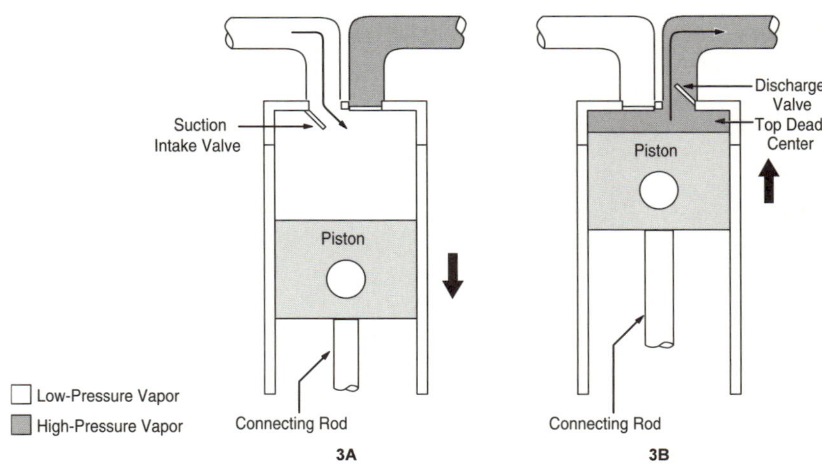

Fig. 10- 3A, B
Reciprocating Compressor Operation

compressor. This type of drive also requires a reliable shaft seal because most reciprocating compressors use refrigerants that operate above atmospheric pressure.

To eliminate the problem of refrigerant leaks around the shaft seal, semi-hermetic and full-hermetic compressors were developed.

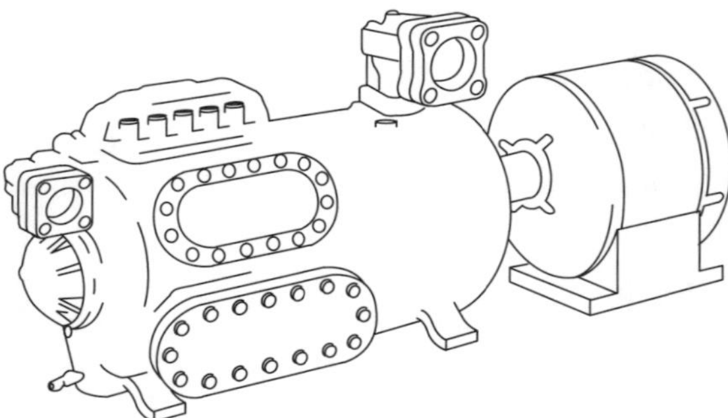

Fig. 10-4 Open Type Compressor

Courtesy of the Trane Company

The semi-hermetic compressor, Figure 10-5, is constructed so the motor and the compressor are contained in the same housing, decreasing the chance of leaking refrigerant at a shaft seal. In fact, the motor of a semi-hermetic compressor will be cooled by the low pressure refrigerant before it is compressed in the cylinders.

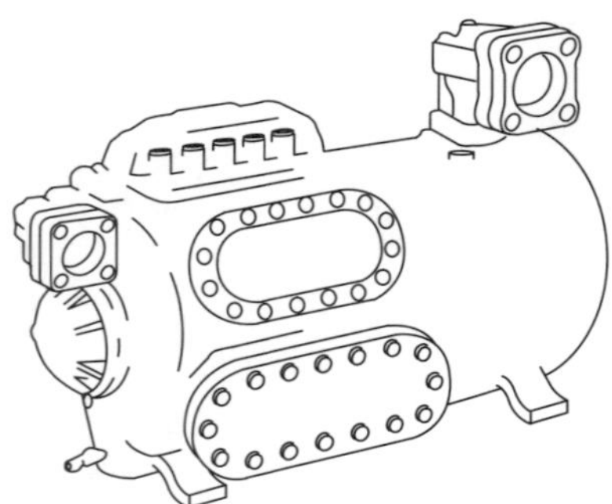

Fig. 10-5 Semi-hermetic Compressor

Courtesy of the Trane Company

The semi-hermetic compressor is manufactured with gaskets and removable cover plates which will allow the compressor to be repaired in the field. The compressor manufacturer makes repair parts such as pistons, bearings and valves for overhauling their compressor.

The full-hermetic compressor shown in Figure 10-6, is manufactured with the electric motor, cylinders and pistons enclosed in a welded steel housing. Because of this, the compressor is not intended to be repaired in the field.

The full-hermetic compressor is very popular in the small size range of 10 tons and below. These compressors are used in refrigerators, residential air conditioning units and small roof top units. Some manufacturers build full hermetic compressors up to 40 tons.

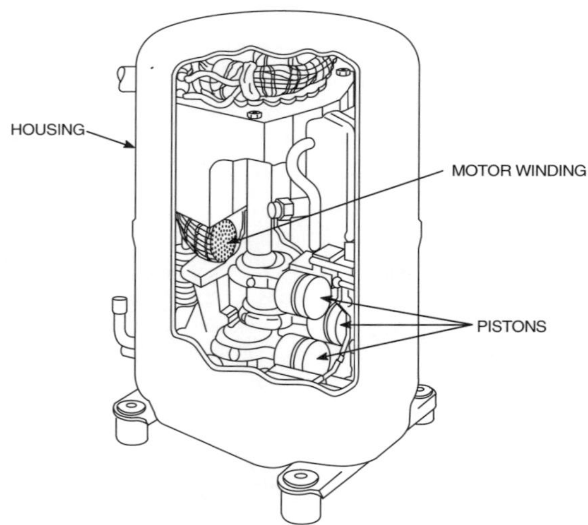

Fig. 10-6 Full-hermetic Compressor

Courtesy of Tecumseh Products Company Company

Rotary Compressors

Like the recriprocating compressors, rotary compressors are also positive displacement compressors. The primary difference is that this compressor uses circular or rotating motion instead of reciprocal motion to increase the pressure of the refrigerant vapor. There are two basic types of rotary compressors:

1. Rotating Blade (Vane) Type
2. Stationary Blade Type

The **rotary blade (vane) compressor**, Figure 10-7, uses two or more blades (vanes) that rotate with the shaft. As the blades rotate past the suction intake, refrigerant vapor is trapped behind the blade. As the blades rotate, the vapor in front of the blade is compressed by decreasing the space between the rotor and the cylinder, increasing the pressure until the vapor is forced out the discharge port.

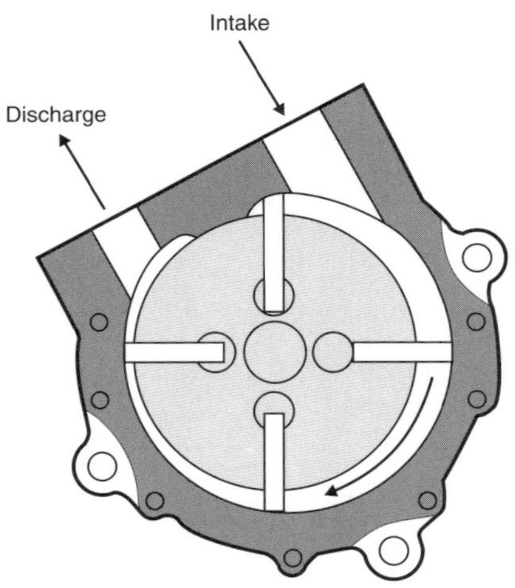

Fig. 10-7
Rotary Blade (Vane) Compressor

Unlike the rotating blade compressor, the stationary blade compressor has one blade which separates the suction and discharge ports. It also has an impeller which rotates by an eccentric shaft constantly sealing the impeller against the cylinder wall, drawing low pressure refrigerant at the suction port and decreasing the space for the vapor, increasing the pressure until it is forced out of the discharge port.

Design requirements for both types of compressor require precision fits and optimum clearance.

Helical Rotary Compressor (Screw Type)

The helical **rotary screw** compressor, Figure 10-8, is a positive displacement compressor, which uses two rotors (screws). The pair of rotors consist of one male and one female rotor which revolve to trap and compress gas in an accurately machined compressor cylinder. The male rotor is connected to the motor and the capacity is controlled by a slide valve. A slide valve covers or uncovers the screws controlling the "breathing" of the screws, thus the capacity of the compressor.

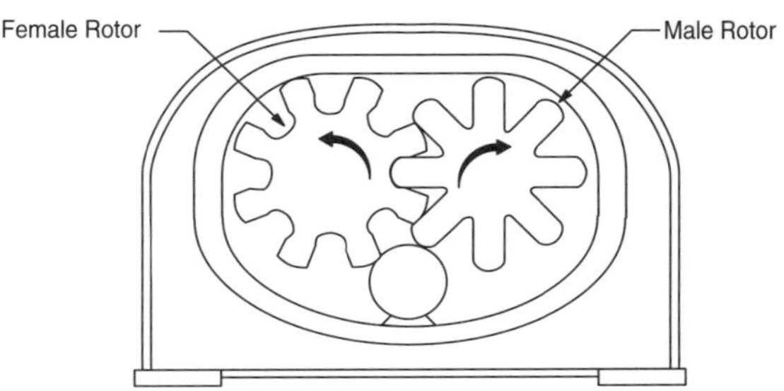

Fig. 10-8
Helical Rotary Compressor (Screw Type)

Unlike the pulsating output the positive displacement compressor delivers, the helical rotor compressor, having both male and female helix rotors, will produce a continuous, flowing output.

The working cycle of compression, Figure 10-9, has these four phases:

1. **Suction:** A pair of lobes unmesh on the suction inlet side and gas flows in the increasing volume formed between the lobes and the housing until the lobes are completely unmeshed.

2. **Transfer/Beginning of Compression:** The trapped pocket of vapor is now isolated from the inlet and outlet, is moved peripherally at a constant pressure. When the vapor is moved into the cylinder between the two lobes, compression begins.

3. **Compression:** As remeshing continues, the volume of trapped vapor is reduced, the charge is gradually moved helically and compressed simultaneously toward the discharge end as the lobes mesh point moves along axially.

4. **Discharge:** It starts when the compressed volume of vapor has been moved to the axial ports on the discharge end of the machine and continues until all the trapped gas is completely squeezed out.

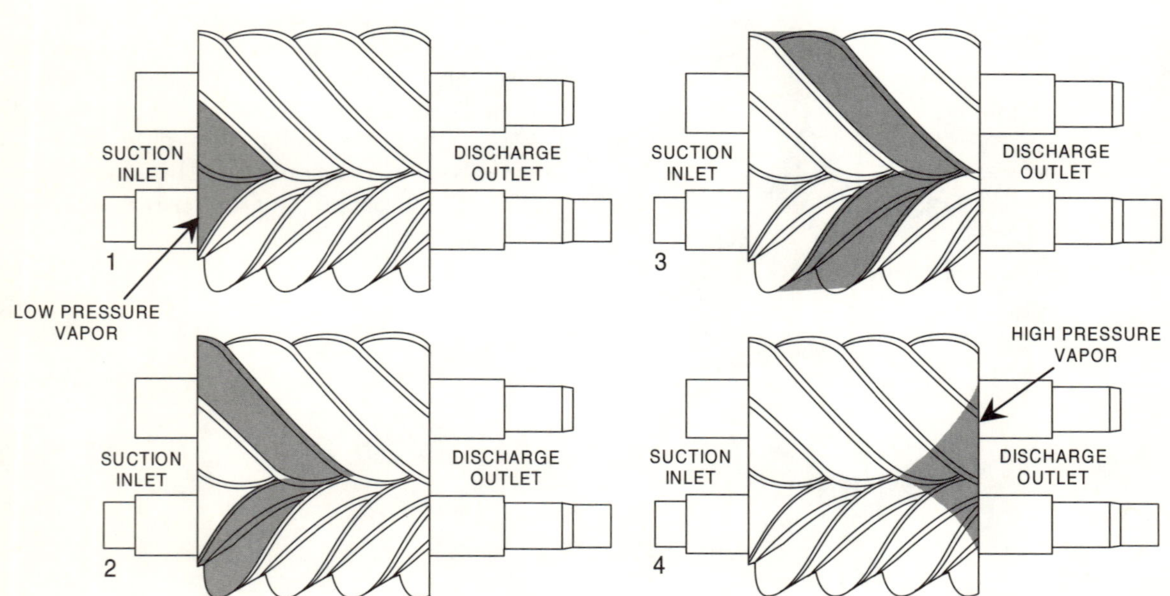

Fig. 10-9 Helical Rotary (Screw Type) Compression Cycle

Illustration courtesy of Dunham-Bush, Inc.

Scroll Compressor

The **scroll compressor** shown in Figure 10-10 is another type of positive displacement compressor like the reciprocating compressor. But unlike the reciprocating compressor which is unloadable, the scroll compressor cannot be unloaded, but the speed of the motor could be varied to control capacity. The cost of building scroll compressors in quantity is low, but designing and tooling for manufacturing is very expensive. The scroll compressor is quickly becoming the compressor of choice, replacing the reciprocating compressor in residential air conditioning systems, and heat pumps as well as rooftop units.

The scroll compressor shown in Figure 10-11 uses two identical scrolls mounted face-to-face. The upper scroll is fixed, while the lower scroll is driven by the compressor. The suction intake takes place around the outer edges of the scroll, while the discharge of the pressurized gas occurs through the discharge port in the center of the stationary scroll.

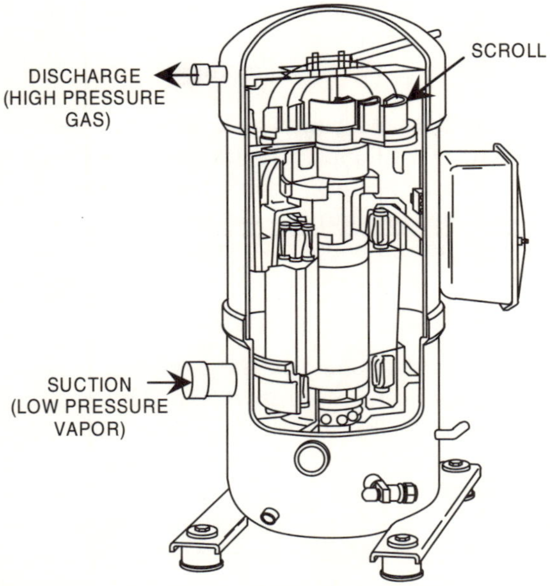

Fig. 10-10 Scroll Compressor

Drawing courtesy of Copeland Corporation

The center of the driven scroll shaft and the motor shaft will be offset by a swing link. This offset creates the eccentric or orbiting motion needed for compression between the two scrolls.

During the first rotation, the intake phase shown in Figure 10-12, the eccentric (orbiting) motion of the scrolls opens the outer portion of the scrolls, allowing gas pockets to form.

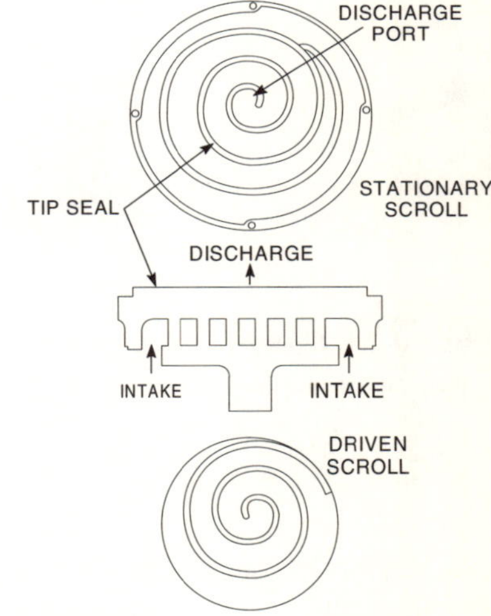

Fig. 10-11 Scroll Compressor Components

Drawing courtesy of Copeland Corporation

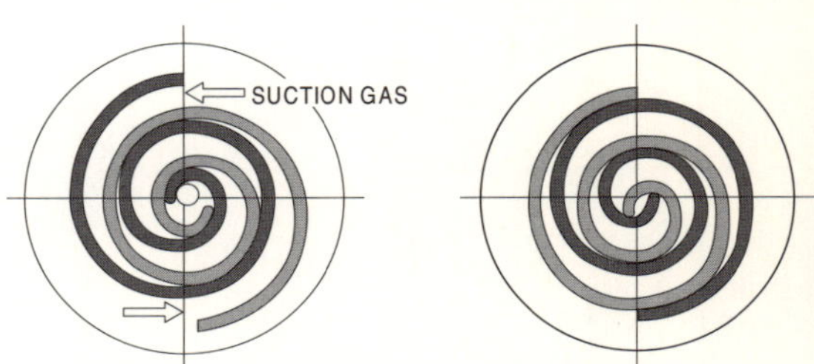

Fig. 10-12 Intake Phase

Drawing courtesy of Copeland Corporation

These gas pockets allow the suction gas to enter. At the end of one revolution, the scrolls re-mate trapping the suction gas inside the scrolls, forming two separate pockets of gas.

During the second rotation of the scrolls, the compression phase shown in Figure 10-13, the volume of the pockets is progressively reduced. As the volume is reduced, the pressure of the gas increases. By the end of the second revolution, the gas is at its maximum compression and, therefore, maximum pressure.

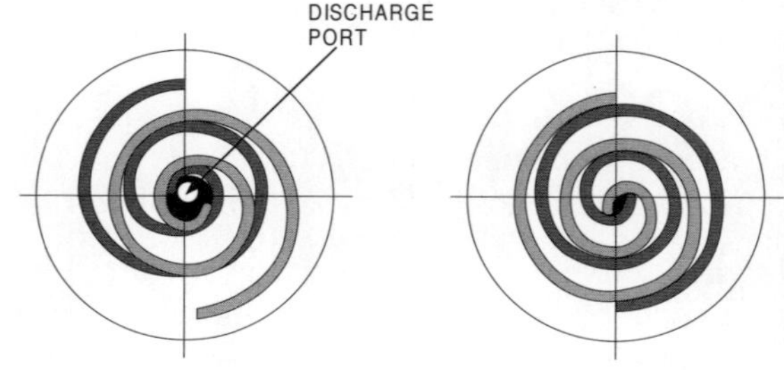

Fig. 10-14 Discharge Phase

Drawing courtesy of Copeland Corporation

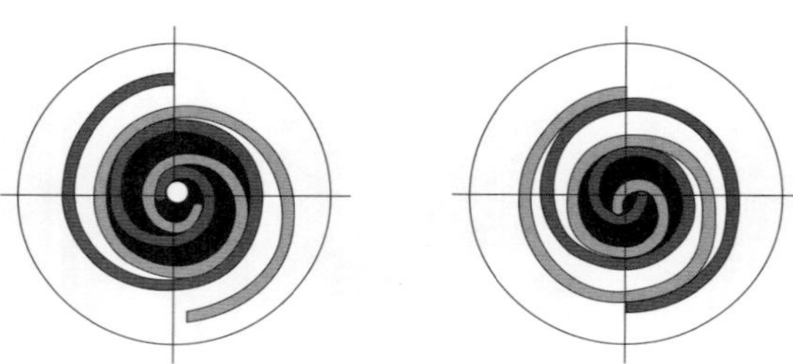

Fig. 10-13 Compression Phase

Drawing courtesy of Copeland Corporation

During the third rotation, the discharge phase shown in Figure 10-14, the scroll ends part, allowing the compressed gas to be discharged through the center hole (discharge port) in the stationary scroll. By the end of the third rotation, the volume between the scrolls is decreased to zero forcing out all the remaining gas.

Centrifugal Compressors

Centrifugal compressors shown in Figure 10-15, or turbocompressors, are members of a family of turbo machines which include fans, propellers and turbines, which rely on centrifugal force for the compression of the refrigerant vapor. Because centrifugal compressors do not have any reciprocating parts (piston, valves or cylinders), very little vibration is produced. The only wearing part is the main bearings.

The centrifugal compressor is not a positive displacement, but rather a continuous flow output compressor. This continuous flow gives these machines a higher volumetric capacity than the positive displacement compressors. They are therefore very suitable in large refrigeration applications.

The main component of the compressor is the impeller shown in Figure 10-16. The impeller draws the vapor into the intake past a set of inlet vanes. Because the impeller is spinning at a constant high RPM, the gas is forced to the outer diameter, thus giving a small increase in pressure. Since centrifugal compressors give only a small pressure increase, the manufacturer may choose to have several impellers feeding each other or connecting them in stages to get the desired increase in pressure.

Because the motors of these compressors run at a constant speed, the inlet vanes control compressor capacity. The inlet vanes open or close to control the amount of gas (vapor) that is allowed into the impeller. The higher the load, the more the inlet vanes are open and vise versa.

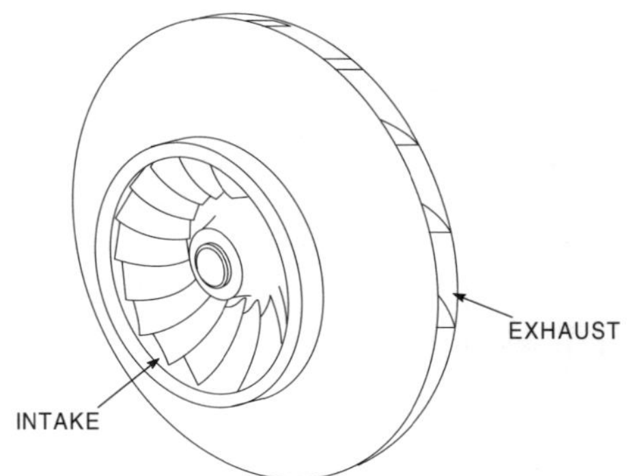

Fig. 10-16 Impeller

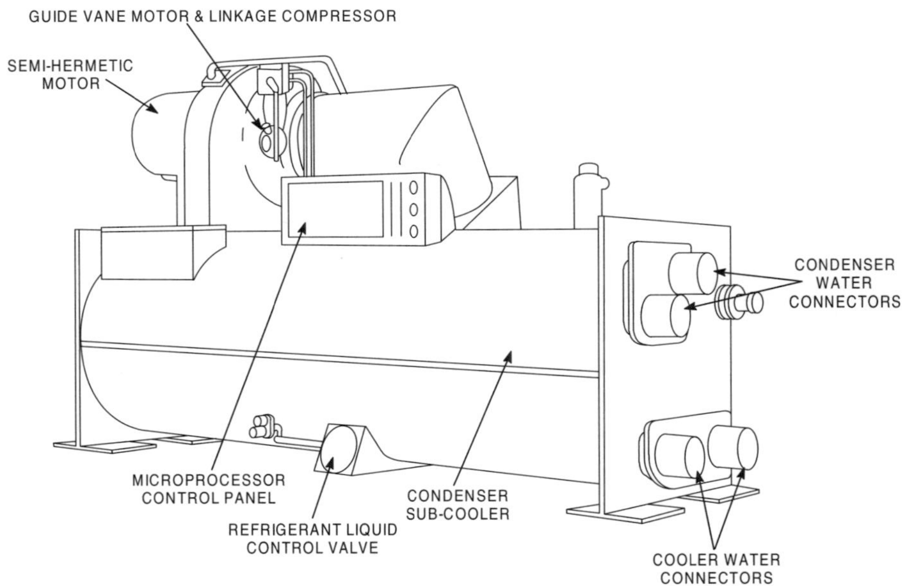

Fig. 10-15 Centrifugal Compressor (shown with centrifugal chiller)

The chart in Figure 10-17 represents all the different type of refrigerant compressors, capacities, refrigerant types and applications. Refer to notes on the chart for advantages and disadvantages of each. Absorption applications will be discussed later in this chapter.

Condensers

A **condenser** removes the heat picked up by the refrigerant in the evaporator and the heat of compression, and transfers the heat from the refrigerant to water or an air stream.

Air-Cooled Condensers

The movement of air across the coil could be caused by natural draft or forced draft, using a fan. The heat transfer process changes the refrigerant from a hot vapor into a cooler liquid. This process is accomplished in three phases:

Phase 1: Desuperheating - This phase is the removal of any superheat that was added in the evaporator. After this phase, the refrigerant vapor would be at its saturation temperature.

Type	Capacity	Application	Refrigerant	Advantage	Disadvantage
Reciprocating	Fractional to 100 tons	Residential & Commercial Air Conditioning & Refrigeration	R-12, R-22, R-134A, R-13, R-500, R-502	Can generate high differential pressures	Many moving parts
Centrifugal	100 tons and up	Commercial Air Conditioning	R-11, R-12, R-113, R-123 R-500, R-134a	Can compress large volumes of refrigerant	Cannot work against large pressure differences
Rotary Screw	75 tons and up	Commercial Air Conditioning Refrigeration	R-12, R-22	Can work with very high pressures	Expensive machine to manufacture machining on lobes
Scroll	Fractional to 30 tons	Residential & Commercial Air Conditioning	R-12, R-22, R-134a, R-13, R-500, R-502	Can handle liquid slugs, very efficient	Slightly higher cost over reciprocating
Absorption	100 tons and up	Commercial Air Conditioning	Water	Very low vibration, low noise	Must operate at extremely low vacuum, must have steam, hot water or direct fired source

Fig. 10-17 Compressor Types And Applications

Phase 2: Condensing - This phase is the removal of the latent heat of vaporization. This removal will cause the vapor refrigerant to change back into a liquid.

Phase 3: Subcooling - This phase is the continuing removal of heat from the liquid refrigerant, cooling the temperature of the refrigerant below its condensing saturation temperature.

Coils are commonly constructed of copper, aluminum or steel tubes ranging from 1/4" to 3/4" in diameter. The most common form is plate fins making a coil bank. Plate fins individually fastened to the tube or a fin is spirally wound onto the tube. Other forms such as plain tube fin extrusions with accordion type fins are also used. The number of fins per inch varies from 4 – 30. The most common range is 8 – 18 fins per inch. An example is shown in Figure 10-18.

Water-Cooled Condensers

This type of condenser is used in large commercial air conditioning units, because they are more efficient than air-cooled condensers. Water cooled condensers use large quantities of water, which must be either discharged to a drain or reclaimed in a cooling tower.

Water-cooled condensers are constructed in three styles, all of which are used in the HVAC industry:

1. Shell-and-Tube Condenser
2. Shell-and-Coil Condenser
3. Tube-Within-A-Tube Condenser

The **shell-and-tube condenser**, Figure 10-19, is the most popular of the condenser styles. The shell portion is constructed of steel and holds the refrigerant vapor. The tubes, made of straight copper tubing, circulate the water through the condenser. The heat from the refrigerant vapor transfers to the condenser water through the tubes. As the tubes get dirty,

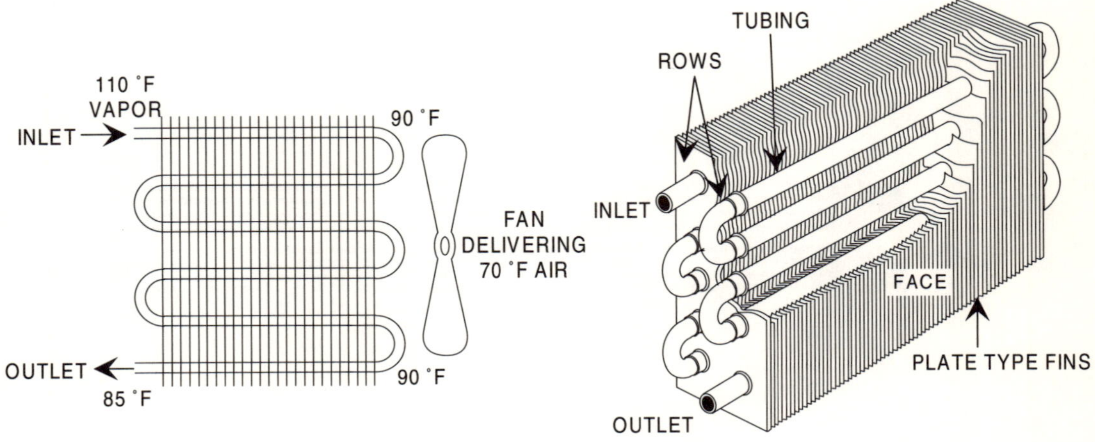

Fig. 10-18 Air Cooled Condenser Construction

the efficiency of the condenser decreases. To maintain efficiency, tubes must be cleaned periodically.

The **shell-and-coil condenser**, shown in Figure 10-20, is very similar to the shell-and-tube condenser. Unlike the shell-and-tube condenser which has straight copper tubes, the shell and coil uses a coil of tubing. They are used in smaller commercial units. They are less expensive than other condensers to manufacture. Unlike the straight tubes of the shell-and-tube condenser which can be cleaned mechanically, the shell-and-coil must be cleaned chemically.

The **tube-within-a-tube** condenser shown in Figure 10-21 uses two tubes. The inner tube has the condensing water running through it, while the outer tube holds the refrigerant vapor. The double cooling effect of the refrigerant vapor coming into contact with both the condensing water and the ambient air around the tube, improves the efficiency of the condenser.

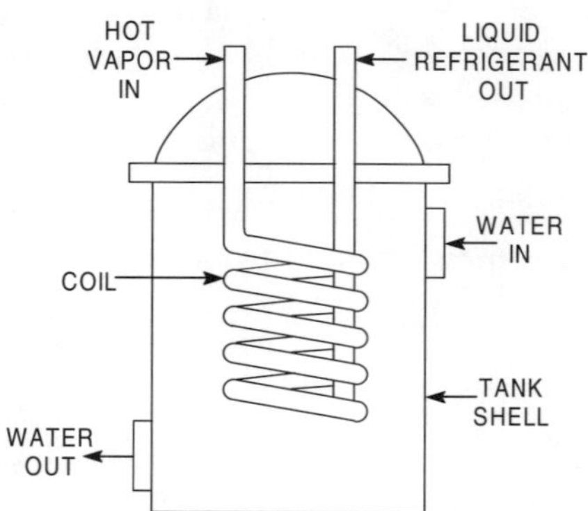

Fig. 10-20 Shell And Coil Condenser

Reprinted with the permission of ASHRAE

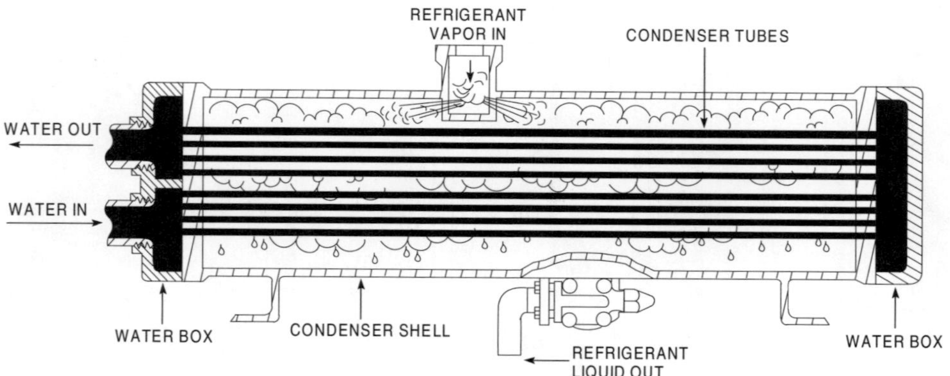

Fig. 10-19 Shell And Tube Condenser

©1977 by TPC Training Systems, a division of Telemedia, Inc. Reprinted with permission

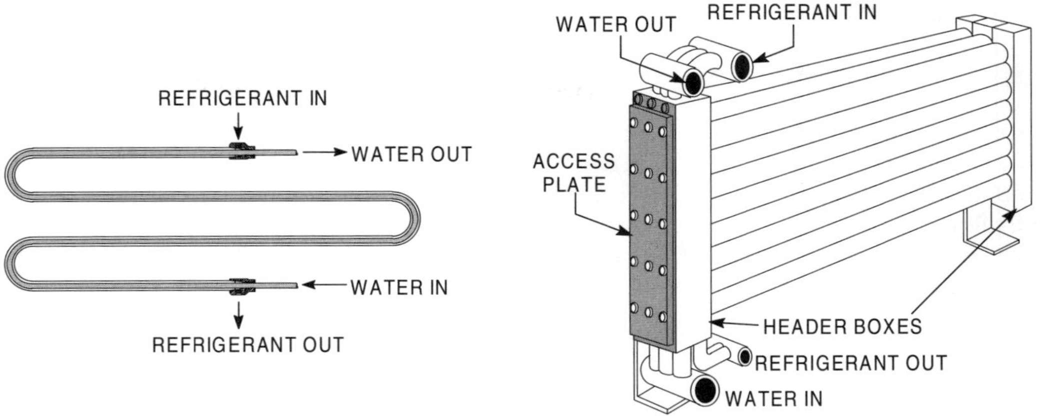

Fig. 10-21 Tube-within-a-tube Condenser

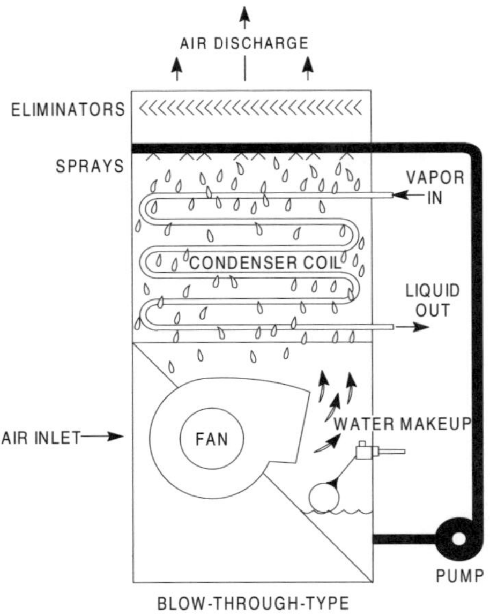

Fig. 10-22 Evaporative Condenser

Evaporative Condenser

The evaporative condenser condenses the refrigerant by evaporative cooling, similar to the way a cooling tower works. These units are usually located outdoors, but they can be located indoors if outdoor air is ducted to the unit and discharge air is vented outside.

This condenser, like the one shown in Figure 10-22, can function as an air-cooled condenser or it can use both air and water for condensing. A thermostat can be used to control the water flow through the unit. When the temperature of the condenser water is below a specific setpoint of the thermostat, only air will be blown through the condensing unit. As the temperature of the condenser water rises above the setpoint, the water pump is turned on and the water can either be dripped or sprayed through the unit. The condensing unit fan will be running whenever the unit is on.

Metering Devices

Refrigerant Metering Devices

In the refrigeration system, the pressure of the refrigerant in the condenser is high, while the pressure in the evaporator is relatively low so that the refrigerant can evaporate at a low temperature. The refrigerant metering device develops the pressure drop necessary for proper operation of the system.

There are five types of refrigerant metering devices commonly used in the HVAC industry:

1. Thermostatic expansion valve
2. Constant pressure expansion valve
3. Capillary tube
4. High side float
5. Orifice plates

Thermostatic Expansion Valve

The **thermostatic expansion valve** controls the flow of refrigerant entering the evaporator in response to the amount of superheat in the refrigerant leaving the evaporator.

The thermostatic expansion valve, Figure 10-23, uses a sensing bulb on the suction line (output of the evaporator). The sensing bulb, valve diaphragm and evaporator pressure together measure the amount of superheat in the refrigerant vapor. The sensing bulb can be charged with either a gas or liquid which may be the same refrigerant used in the system. Cross-charged sensing bulbs are more common and are filled with a fluid which is different from the refrigerant in the system.

As the load increases, Figure 10-23, more liquid in the sensing bulb changes to gas which increases the pressure above the diaphragm. The pressure increase modulates the valve open, allowing more refrigerant to enter the

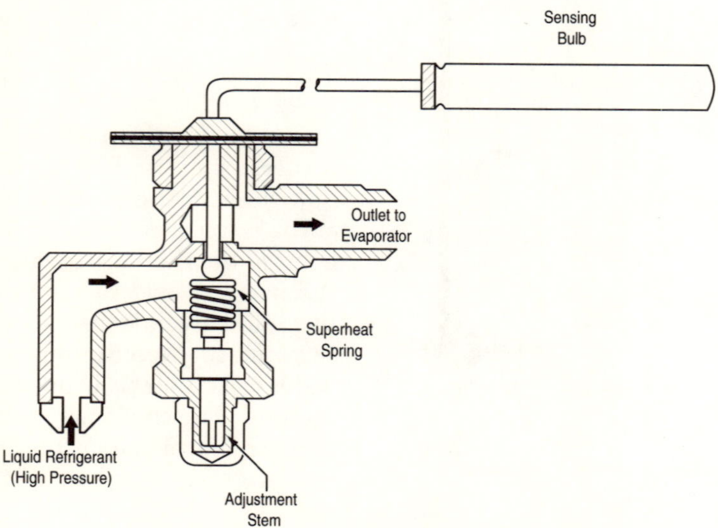

Fig. 10-23 Thermostatic Expansion Valve

evaporator. The opposite sequence would occur when the load decreases.

Also affecting the amount of refrigerant entering the evaporator would be the pressure of the evaporator which will be established by the compressor's pumping action. The thermostatic expansion valve provides a high flow rate when the evaporator is under a high load, and a lower flow rate when the evaporator starts to fill with liquid refrigerant due to a decrease in the load. Because the thermostatic expansion valve controls the flow rate dependent on the load by sensing superheat, it ensures that only vapor gets sent to the compressor.

Thermostatic expansion valves are used to control refrigerant flow to all types of evaporators in air conditioning and low temperature systems. Their use is not limited to constant capacity applications. They are widely used because they can maintain a preset level of superheat and prevent liquid slugging by ensuring that all the liquid refrigerant is turned into vapor.

Constant Pressure Expansion Valve

The constant pressure expansion valve, Figure 10-24, is operated by the pressure in the evaporator to regulate the flow of liquid refrigerant entering the evaporator keeping the pressure constant.

The valve has an adjustable spring which exerts force on the top of the diaphragm in the open direction. In the closing direction of the valve, there is a combination of evaporator pressure applied to the diaphragm along with the pressure of the closing spring. This pressure moves the valve more closed, decreasing the flow rate of refrigerant.

The adjustable spring, which is controlled by the adjusting screw, allows the constant pressure expansion valve to be adjusted to work with the different refrigerants and under different atmospheric pressures.

When the valve is set and feeding refrigerant at a given pressure, a small increase in the evaporator pressure due to a increase in load will act beneath the diaphragm forcing it upward. The upward movement causes the valve

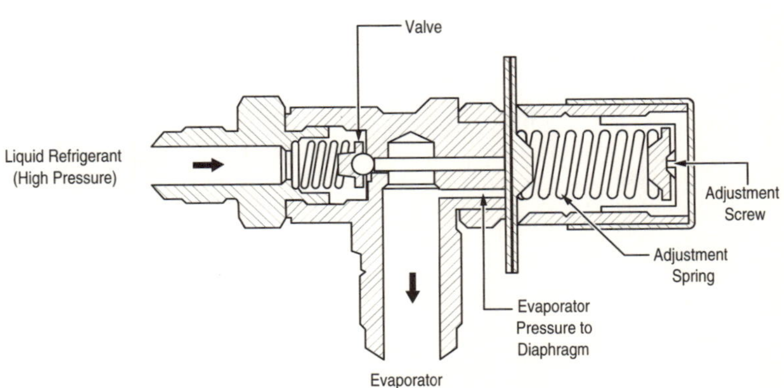

Fig. 10-24 Constant Pressure Expansion Valve

pin to move in a closing direction, restricting the refrigerant flow and limiting the evaporator pressure.

When the evaporator pressure drops below the valve setting because of a load decrease, the top spring pressure moves the valve pin more open, increasing the refrigerant flow in an effort to raise the evaporator pressure to the balanced setting.

Constant pressure expansion valves are used with systems that have a constant load such as air dryers.

Capillary Tubes

The **capillary tube** shown in Figure 10-25 is a metering device that consists of a length of seamless tubing having a small inside diameter. It usually has a fine filter or filter dryer on the inlet to prevent the tube from plugging up. It has no moving parts.

This metering device is primarily used in household refrigerators and air conditioners. Due to its limited use in larger tonnage equipment, further discussion is not necessary for this text.

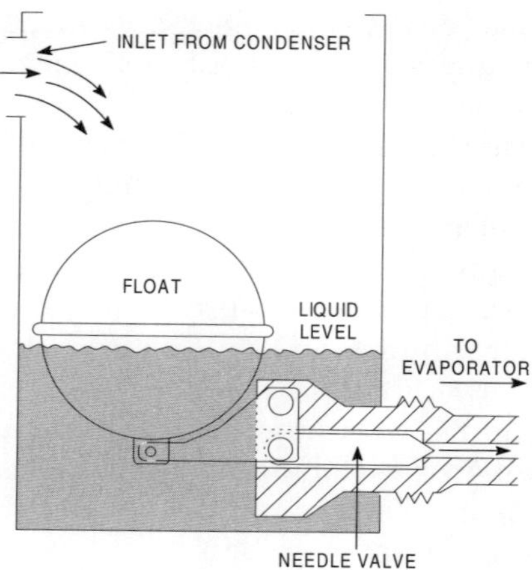

Fig. 10-26 High Side Float Valve

*©1977 by TPC Training Systems,
a division of Telemedia, Inc. Reprinted with permission*

High Side Float Valve

The **high side float valve**, Figure 10-26, controls the flow rate of refrigerant entering the evaporator at the same rate as the vapor is pumped out of the evaporator. As the name denotes, this valve is in the high pressure side of the system.

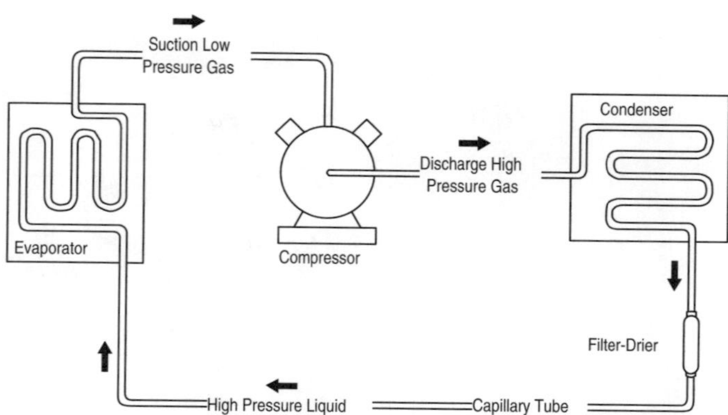

Fig. 10-25 Capillary Tubing

As the load on the system increases, more refrigerant changes from a liquid to a vapor in the evaporator. As a result, the amount of liquid refrigerant in the condenser increases. This increase raises the level in the high side float valve body which raises the float. When the float starts to rise, it moves the needle valve to the open position, allowing more liquid refrigerant to enter the evaporator. When the load decreases, the level of liquid in the valve body drops. As a result, the float moves down, moving the needle valve towards it closed position.

Orifice Plates

The **orifice plate** metering system is yet another method of metering liquid refrigerant into the evaporator. The orifice plate metering system balances the refrigerant flow rate with load condition in the evaporator. This method used in centrifugal chillers, consists of two orifice plates spaced a specific distance apart, giving successive pressure drops. After the pressure drops, some of the refrigerant turns to vapor, while some stays as liquid. The first orifice plate will pass *only* liquid refrigerant due to the pressure exerted by the column of liquid in the liquid line. The second orifice plate can pass either liquid or a combination of liquid and vapor, allowing liquid refrigerant to pass easier than vapor. This configuration allows the correct amount of liquid refrigerant to be metered into the evaporator for changing load conditions.

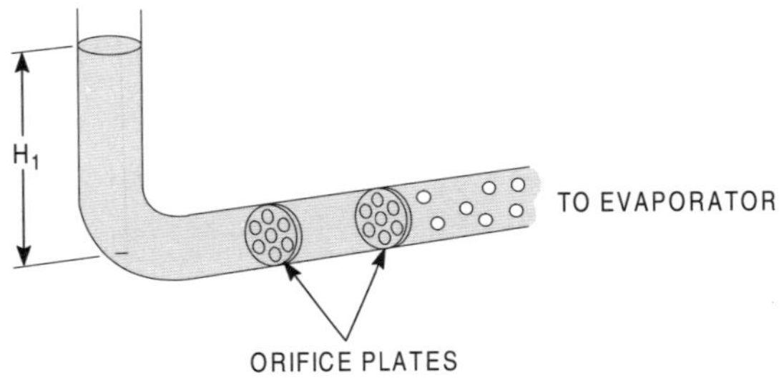

Fig. 10-27a Orifice Plates

Courtesy of the Trane Company

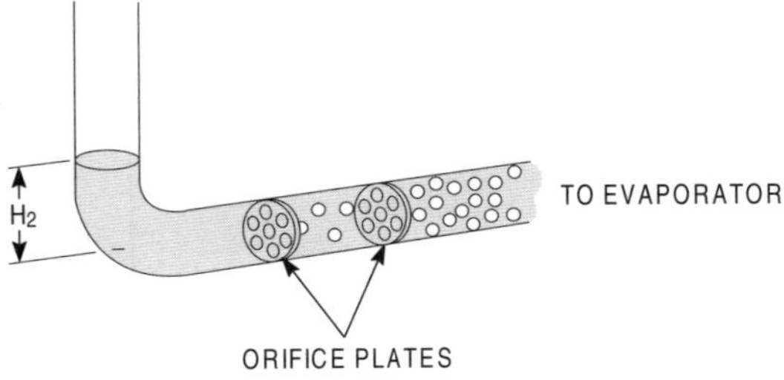

Fig. 10-27b Orifice Plates

Courtesy of the Trane Company

Given an increase in load, a large amount of refrigerant is moved within the chiller, building a liquid head (H_1) as shown in Figure 10-27a. This head pressurizes the liquid at the base of the column. This head pressure is great enough to push the refrigerant as a liquid through both orifice plates. Because both orifice plates have liquid flowing freely through them, the quantity of refrigerant satisfies the load of the chiller.

As the load decreases, less refrigerant is moved by the chiller and the head decreases as shown in Figure 10-27b. This decreased head results in a lower pressure across the orifice plates. As a result of the lower pressure, some refrigerant will flash into a vapor, limiting the flow of refrigerant through the orifice plates. Vapor bubbles impede the flow. The metered quantity of refrigerant is less, thus satisfying the decreased load on the chiller.

Direct Expansion Coils

The direct expansion coil (evaporator) allows the air stream to be cooled by coming into direct contact with a coil of refrigerant, transferring the Btus from the air directly into the refrigerant in order for the refrigerant to change from liquid to a vapor.

Figure 10-28 shows the flow of refrigerant beginning at the liquid receiver (A). The refrigerant then flows through the metering device (B), removes heat from an air stream vaporizing the refrigerant (C), then leaves the evaporator at a superheated condition (D).

The metering device most commonly used with direct expansion coils is the thermostatic expansion valve. The capillary tube may be used for some applications.

The thermostatic expansion valve is preferred because of its ability to operate over a wide range of load changes, along with ensuring that all the liquid refrigerant is turned to vapor to prevent liquid slugging. The thermostatic expansion valve, Figure 10-29, controls for a constant value of superheat at the outlet (suction side) of the evaporator. As the temperature rises on the outlet pipe, the thermostatic expansion valve opens allowing more refrigerant to enter the evaporator. On the average, the thermostatic expansion valve will control for 10°F of superheat, but that value can go as high as 20°F.

Evaporators for Liquid Chillers

Evaporators for liquid chillers (also referred to as coolers) use the evaporation of refrigerant to produce a cooling effect on a fluid, usually water. There are two types of liquid chillers:

1. Direct expansion chilled water evaporator
2. Flooded shell-and-tube

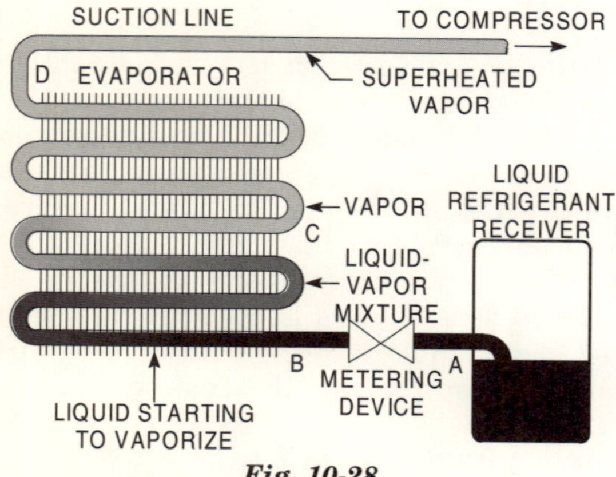

Fig. 10-28
Direct Expansion Coil (Evaporator)

©1977 by TPC Training Systems, a division of Telemedia, Inc. Reprinted with permission

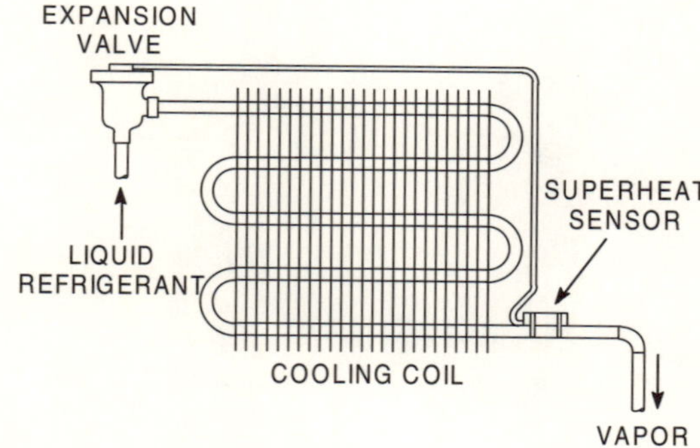

Fig. 10-29 Direct Expansion Coil With Thermostatic Expansion Valve

Courtesy of the Trane Company

Direct Expansion in Chilled Water Evaporator

In a **direct expansion** chilled water evaporator, the refrigerant evaporates inside the tubes, Figure 10-30, while the water that is being chilled is in the shell, around the tubes filled with refrigerant. The performance of the chiller depends mainly on the distribution of refrigerant through the tubes. Most direct expansion evaporator tubes come with a twisted aluminum spine in the center of the tube which causes the refrigerant to spin outward against the copper tube wall, increasing the amount of heat transfer.

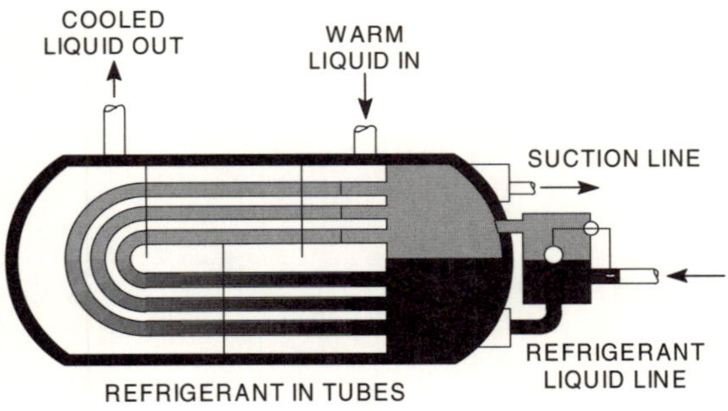

Fig. 10-30 Direct Expansion Chilled Water Evaporator Shell And Tube Type

©1977 by TPC Training Systems, a division of Telemedia, Inc. Reprinted with permission

Direct expansion chilled water evaporators are used with positive displacement compressors, such as reciprocating, rotary and rotary screw compressors.

Flooded Shell-and-Tube

In a **flooded shell and tube** chilled water evaporator, Figure 10-31, the water to be chilled is in the tubes, and the refrigerant is in the shell. The tubes of water are submerged in liquid refrigerant. The warm water in the tube transfers its heat into the refrigerant, causing it to boil. As the refrigerant boils, a mist forms above the tube bundle. This mist is a combination of vapor and liquid. The liquid must be separated from the vapor before it enters the compressor. They are separated by having a drop-out area or eliminators (fins that capture liquids from vapor) above the suction pipe of the compressor.

The flooded chilled water evaporators are commonly used with rotary screw or centrifugal compressors.

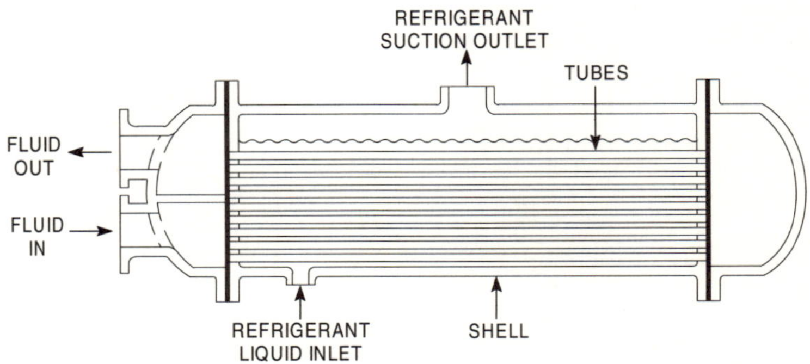

Fig. 10-31 Flooded Shell-and-tube

Reprinted with the permission of ASHRAE

Introduction to Absorption Refrigeration Cycle

The **absorption chiller** does not use the mechanical refrigeration cycle principle to operate. Instead, it uses physical properties of an absorbent to attract a refrigerant vapor and hold it in solution. To better understand this process, use common table salt and water as examples. Visualize the following sequence: First, cover the bottom of a sauce pan with a layer of table salt (absorbent). Begin to add water (refrigerant) until all the salt has dissolved. You now have a strong salt solution. The salt or the water have not chemically changed; they are just in solution with each other. Now add more water. This will give you a weak salt solution. Now place the pan on a kitchen stove and heat. Eventually, when the solution becomes hot enough, some of the water will boil out of the solution as a vapor leaving the salt behind which makes the remaining solution a strong salt solution again. You can add water again to create a weak solution and this whole procedure can be repeated. These are the principles behind the absorption refrigeration cycle.

Two combinations of absorbent and refrigerants are used today in absorption air conditioning units. One combination is water as the absorbent and ammonia as the refrigerant. The second is lithium bromide as the absorbent and water as the refrigerant. For this discussion, we will assume that we have a unit which utilizes lithium bromide and water.

Typical Absorption Unit

The following unit, Figure 10-32, is used primarily in air conditioning applications for producing chilled water. They are available with capacities from 100 – 1600 tons. Small in size, relatively light in weight and vibration free, these units can be located wherever a source of steam or high temperature water is available. Direct-fired units are gaining in popularity. These units use fuel oil or natural gas type burners as the heat source.

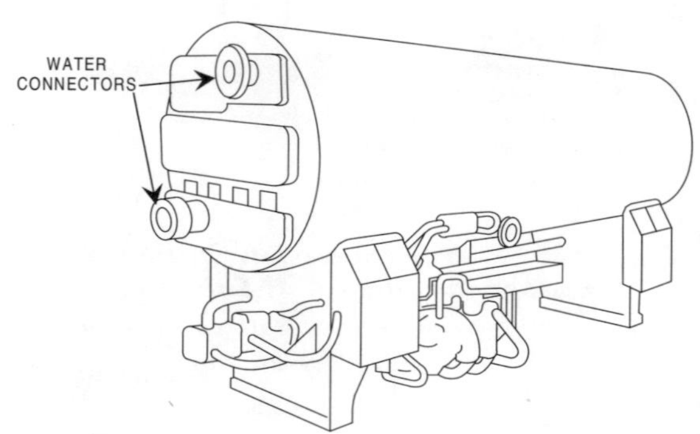

Fig. 10-32 Typical Absorption Unit

Absorption Refrigeration Cycle Components

The absorption refrigeration cycle consists of four major components (Figure 10-33):

1. Evaporator
2. Absorber
3. Concentrator
4. Condenser

Evaporator

The evaporator operates at a very low pressure such as 0.25 psia. The saturation temperature for this condition is 40°F.

The refrigerant pump circulates the refrigerant (water) to the spray trees. In order to utilize the maximum surface for evaporation, the refrigerant is sprayed over the evaporator tubes. As the spray contacts the relatively warm surface of the tubes carrying the water to be chilled, a vapor is created. In this manner, heat is extracted from the tube surfaces, chilling the fluid in the tubes. The vapor created in this process passes through the eliminators to the absorber.

Absorber

The concentration of lithium bromide (absorbent) and water (refrigerant) delivered to the absorber section determines the operating pressure in the evaporator section, thus controlling the chilled water temperature. When the solution in the absorber has more lithium bromide than water, it can absorb a lot of water vapor out of the evaporator. This is due

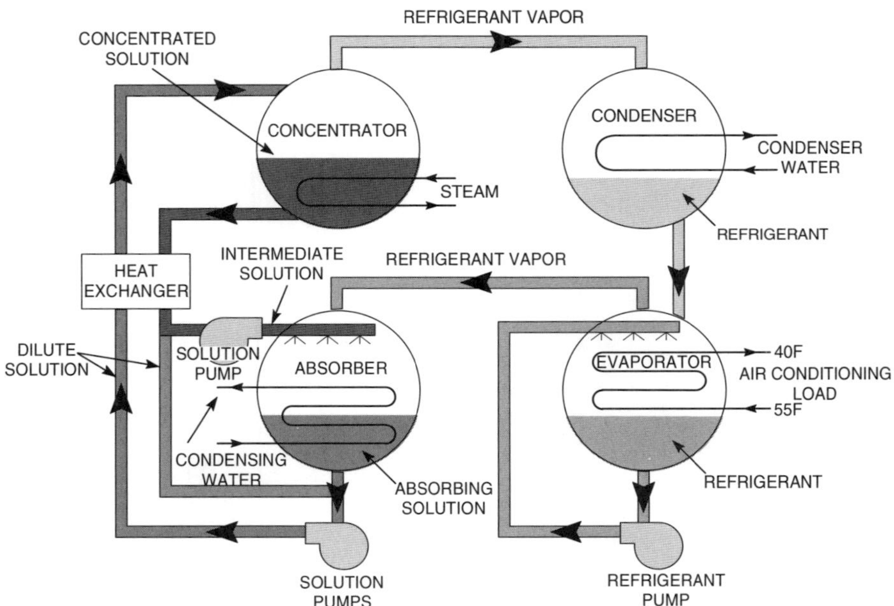

Fig. 10-33 Absorption Refrigeration Components

to the high affinity that lithium bromide has for water. The more water vapor the lithium bromide draws out of the evaporator, the lower the pressure gets in the evaporator, resulting in lower chilled water temperatures.

Just the opposite would take place when the solution had more water than lithium bromide. The solution could not absorb as much water vapor out of the evaporator, raising the pressure in the evaporator, thus raising the chilled water temperature. The evaporator and absorber section will typically operate at a pressure of 1/100th of 1 atmosphere.

Concentrator

The dilute absorbent solution (more water than lithium bromide) will be pumped from the absorber section to the concentrator section. In the concentrator, the refrigerant (water) will be driven out of the solution with the heat from the steam, high temperature hot water or direct fired burner. When the steam or hot water is applied to the solution of lithium bromide and water, the water boils first because its saturation temperature is lower than that of the lithium bromide.

The temperature of the chilled water supplied to the system controls a steam, hot water or gas valve. As the chilled water temperature drops, the valve is modulated more towards its closed position. When the valve modulates closed, less refrigerant is boiled out of the dilute solution, resulting in a concentrated solution having more water than lithium bromide. Temperature in this section may be as high as 210°F.

When the chilled water temperature increases, the steam, high temperature hot water or gas valve will be modulated more open. When the valve modulates open, more refrigerant is boiled off, increasing the concentration of lithium bromide (more lithium bromide than water) to the absorber section.

Condenser

The refrigerant vapor from the concentrator is condensed on the tube surface of the condenser and falls into the pan below the tube bundle. The supply of water to the condenser is coming from a cooling tower. Now the refrigerant is ready to meter back into the evaporator.

Summary

Mechanical refrigeration performs the sensible cooling and dehumidification requirements in the HVAC systems either by using the refrigerant or chilled water to condition the air stream.

In this chapter you learned:

- The general principles of refrigeration, how the capacity (tons) of refrigeration machines is determined, and the different types of refrigerants along with the temperature pressure relationships of refrigerants.
- Basic components of the mechanical refrigeration: the compressor, condenser, evaporator and metering device
- The different types of refrigerant compressors and their operating characteristics and applications.
- The different types of condensing units which includes air-cooled, water-cooled and evaporative cooled condenser.
- The different types of refrigerant metering devices, their operation and applications.
- Direct expansion coils (evaporators) operation, types and applications.
- Evaporators for liquid chillers (chilled water systems) operation and application.
- Absorption refrigeration cycle including its, components, refrigerant, absorbent and operation.

11

Evaporative Cooling and Cooling Towers

Evaporative cooling occurs in nature whenever water comes in contact with air. Near water falls, under summer showers and even upon wet perspiring skin. The process of evaporative cooling has been known for centuries dating back to the ancient Egyptians. Early paintings show servants fanning jars, presumably porous enough to maintain wet surfaces to facilitate the evaporative cooling process. The low desert DB temperatures combined with low WB temperatures allowed ice to be made in this way.

This chapter will acquaint you with the processes and equipment used to provide evaporative cooling. Upon completion you will be able to:

1. Distinguish the difference between direct and indirect evaporative cooling processes and equipment.
2. Understand the psychrometric processes that occur during the evaporative cooling processes.
3. Explain the terms, range and approach as related to cooling towers and evaporative coolers.
4. State the different arrangements of airflow, fans, and sprays, within a cooling tower.
5. Understand the design and operational differences between a cooling tower and evaporative condensers.

Theory of Evaporative Cooling

Evaporative Cooling

Whenever water and non-saturated air are in contact, isolated from other thermal influences, heat transfer will occur. **Evaporative cooling** takes place whenever water and air come into contact and sensible heat is exchanged for latent heat. Evaporative cooling changes the physical state of water into vapor, resulting in cooling of the air or cooling of the water, or both.

The cool air felt near a waterfall is an example of this effect. The process works in reverse as well. Whenever the water is significantly warmer than the air, the water is cooled by the air. Whenever the water is warmer than the air, two cooling processes take place, sensible cooling as well as latent cooling.

There are two types of evaporative cooling. **Direct contact evaporative cooling** takes place whenever air comes in direct contact with the water being cooled. **Indirect contact evaporative cooling** takes place whenever a *wetted* surface is used to facilitate heat transfer to another medium. The equipment required to accomplish these processes will be discussed later in this chapter.

The underlying principle of evaporative cooling deals with the exchange of sensible heat for latent heat. The psychrometric representation of this process is shown in Figure 11-1.

Air enters at point A, absorbs heat and mass (moisture) from the water, and exits at point B in a saturated condition. Vector AB may be separated into components AC (sensible air heating/sensible water cooling) and CB (latent air heating/latent water cooling). The total heat removed is the enthalpy (h) change of hB-hA. Since the wet bulb and enthalpy lines

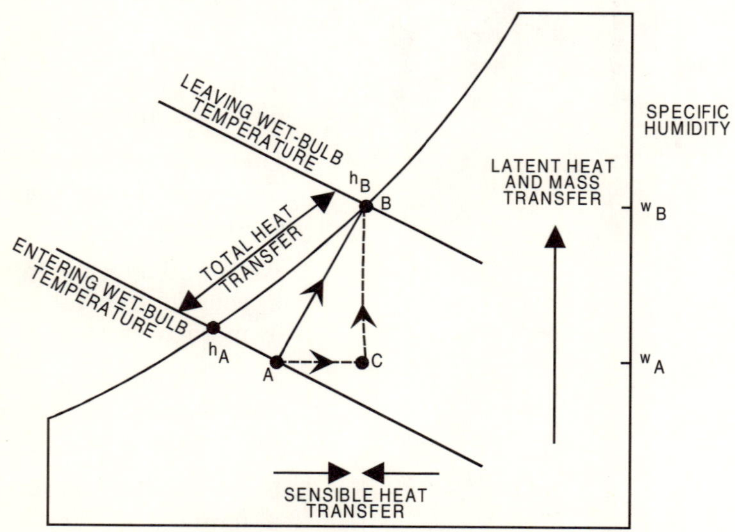

Fig. 11-1 Psychrometric Analysis of Air Passing

Reprinted with the permission of ASHRAE

closely approximate one another, the change in enthalpy can be determined by the change in wet bulb in the air. The change in enthalpy of the air, an increase in this case, reflects the amount of heat removed from the water. As a result, the water has been cooled.

The evaporative cooling (refrigerating effect) of water at 75° is 1050 Btu/lb of H_2O evaporated. To evaporate water, Btus are utilized (thus removed) from the air source providing the heat needed to change water into a vapor, cooling the air. In most cases, the water to be cooled is considerably warmer than the wet bulb temperature of the air, and the process cools both the air and the water. Air and water are jointly cooled, not just the air alone. The primary benefit is the cooling of the condenser water which serves the condenser of the chiller.

Adiabatic Saturation

Adiabatic saturation is the process of exchanging sensible heat for latent heat without the involvement of any external heat. In a once-through process, the temperature of the air and water approach one another. If the air and water were left to cycle through time and time again, eventually the air and water would achieve the same temperature. Given time, both the air and water temperatures and the vapor pressures would equalize, ending heat transfer. This equilibrium occurs when the air is saturated with water vapor. See Chapter 5 Pyschrometrics.

There are limits to the cooling achieved by adiabatic saturation. In any case, the amount of sensible heat removed cannot exceed the latent heat required to saturate the air.

Typical System

A typical cooling system where cooling towers are used with a chiller is shown in Figure 11-2. Here the typical temperatures at design load conditions are shown. Note the temperature of the condenser water leaving the chiller, 95°F compared to the temperature of the condenser water entering the chiller, 85°F. The cooling tower has rejected the heat carried in the condenser water through the evaporative cooling process.

Cooling Tower Design

Cooling tower design is a function of many variables. The amount of heat that the tower must reject is determined by the process being served. The heat to be rejected is equal to the refrigeration load *plus* the heat of compression from the compressor of the chiller. This amount of heat has to be rejected by the cooling tower.

For a typical chilled water system with a 10°ΔT, this relationship amounts to a flow rate of 3 GPM of condenser water per ton of refrigeration at design load.

Analysis of a Cooling Towers Performance

The **design DB** and its **coincident WB** must be considered for the local climate. (See Selecting Design Weather Conditions in Chapter 4, Determining the Loads on the HVAC System). The cooling tower's thermal performance is

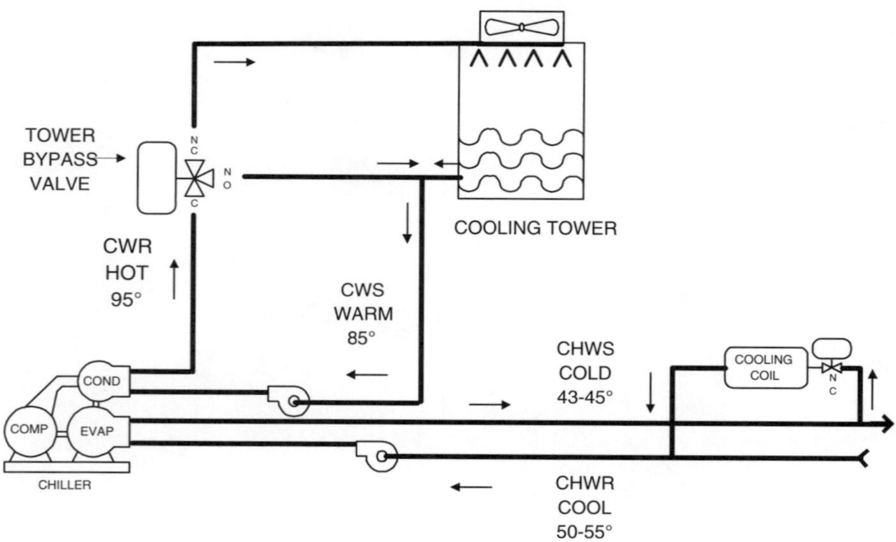

Fig. 11-2 Typical Cooling System

affected by the WB temperature of the air entering the tower. The measurement of this performance on the water being cooled is often expressed in two ways, range and approach. These values are typically stated for design conditions, full tower load. The **range** is the change in the water temperature through the cooling tower. **Approach** is the difference between the wet bulb temperature of the entering air and the temperature of the leaving water.

The relationship between WB temperature, approach temperature and water temperature (leaving the tower) is **Wet bulb temperature + Approach = Water Temperature**. Most cooling towers can achieve an approach as low as 7 - 9°. See Example 11-1.

Example 11-1: A tower has a design approach temperature of 7°. The ambient wet bulb temperature on a given day is 68°. Assuming full tower load, what is the coldest temperature that the tower can produce?

Solution:

Coldest temperature = 68° + 7°
= 75°

The coldest temperature that can leave the tower given the 7° approach on a 68°WB day is 75°. The limiting factors are the outdoor WB and the cooling tower's approach.

On a "design day", in summer when the humidity is high, the maximum energy would be consumed at the chiller. An example day when the WB is 75°, a tower with a 9° approach could produce a condenser water temperature as cool as 84°.

Given a "low ambient" day, a cool dry day in fall, a minimum level of energy would be consumed at the chiller. Due to the lower load at this time of the year, fewer Btus are being removed from the building. At this light load, few Btus are being rejected from the tower. At this time, the tower is in fact oversized for the current load. As a result the tower can drop the condenser water temperature lower than desired (closer to the ambient WB temperature). Controls are used to resolve this problem of condenser water temperatures that are too low. These controls play a large part in maintaining minimum condenser water temperatures that are acceptable to the chiller. Temperatures that are too low can result in damage to the chiller.

Cooling Tower Classification and Construction

Cooling towers can be classified on the basis of air flow in relation to water flow through the fill section of the tower.

Parallel Flow

In a parallel flow tower, air and water flow is in the same direction while traveling through the tower. Air velocities are kept relatively low. Low air velocities make such towers susceptible to cooling from adverse wind. Therefore, these towers are specified to satisfy low first costs for those systems where operating temperatures are not critical. See Figure 11-3.

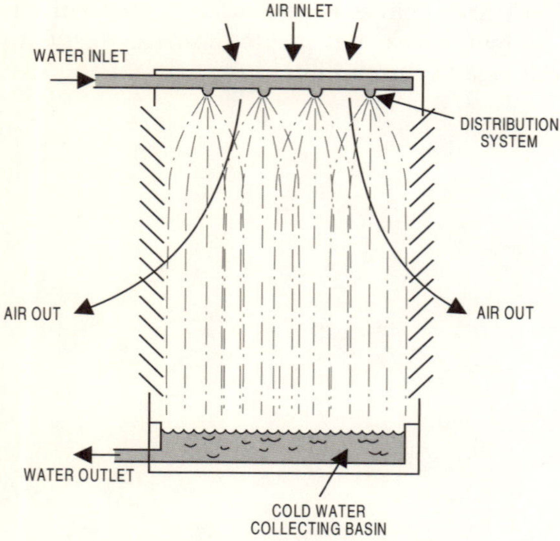

Fig. 11-3 Parallel Arrangement

Reprinted with the permission of ASHRAE

Cross Flow

In a cross flow system, the water flows from the top to the bottom of the tower. Fans draw in air from the sides of the tower and discharge it from the top. Therefore the air and water cross each other in the fill section. This system is inexpensive to manufacture and is popular for small installations. Figure 11-4 is an example of this configuration.

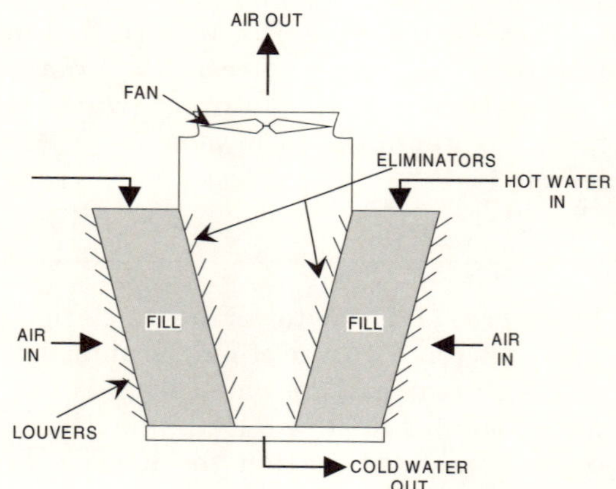

Fig. 11-4
Cross Flow (Induced Draft) Arrangement

Reprinted with the permission of ASHRAE

Counterflow

In a counterflow system as shown in Figure 11-5, the flow of the water is opposite the flow of air as they travel through the fill section. Air is drawn through the louvers at the base and flows upward. Water is pumped into the top of the tower and sprayed downwards. Counterflow towers can be used in places where equipment space is limited.

Atmospheric and Mechanical Draft

There are two major styles of cooling towers used today, the atmospheric and the mechanical draft tower. Few towers in the HVAC industry make use of the atmospheric or natural draft style. They are used occasionally in small applications, but they are susceptible to wind currents. For this reason this chapter will omit any further discussion. Due to better control and less dependence on atmospheric conditions, most towers used in the HVAC industry are the mechanical draft type.

Air can be moved through a tower by a fan in one of two ways. Whenever air is moved by pulling it through the tower it is called **induced draft**. Once again see Figure 11-4. The other method is by blowing it into the tower, in which case the tower is called **forced draft**. See Figure 11-5.

Other Tower Components

In most towers, the water enters the top of the tower and is distributed in one of two ways. In one way, water enters a pan or shallow tank called a **distribution basin** and drops through small holes **(orifices)** allowing the water to pass into the tower. See Figure 11-6.

As the water continues to fall, it comes into contact with a material called **fill or packing.** Splash type packings are shown in Figure 11-6. The fill allows the water to stay in contact with the air for a longer time and for the water droplets to break into even more droplets before finally recombining back into water. This breaking into droplets and slowing of the water allows more cooling to take place. Another type of packing, the film type, is yet another way to slow the water, causing the water to spread over large areas, exposing the water to air flow.

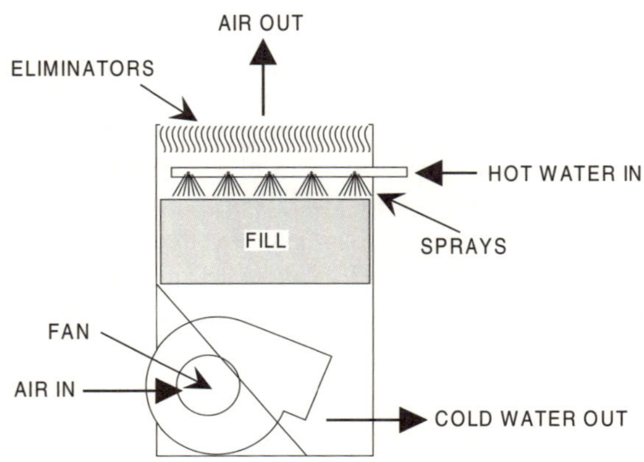

Fig. 11-5
Counterflow (Forced Draft) Arrangement
Reprinted with the permission of ASHRAE

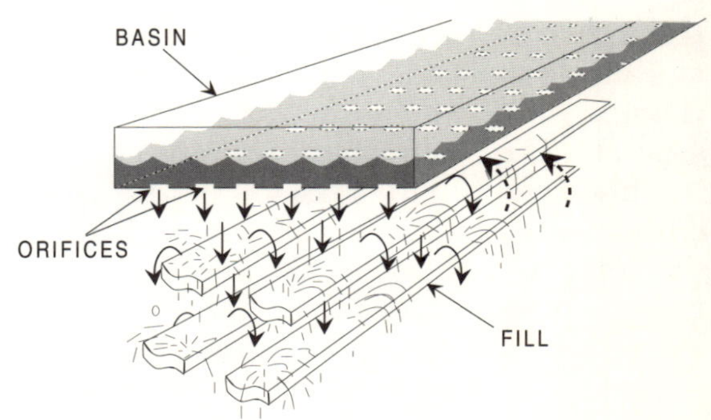

Fig. 11-6
Distribution Basin and Fill-Packing Arrangements

In another arrangement, water is supplied to a pipe header that has **spray nozzles** mounted on it. See Figure 11-7. In either case, sprays or distribution basin, as the water leaves the holes or the nozzles, the water is broken into droplets. Each droplet has a given surface area. When all the droplets are considered, the droplets have created a very high air-to-water contact area. Since the tower cools by evaporating the outside layer of the water droplet, this high amount of surface area significantly increases the amount of cooling done in the tower.

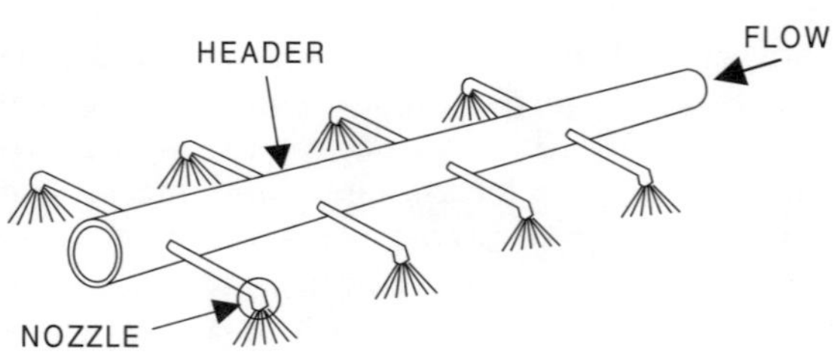

Fig. 11-7
Header and Nozzle Distribution System

Air tends to carry water broken into droplets out of the tower. **Eliminators** help prevent this from occurring. Eliminators are designed to cause the air to take several direction changes which cause the air and water to separate. These eliminators prevent the droplets from drifting out of the tower, thus eliminating the water from the air stream. Because the droplets are heavier than air, they cannot react to sudden direction changes. These droplets either continue to the sump or are captured by the eliminators and returned to the sump. Water droplets that carry over can spot cars and windows as well as increase the demands for make-up water.

Eventually the water collects in a **sump** or water basin connected to the suction of the condenser water pump. The sump may be outdoors as an integral part of the tower or indoors as a separate unit. Depending on the seasonal operation, the designer may choose to locate the tower indoors to prevent freezing. The sump is also an excellent place to introduce makeup water to replenish evaporated water. Also silt, sludge and solids are allowed to settle here for removal later.

Water treatment system components are added to tower systems to combat the growth of bacteria and to inhibit the formation of corrosion and mineral deposits (scale). Slime, algae, scale and corrosion can result in poor heat transfer, system fouling and unsafe operating conditions.

It is generally accepted that hot humid weather facilitates the growth of bacteria such as legionella pneumophilia which is known to cause Legionnaires' Disease, so named after an outbreak at a convention in Philadelphia in 1976. Oxidizing biocides, generally either clorines, bromines or ozones in the proper amounts help prevent the breeding and spreading of bacteria.

The **blowdown cycle** helps reduce the increased mineralization of tower water that results as the water is evaporated. The blowdown valve opens, allowing tower water to drain away. Makeup water is introduced to replace the water just drained thereby diluting the mineral content of the tower water. This cycle is initiated whenever a given quantity of makeup water has been introduced, as measured by a water meter. Overflow protection at the sump is also provided in the event of a make-up valve failure.

Inhibitors put a coating on metal surfaces to prevent oxidation of pipes and condensor tubes. Certain additives also help keep minerals suspended, preventing the buildup of scale. These inhibitors are added automatically fed into the system by special feeder systems calibrated and adjusted by firms specializing in water treatment service.

Strainers downstream from the cooling tower help control foreign materials that may enter the tower water system. Most of this material is brought in through the air. Leaves, seeds, twigs, bugs, even an occasional pigeon, are potential contaminants which can reduce the efficiency of the condensing system. Clogged strainers can reduce flow rates, causing problems elsewhere in the system.

Tower Control

From design to low ambient conditions, the cooling tower controls must optimize the water temperatures to the condenser of the chiller. This temperature is recommended by the chiller manufacturer and is often stated as a minimum acceptable temperature. This minimum temperature keeps the operating pressures within a chiller at acceptable levels. Towers often have condenser water bypass valves, fans or dampers, or some combination to maintain a specified condenser water temperature. Follow the tower manufacturer's recommendation for control.

Much energy can be conserved at the chiller with optimum condenser water temperatures. The tower must perform over a range of conditions. From design conditions to low ambient conditions, the cooling tower must produce the lowest possible temperature acceptable to the chiller. To achieve the desired end result, an optimum temperature to the chiller, some form of control must be applied.

Fan Control

Fan control involves simply turning a fan on or off, from off to low to high or varying fan speed or pitch. In any case, the volume of air through the tower is varied. When the volume changes, a corresponding change in heat transfer results. Variable speed drives are a popular choice because of their infinite speed control capabilities.

Valve control

A bypass valve is used to proportion the amount of water flowing through the tower. Since the condenser water circuit is constant volume, valves must be used to modulate the flow to the tower. A controller senses the leaving temperature of the tower as the controlled variable (generally placed where the water enters the chiller). This controller then proportions the flow to the cooling tower, bypassing as required to maintain the desired temperature. A system diagram is shown in Figure 11-8.

This type of control is used for towers, with or without fans, as a means of controlling the cooling tower at low ambient conditions.

A typical sequence of operation is as follows: when the condenser water temperature increases, the bypass valve modulates open to maximum flow to the tower at which time one or more fan stages start. Generally, the water through the tower will be at full flow before the first stage energizes. Full water to the tower allows the system to make use of as much cooling from natural drafts as possible before utilizing fan energy. Full flow continues as successive stages of fans are energized. If the temperature decreases, the tower fans are de-energized in sequence.

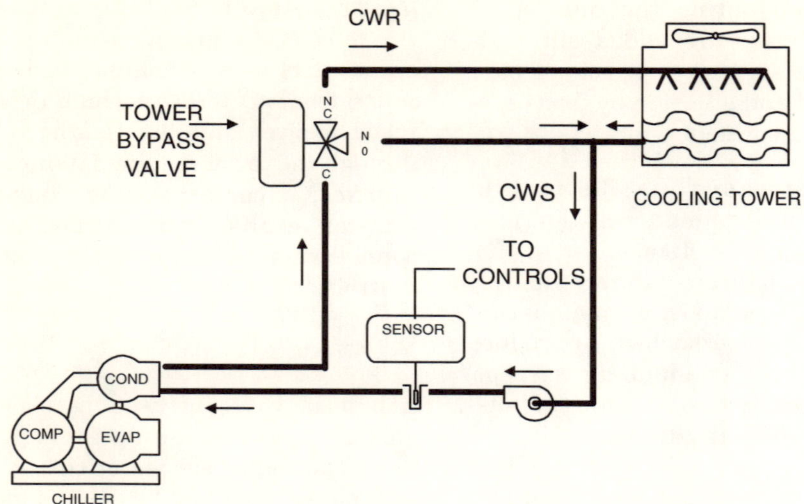

Fig. 11-8 Valve Control

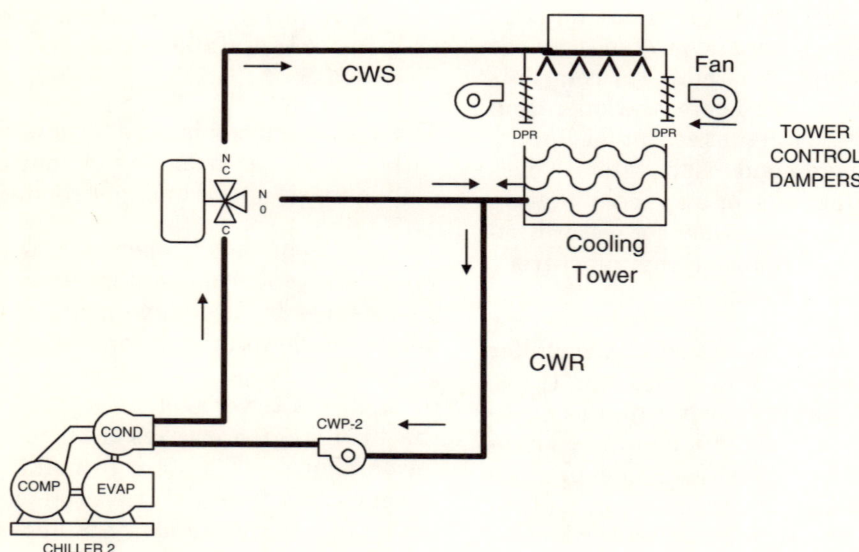

Fig. 11-9 Damper Control

Damper control

Damper control is often used for towers operating at low ambient conditions. Given staged fan operation with the fan on low speed, the quantity of air moving through the tower may cool the condenser water to an undesirable low temperature. In forced draft towers, these dampers provide yet another means of controlling the condenser water temperature by limiting the quantity of air through the tower. The dampers would modulate the airflow to achieve the desired condenser water temperature. See Figure 11-9.

Evaporative Condensers

Another evaporative cooling device is the evaporative condenser. The purpose of an evaporative condenser is the same as a cooling tower, that of rejecting heat. The advantage is that the evaporative condenser eliminates pumping large quantities of water and reduces the amount of water treatment required by cooling towers. In this type of arrangement the refrigeration components are *split* apart, often referred to as a split system. The compressor and the condenser are split apart with the condenser generally located on a roof or on the side of a building. Refrigerant gas having just left the compressor at a higher pressure naturally migrates toward the cold condensing surfaces within the condenser.

Recall the following definition from earlier in the chapter: **indirect evaporative cooling** is whenever a *wetted* surface is employed to facilitate heat transfer from or to another medium. An evaporative condenser consists of a condensing coil, a water spray bank, a fan, a circulating pump, a water basin and a housing. Water is sprayed directly over the outside surface of the condensing coil. Refrigerant gas leaving the compressor enters the condensing coil, rejecting its heat to the cooler coil. This heat causes evaporation to take place. In this way, the heat from the refrigerant is extracted through the wet surface of the coil, condensing the refrigerant. See Figure 11-10.

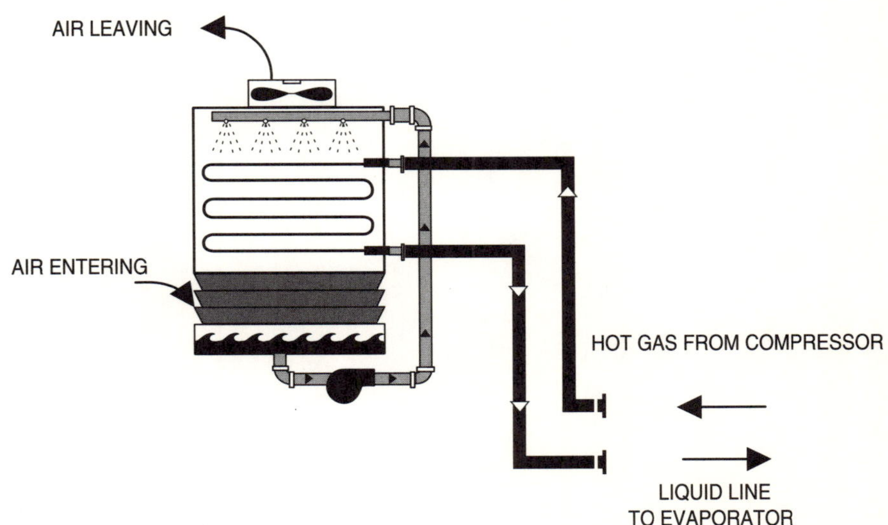

Fig. 11-10 Evaporative Condenser

The materials used for these condensing coils are generally copper, galvanized steel or stainless steel.

Summary

In Chapter 11 you learned how the evaporative cooling process rejects heat from a building. You also learned the construction and operation of equipment, cooling towers and evaporative coolers. The following summary will recap what you have learned in this chapter.

- Evaporative cooling takes place whenever water and air come into contact and an exchange of sensible for latent heat takes place.
- Indirect evaporative cooling occurs whenever a *wetted* surface is employed to facilitate heat transfer to another medium.
- The range is the temperature difference between the water entering and the water leaving the tower.
- Approach is the difference between the wet bulb temperature of the entering air and the temperature of the leaving water.
- Cooling towers can be classified on the basis of air flow in relation to water flow, and by natural or mechanical draft.
- Tower controls maintain temperatures acceptable to the chiller as well as optimize energy savings.
- Evaporative condensers perform both the evaporative cooling and condensing function for the chilled water system.

12

Centrifugal Pumps and Hydronic Systems

Centrifugal pumps do several jobs in HVAC systems. They do more than simply circulate hot and chilled water in hydronic systems. They supply water to boilers. They pump condensate back to steam boilers. And they circulate water from refrigerant condensers to cooling towers and heat exchangers. After reading and studying this chapter, you will be able to:

1. Describe how a centrifugal pump works.
2. Identify the primary parts of a centrifugal pump.
3. Describe how a centrifugal pump adds pressure to a fluid.
4. Recognize several types of the centrifugal pumps.
5. Explain pump performance characteristics, including head, flow rate, horsepower and efficiency.
6. Use pump performance curves to evaluate a pump's capability.
7. Evaluate the flow characteristics of a hydronic system.
8. Calculate the flow requirement for a hydronic system.

How A Centrifugal Pump Works

A centrifugal pump uses centrifugal force to move and pressurize water. Every time you drive around a corner, you experience centrifugal force, the force that pushes you outward. The faster you drive around the corner, the greater the centrifugal force you feel.

The part of a centrifugal pump that creates the centrifugal force is the **impeller**. See Figure 12-1. The impeller spins rapidly as it is driven by the pump motor. Water enters the impeller eye and is then directed outward by the spinning impeller vanes. The faster the impeller spins, the greater is the centrifugal force imparted to the water. Once flung free of the impeller, the water is constrained by the **volute**, the casing surrounding the impeller.

Look at the axial view of the pump in Figure 12-1. Notice that the size of the volute increases away from the center of the pump. This expanding cross-sectional area slows the wildly flowing water. This slowing also converts the water's velocity (kinetic energy) into static pressure (potential energy).

Types Of Centrifugal Pumps

Centrifugal pumps come in many designs and sizes. Most designs meet specific performance goals, while others meet the needs of special uses or size constraints. The following discussion is only an introduction to the diversity of these pumps.

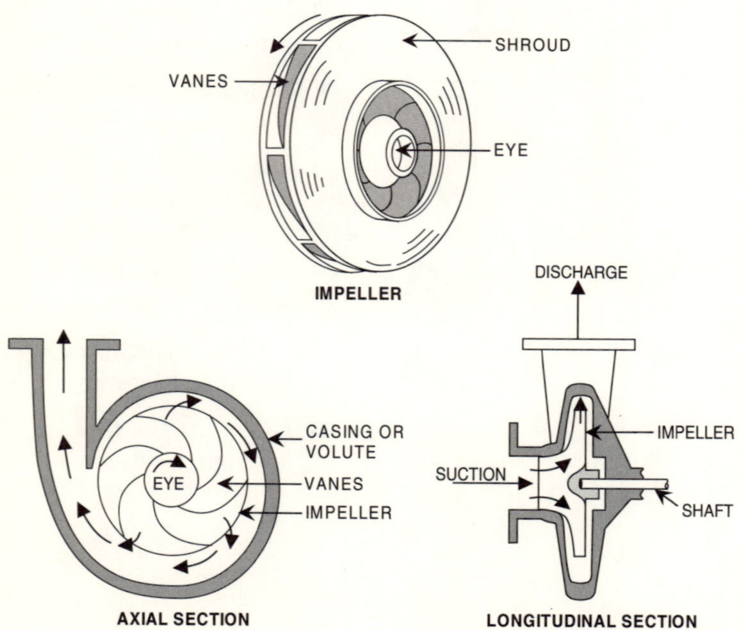

Fig. 12-1 Internal Views of a Centrifugal Pump

Most centrifugal pumps used in HVAC applications are driven by electrical motors. Others are driven by steam turbines or engines. Although only motor-driven pumps are discussed here, remember all centrifugal pumps are similar, regardless of their driving device.

Pumps may be classified in several ways. A major classification is by the motor size. Pumps with motors less than one horsepower are nearly all of in-line or circulator design. Pumps with motors larger than one horsepower are usually classified on the basis of other mechanical or installation details.

A common centrifugal pump is the in-line pump. See Figure 12-2. This low-flow, low-pressure pump is found in small hydronic systems. The **in-line pump** takes its name from its distinguishing feature: The inlet and discharge are in a straight line, or in-line. Because the pump and motor share the same housing, the entire pump assembly is small and lighter weight. Its small size and low weight allow the in-line pump to be installed on and be directly supported by the piping system. The centrifugal pump is also known as a booster pump or circulator. As a **booster pump** it raises fluid pressure within the piping system. As a **circulator pump** it recirculates water through coil units to transfer more heat from the water.

Another method of classification is by the method used to connect pumps to their motors. Pumps may be closed-coupled or flexible-coupled.

The impellers of **close-coupled** pumps are mounted directly on the motor shaft. See Figure 12-3. This works well for smaller pumps because pump vibration is not likely to damage the motor or its bearings.

The impellers of larger, **flexible-coupled** pumps are mounted on an impeller shaft connected to the motor by a flexible coupling. See Figure 12-4. The coupling isolates the motor from the pump vibration. The flexible coupling also allows the motors and pumps to be installed and serviced independent of one another. The pump and motor of Figure 12-4 are mounted on a common baseplate, a normal arrangement for small pumps and motors. Large pumps and their motors are usually mounted on separate bases.

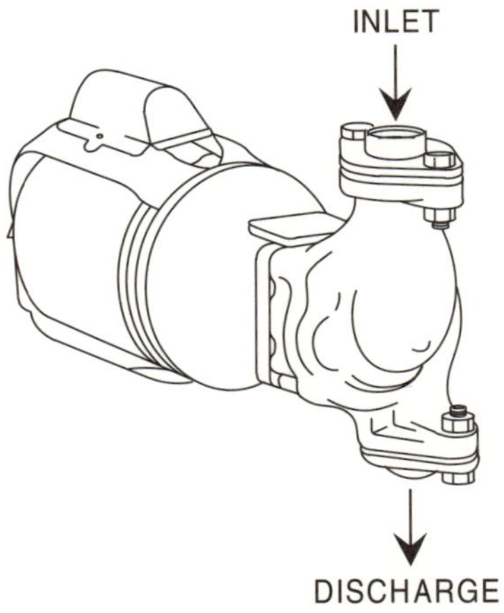

INLET

DISCHARGE

Fig. 12-2 In-line or Circulator Pump

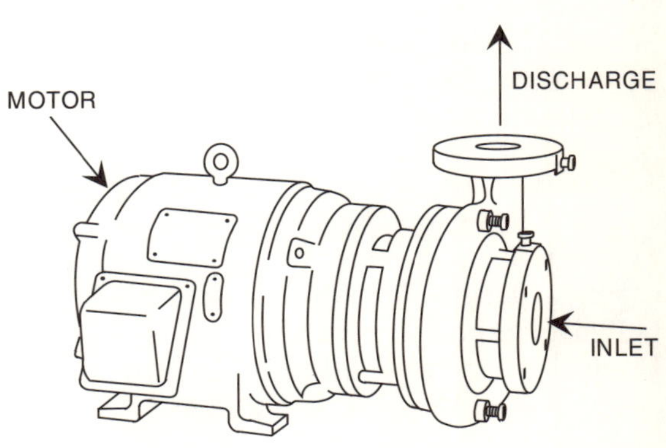

Fig. 12-3 Close Coupled Pump

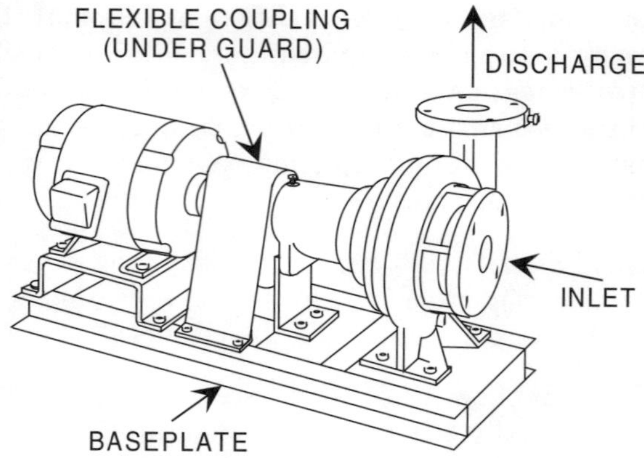

Fig. 12-4 Flexible Coupled Pump on Common Baseplate

Another method of classifying centrifugal pumps is by type of suction. Pumps are designed with either single or double suction. In **single suction** pumps the water enters the pump casing from only one side. In **double suction** pumps the water enters the pump from both sides of the casing.

The direction from which the water enters the single suction pump is the basis for yet another classification. In the **in-line** design water enters the pump on the same line as the outlet. See Figure 12-2. If water enters from the end of the motor shaft, the pump is called an **end suction** pump. The centrifugal pumps shown in Figures 12-3 and 12-4 are examples of the end suction arrangement.

A final method of classification is by type of casing. The casings of small pumps are often cast as an entire unit. In larger pumps, the casing is designed in two halves that are then bolted together to form the complete casing. Pump casings may be split as **horizontal split case** or **vertical split case**.

Characteristics of Centrifugal Pumps

Pump performance is rated with six **pump characteristics**:

1. Head
2. Flow rate
3. Horsepower
4. Speed
5. Efficiency
6. Impeller diameter

Head

Head is a measure of the pressure a pump develops. It is usually expressed in **feet** and you can think of it as a column of water of equivalent height. For example, 100 feet of head would be a column of water 100 feet high. Head is also expressed in **pounds per square inch** or **psi**. One foot of water column is equivalent to 0.433 psi or 100 feet of water equals 43.3 psi. Looking at head from another perspective, 2.3 feet of head is equal to 1 psi. Both examples are shown in Figure 12-5.

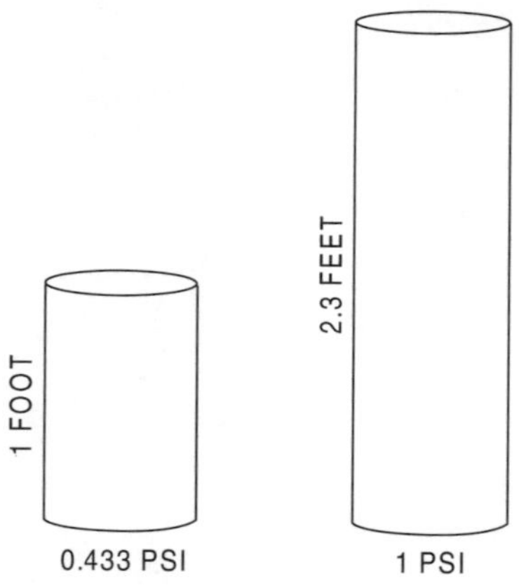

Fig. 12-5 Feet of Head

Flow Rate

Another measure of pump performance is its **flow rate** or **capacity**, which describes the amount of water the pump pumps. This capacity is expressed in **gallons per minute** or **gpm**. The capacity of a pump varies, too. Pumping against a high system resistance reduces the flow rate, while pumping against a lower system resistance increases the flow rate.

Horsepower

The output of these motors and engines is rated in **brake horsepower** or **bhp**. Rating a pump's horsepower requirement is important because pumping more water requires more horsepower and more horsepower uses more energy.

Speed

Since a centrifugal pump connects directly to its motor or engine, it rotates at the same speed as its motor or engine. Horsepower requirements are measured at these specific motor or engine speeds. The two common **pump speeds** driven by induction type electrical motors on 60 hz power are 1725 and 3450 revolutions per minute or **rpm**. Turbine and engine driven speeds vary over a much wider range.

Efficiency

The power input to a pump is always greater than the power output of the pump because of friction losses in the system. The ratio of input to output horsepower is called **efficiency**.

Impeller Diameter

Finally, pump casings are often fitted with impellers of different sizes. For example, a casing might accept impeller diameters of 5, 6 or 7 inches. When switching impellers the manufacturer also matches a motor of the right horsepower to the impeller. Now, the manufacturer can sell a specific pump casing as several different pumps. Each impeller/motor combination creates a new pump with different head, flow rate, horsepower, speed and efficiency characteristics.

Performance Curves of Centrifugal Pumps

Manufacturers test their pumps for these characteristics and publish the results as **pump performance curves**. The three basic pump performance curves are:

1. Head-capacity
2. Horsepower-capacity
3. Efficiency-capacity

As you will see, each curve has a typical shape and all centrifugal pumps have similarly shaped curves. Remember that pumps operate at different speeds and a performance curve could be plotted for each speed.

Head-Capacity Curve

A **head-capacity curve** is a plot of the pump's head on the vertical axis versus its flow rate on the horizontal axis. See Figure 12-6. Notice that at zero flow the pump develops its greatest pressure. This is known as the **shut-off head**. Notice too that as the pump output increases, its ability to develop pressure decreases.

Many casings can be fitted with impellers of different diameters and manufacturers will plot a "family" of head-capacity curves for the same casing on the same graph. See Figure 12-7.

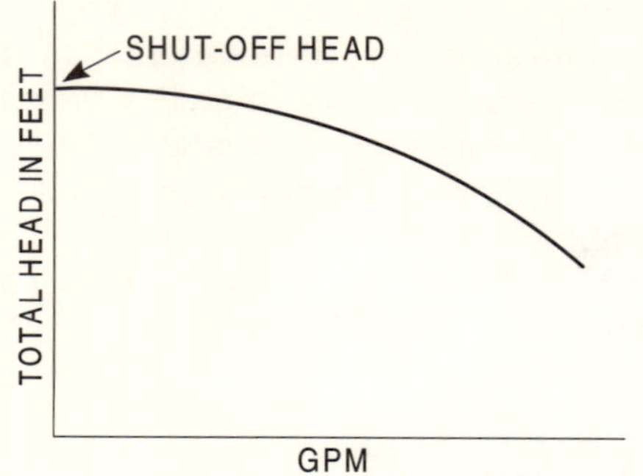

Fig. 12-6 Head vs. Capacity Curve

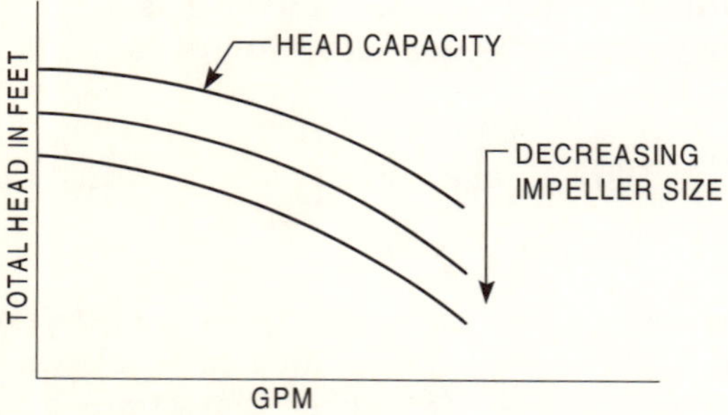

Fig. 12-7 Head-Capacities for One Pump with Different Impellers

Head-capacity curves are also described as being flat or steep. See Figure 12-8. A curve is a **flat curve** if the pump's shutoff head is about 1.1 to 1.2 times its pressure at its peak efficiency. If the ratio of shutoff to peak efficiency pressure is greater than 1.2 the pump curve is a **steep curve**.

The flat-curved pump on the left is used with systems that experience widely varying loads and flow rates but that need nearly constant pressures. In general, flat-curved pumps are more costly to make but they are also practical because they are more easily matched to system operating requirements.

The steep-curved pump develops high pressures and at low flows and low pressures at high flows. This pressure variance can affect the control of valves and other system components. This problem can be corrected by the addition of a pressure bypass to the system.

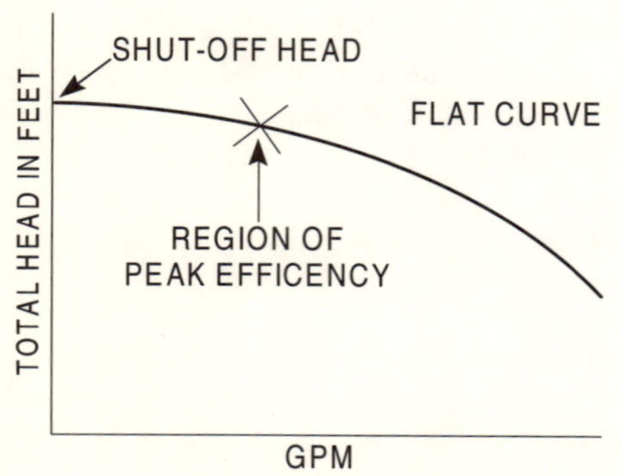

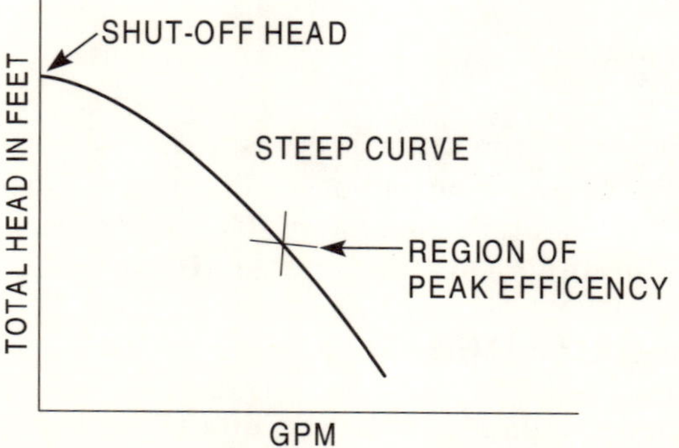

Fig. 12-8 Flat and Steep Head-Capacity Curves

Horsepower-Capacity Curve

A **horsepower-capacity curve** is a plot of the pump's horsepower requirements on the vertical axis versus its flow rate on the horizontal axis. See Figure 12-9. Notice that as the pumping rate increases so must the horsepower.

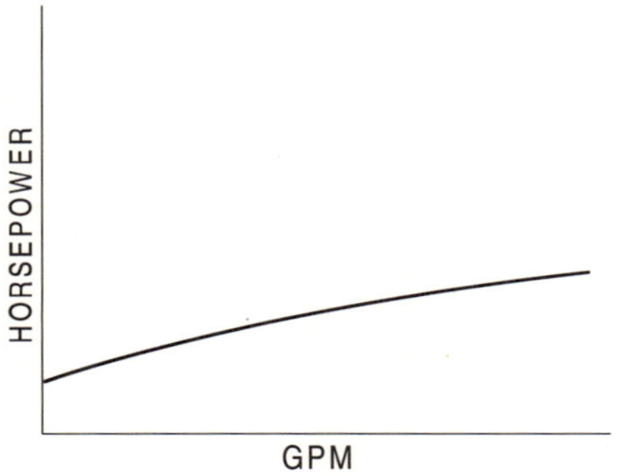

Fig. 12-9 Horsepower-Capacity Curve

Efficiency-Capacity Curve

Every pump operates at less than perfect efficiency because of power losses due to bearing friction, water friction and recirculation within the pump casing. Efficiency is defined as the ratio of input horsepower to output horsepower. Water horsepower is a measure of the energy content of water leaving a pump. Brake horsepower is the energy required to run the motor. As an equation, efficiency looks like this:

$$\textit{Efficiency} = \frac{\textit{Water Horsepower}}{\textit{Motor Brake Horsepower}} \times 100$$

An **efficiency-capacity curve** is a plot of the pump's efficiency on the vertical axis versus its flow rate on the horizontal axis. See Figure 12-10. Notice that this curve begins with a value of zero efficiency at zero flow. As the flow increases so does the efficiency. Unfortunately, the pump reaches its peak efficiency before its peak flow rate. The peak efficiency for most pumps is in a range of 60-80%.

Now, by combining the head-capacity and efficiency-capacity curves, you can determine the pump's head when operating at peak efficiency. See Figure 12-11. Note that even though a pump may have a wide range of flow rates, it is inefficient to use much of this capability.

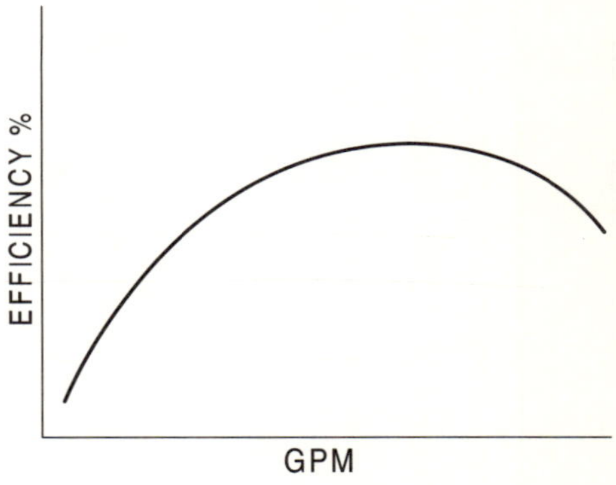

Fig. 12-10 Efficiency-Capacity Curve

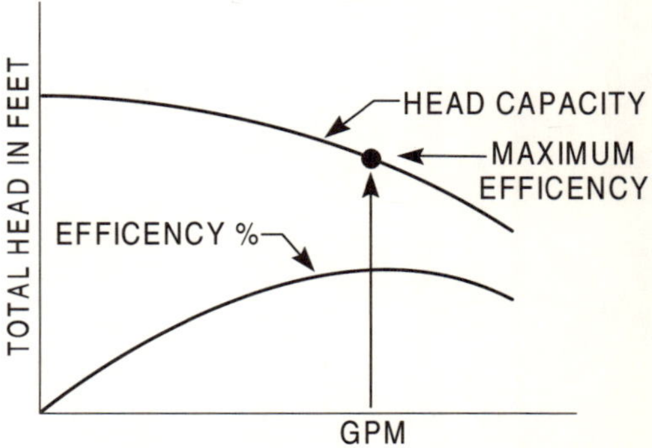

Fig. 12-11
Head, Efficiency and Capacity Curves

Despite the designer's careful selection, pumps rarely operate at only their point of peak efficiency. The designer must therefore look at operating the pump over a range of efficiencies near the peak efficiency. To aid designers, manufacturers plot the curves of efficiency ranges that develop around the point of peak efficiency. See Figure 12-12. The efficiency values shown in the figure vary depending on the pump design.

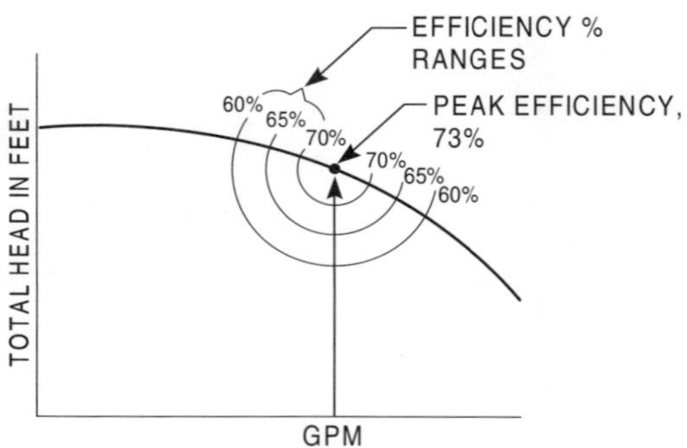

Fig. 12-12
Typical Pump Efficiency Curves

Finally, manufacturers do not furnish the designer with all these separate curves. Instead, they combine the data on one graph like the one shown in Figure 12-13. This is a graph for the same pump fitted with different impellers and motors. With this graph, the designer can simultaneously check head, capacity, horsepower requirements and efficiency for several impeller sizes.

> ***Example 12-1***: The preliminary evaluation of a hydronic system suggests that it needs a flow of 90 GPM and its total head is about 25 ft. Using Figure 12-13, which pump impeller meets these criteria? What size motor should be selected?
>
> ***Solution***: Locate 25 ft. on the head axis and read across to the pump curves. At this head the 5½ in. impeller pumps 90 GPM. This pump will operate at an efficiency of about 65%. Since a ¾ Hp motor is undersized for this application, a 1 Hp motor will be selected.

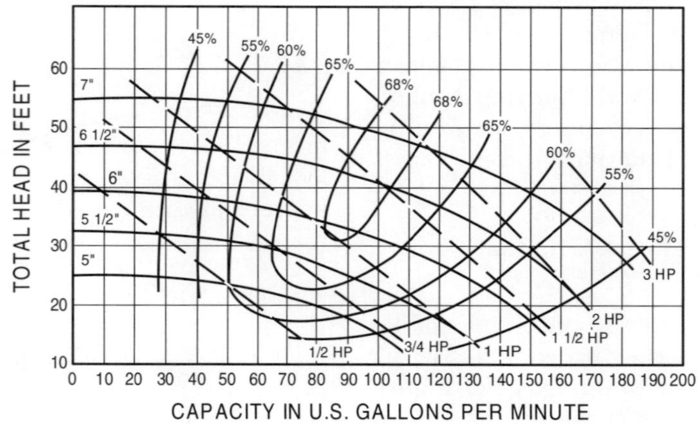

Fig. 12-13 Performance Curves for a Pump Series

Piping Characteristics of Hydronic Systems

Now that you understand how a centrifugal pump works and how to rate its performance, you will learn how to match that performance to the needs of the piping system. To understand piping characteristics, you need to know:

- Whether the piping system is open or closed,
- What flow rate is required, and
- What the system pressure loss is.

An **open piping system** opens to the atmosphere at some point. When you open a faucet to fill a sink you have changed the piping system into an open one. Or, a cooling system's condenser might be in the basement and its cooling tower on the roof. Water entering the tower is in effect poured into the atmosphere. A **closed piping system** has no openings. Its water has no chance to escape and is recirculated. Most chilled water and hot water systems are closed systems.

Calculating the Flow Rate

The required flow rate in a piping system is determined by the type of system. A hydronic system, for example, delivers hot or cold water and collects return water. Its flow rate is based on the amount of heat the system must add or remove from its zones.

The flow rate necessary to condition its zones can be calculated by the formula:

$$GPM = \frac{Q}{500 \times \Delta T}$$

where: GPM = gallons per minute

Q = rate of heat to be added or removed, btu/hr

500 = constant (derived by water 8.34 lb per gallon and 60 min per hour) 8.34 x 60 = 500

ΔT = difference between water entering heat exchange device in °F and water leaving in °F.

The flow rates in condensate and boiler feed systems are a response to maintaining the water in their tanks and boilers at preset levels. Pumps applies to these applications are controlled in an ON/OFF fashion. Their flow rate is selected much differently.

When water flows in a piping system, either open or closed, its movement is resisted by the pipe walls and by every fitting and device that is a part of the system. Elbows, tees, valves, coils, gauges and vents all resist the flow of water. Collectively, these resistances are called **friction head loss**. The amount of head loss due to friction depends on the characteristics of the piping system and the volume of water being pumped. Friction losses increase as the length of pipe increases and as the number of fittings and devices increase. Each resistance has been measured in feet of head. Tables of their values are readily available from manufacturers.

The amount of head loss due to friction also depends on the amount of flow in the system. As the flow increases, so does the friction loss. Because the friction head increases at the *square* of the flow rate, the increase is dramatic. When plotted on a graph the results look like Figure 12-14. This curve is called the **system characteristic curve**.

The other part of system head loss is the water's elevation change. This is called **elevation** or **static head loss**. Specifically, how high must the water be pumped? In an open system the pump must push the water to the system's highest level. If the cooling tower is on the roof and the building is 10 stories high,

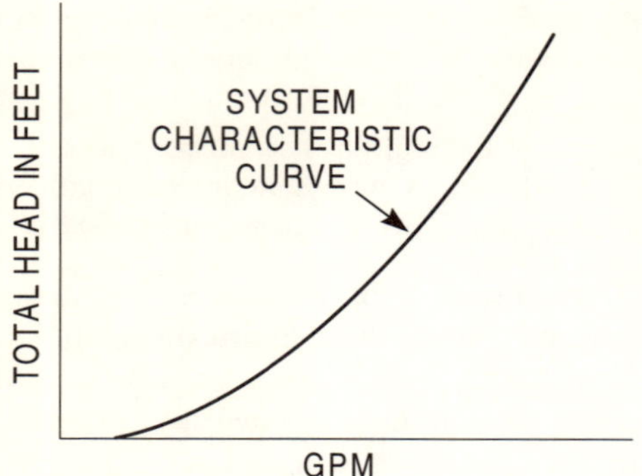

Fig. 12-14 System Characteristic Curve

the static head is about 120 feet (assuming the pump is in the basement, each floor is 10 ft high and the tower is 10 ft tall).

Static head loss is a concern only in open systems. Every time the pump is pumping, the system loses all its pressure because of its opening to the atmosphere. In an open system the pump must therefore develop enough pressure to overcome both elevation and frictional head losses.

In a closed system however, the return line connects directly to the pump inlet. Regardless of how high the water flows in the building, the static pressure from that elevation change is returned to the pump. Since static head is preserved in a closed system the pump only has to overcome friction head loss.

Figure 12-14 shows the system head curve for a closed system. For open systems the static head is added to the system characteristic curve, as shown in Figure 12-15.

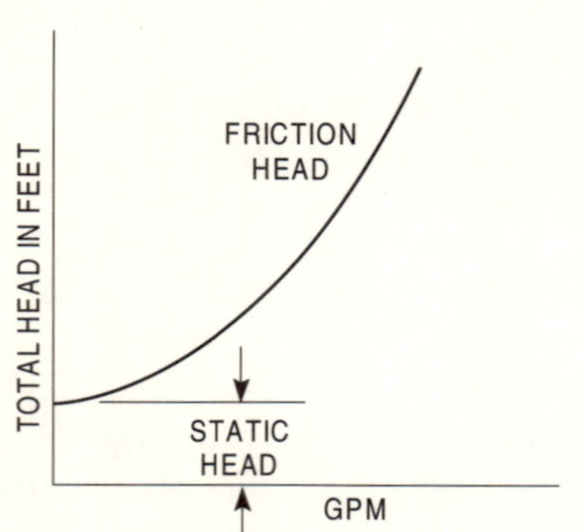

Fig. 12-15 Characteristic Curve

Operating Characteristics of Hydronic Systems

What is the flow rate and pressure within a hydronic system? The answer to this question is found by combining the pump head-capacity curve and the system characteristic curve. See Figure 12-16.

Regardless of its maximum capacity, a pump's output is always determined by the system head. Said another way, the **system operating point** is always at the intersection of the pump's head-capacity curve and the piping system's characteristic curve. For example, as the piping system ages it becomes more resistant to flow due to a buildup of scale. This increase in the friction head shifts the entire system characteristic curve to the left as in Figure 12-17. The pump will respond to this change by developing greater head, but less flow.

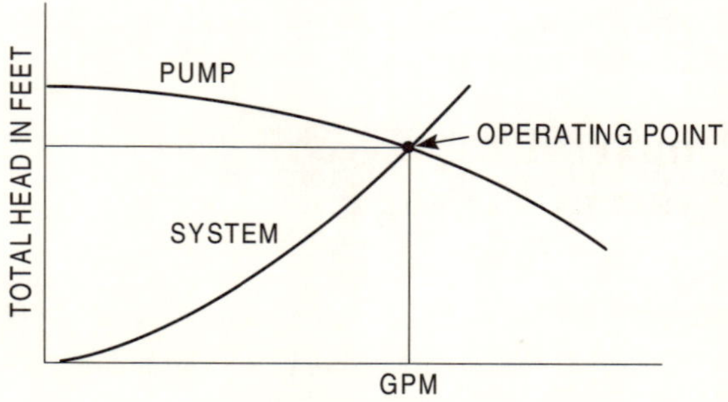

Fig. 12-16
System Operating Point

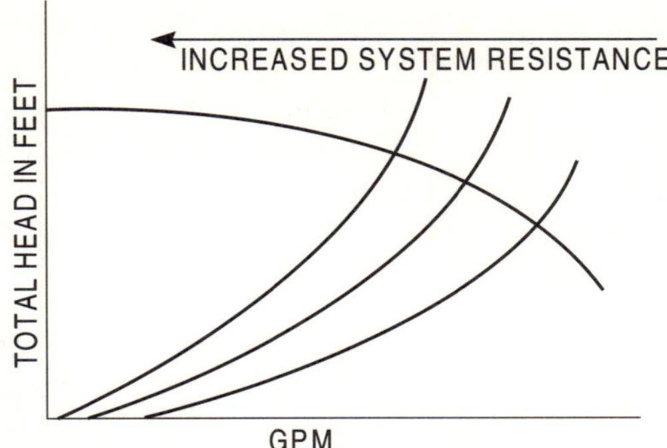

Fig. 12-17
Increasing System Head

> ***Example 12-2*:** After 10 years of operation, the system head in Example 12-1 now measures 30 ft. What is the present output from a pump fitted with the 5½ in. impeller?
>
> ***Solution*:** Locate on the pump curve, Figure 12-13, 30 ft. of head. Read across to the 5½ in. impeller curve and down to the GPM scale. At 30 ft. of head this pump is now capable of pumping about 60 GPM. If this is a hydronic system, the system can no longer meet the design heating and cooling requirements as well as it once did.

Controlling Hydronic System Pressure

The pressure of a hydronic system must be maintained within a narrow range. If the pressure is too high, it can prevent flow control valves from seating properly, it can weaken pipe joints and cause leaks, and it can damage boilers, coils and gauges. On the other hand, if the pressure is too low, it can cause the vapor pressure of water to fall below its saturation temperature. The result is, instant formation of steam. If the steam forms in a pump, its explosive force will damage the pump. And, if the water pressure falls below atmospheric pressure, air may enter the piping system.

The centrifugal pump is not the only source of system pressure. The volume of the water in the system, the air trapped in the piping and the position of the flow control valves on each terminal unit also increase system pressure. Two devices and one piping arrangement help the operator control and maintain system pressure. They are the expansion tank, air venting valves and the pressure bypass.

Expansion Tank

Water expands as it is heated, and contracts as it cools. In a closed pipe system, hot water increases the system pressure. Conversely, cool water lowers system pressure. The **expansion** tank is used to control this cyclical expansion and contraction of water volume.

A typical closed expansion tank is shown in Figure 12-18. The expansion tank also comes in an open version that is vented to the atmosphere, but it is used infrequently today.

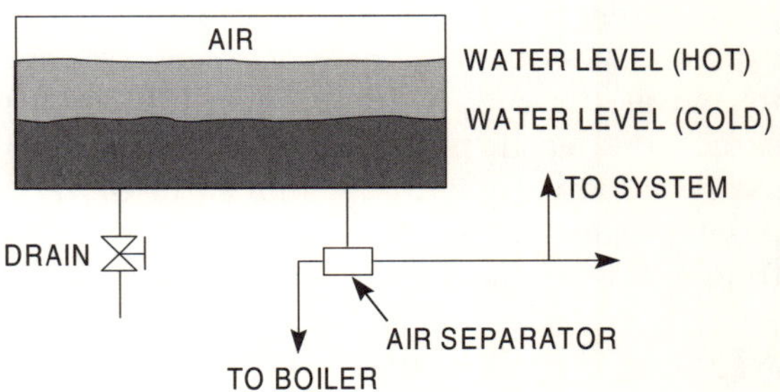

Fig. 12-18 Closed Expansion Tank
AIR CONDITIONING PRINCIPLES AND SYSTEMS, 2nd ed. by Pita, Edward G., ©1989 Adapted by permission of Prentice-Hall Inc., Upper Saddle River, NJ

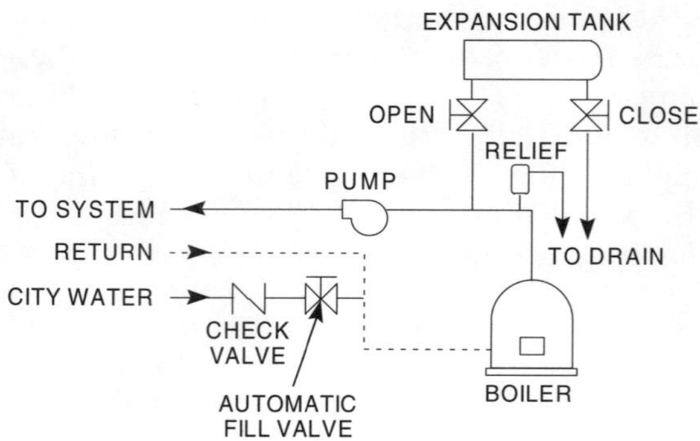

Fig. 12-19
Best Compression Tank Location

An expansion tank acts as a cushion. The extra hot water volume flows into the tank. Then, as the water cools between cycles, it flows back into the system. Expansion tanks also help control system air by storing some of the air trapped in the system.

The best location for the expansion tank in a hot water system is between the boiler and the pump. See Figure 12-19. The tank location then becomes a point of no pressure change within the system. By connecting the expansion tank piping to the pump suction, the suction pressure is always higher than atmospheric pressure. This prevents the pump from drawing air into the system. A gauge connected at the suction of the pump reads the same whether the pump is running or not. The gauge will always read the sum of the highest point in the system and the pressure added from city water source.

Air Venting Valves

Air trapped in the piping displaces water volume. Trapped air collects at the higher points in the system. The rising air displaces water from these regions and blocks water circulation. These air locks reduce heating or cooling effectiveness. Air in the system may also cause undesirable noise as moving water tries to dislodge the air pockets. Furthermore, the trapped air is corrosive to the metal piping and components of the system.

Air can enter the system in four ways. First, some is trapped when the system is filled with water. Second, when the water is heated the first time, dissolved air is liberated as a gas. Third, makeup water added to the system contains more dissolved air. Fourth, normal maintenance and repair often introduce additional air.

To combat these problems, **air venting valves** can be added to the high points of the system and to the terminal units. The small valves, like those shown in Figure 12-20, will open automatically, or may be manually operated, to release trapped air and its pressure. Inside the automatic venting valve is hygroscopic material which, when wet, forms a tight seal. If air collects in the valve, the hygroscopic material dries out, allowing air to bleed out of the valve.

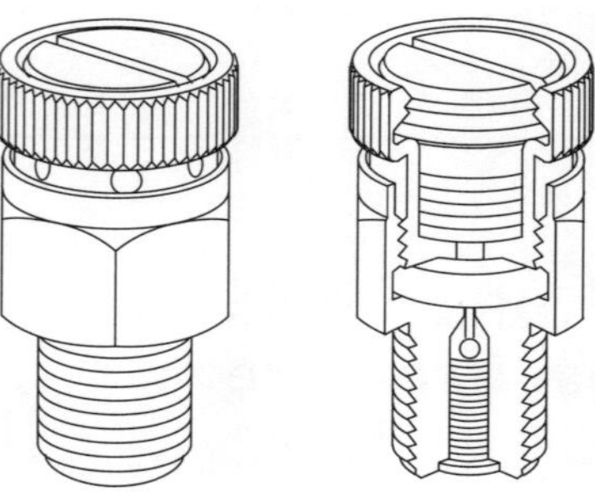

Fig. 12-20
Automatic Air Venting Valves

Pressure Bypass

The zone thermostat controls the flow of water through its terminal units. When the demand for heating or cooling is high, the thermostat signals the valves to open fully, letting the maximum flow into each unit. As the demand lessens, the thermostat signals the valves to begin closing. The closing valves add resistance or head to the system. As the system head builds, the centrifugal pump maintains the new head by moving to the left along its performance curve. This pressure buildup can cause flow control valves to lift off their seats, and it can easily exceed the volume of the compression tank to absorb it.

The solution to the problem is the pressure bypass piping arrangement shown in Figure 12-21. The **pressure differential control valve** monitors the difference between the pump discharge pressure and the return line pressure. When the difference exceeds the acceptable limit, the control valve opens, letting some water flow from the system's supply line directly into the return line. This bypassing reduces the system head.

The effect of the pressure bypass on the pump's performance curve is shown in Figure 12-22. Here you can see that the pump's head capacity is effectively shaved off and kept below an acceptable maximum pressure.

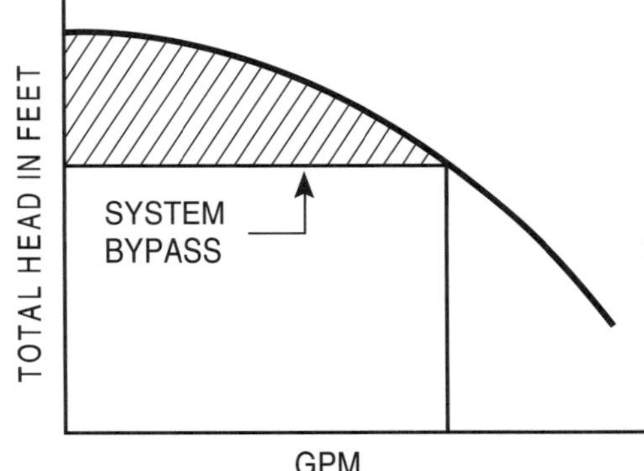

Fig. 12-22 Effect of Pressure Bypass on Head-Capacity Curve

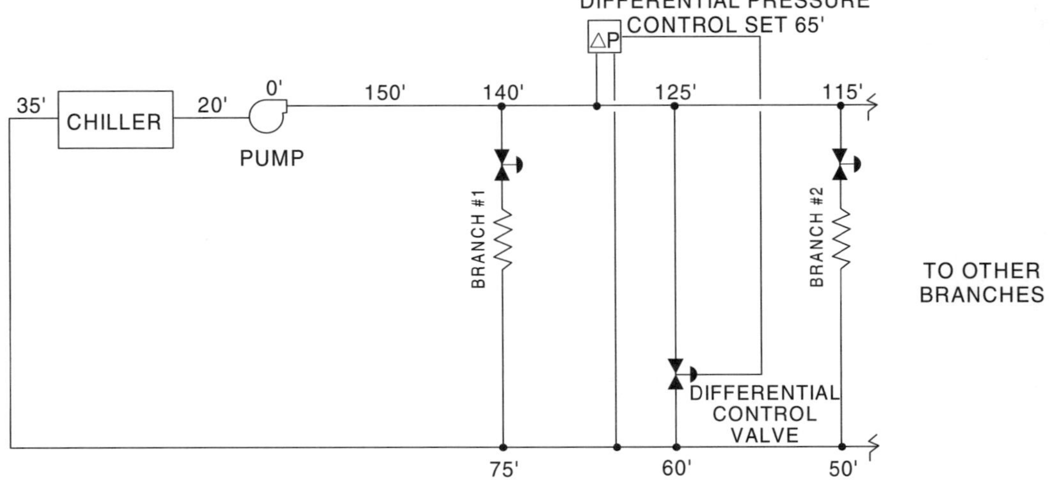

Fig. 12-21 Pressure Bypass

Summary

Centrifugal pumps perform several duties in the HVAC system. They circulate heated and chilled water, they supply boiler feed water, they pump condensate return water and they maintain water flow in refrigerant condenser systems. In this chapter you learned that:

- A centrifugal pump raises water pressure by adding to the moving or kinetic energy of the water. When the water velocity slows, the conservation of energy principle transforms the kinetic energy into the potential energy of pressure.
- When rating pump performance use the head-capacity, horsepower-capacity and efficiency-capacity curves supplied by the manufacturer.
- An open piping system has to produce pressure to overcome elevation head and friction head. A closed piping system has only friction head.
- The pressure in a hydronic system is controlled by the centrifugal pump, a compression tank and a differential controlled-system bypass.

13

Air Cleaning Equipment

People today are concerned about indoor air quality. In fact, people's health and comfort, the efficient operation of equipment, and a building's cleanliness all depend on it. Everyone assumes that the air inside his or her building will be clean. However, not all building air is clean and odor free. Today, it's the job of the HVAC system's air cleaning equipment to ensure the indoor atmosphere is made as clean and odor free as possible. After reading and studying this chapter, you will be able to:

1. List the common components of outdoor and indoor air pollution.
2. Explain why clean HVAC air is important.
3. Describe the general classes of airborne pollutants.
4. Explain the three primary methods of removing particulate matter from air.
5. Describe the several types of air filters, their media and their uses.
6. Evaluate a filter's operating performance.
7. Understand the unique problems of defining and reducing gas and odor pollution.

What is Dirty Air?

Dirty air contains many substances, ranging from eye-burning smog to sneeze-provoking pollen to headache-causing industrial fumes to irritating cigarette smoke from the next office. Depending upon where they are taken, samples of outdoor air may contain soot and smoke, ozone, silica, clay, decayed animal and vegetable matter, lint and plant fibers, metallic fragments, mold spores, bacteria, pollen, and a range of gaseous emissions from business and industrial processes. Unless properly filtered and cleaned, outdoor air drawn into the HVAC system may introduce some of these same substances to the indoor air.

Samples of indoor air may contain some of the same outdoor substances. They may also contain tobacco smoke; gaseous emissions from fabrics, furniture and adhesives; and offensive odors from equipment, solvents, and people. Indoor air should not contain contaminants that exceed concentrations known to adversely affect health or cause discomfort to a building's occupants.

Why Clean HVAC Air?

There are least five reasons for maintaining high indoor air quality.

1. Protect everyone's health. Many people suffer from respiratory ailments that include allergies and asthma. Medical research increasingly demonstrates that all forms of airborne pollution aggravate these conditions. Clean air protects people's health and reduces medical costs.
2. Improve employee productivity. Although some employees pay little heed to airborne contaminants, all employees work more comfortably in clean air.
3. Protect sensitive equipment. Today's office, school, hospital and businesses depend on electronic equipment. Clean air protects this equipment and adds to its longevity.
4. Preserve furnishings. Clean air reduces the dust, gaseous pollutants and odors that often cling to wall coverings, window treatments and furnishings.
5. Protect air conditioning equipment. Dust and dirt collects on air conditioning coils, fans and ducts, and reduces their conditioning efficiency. Cleaning the air before it flows through this equipment adds to efficient system operation.

Air Cleaning Methods

The range of potential airborne substances is certainly large. To understand the equipment and processes that have been designed for removing these substances, the contaminants can be grouped into four broad categories:

- Inert particles
- Gaseous pollutants and odors
- Microorganisms
- Liquid particles, such as mist and fog

A variety of methods is used to remove these contaminants. Large and medium-sized inert particles are typically removed with mid efficiency filtration, and the smaller particles are removed with high efficiency filtration or by electrostatic precipitation. Gaseous pollutants and odors may be removed with air washers or activated charcoal filters. Microorganisms and liquid particles are controlled by ventilation, filtration and proper maintenance. The point to remember is this: while filtration is the most common means of removing dust and dirt, it is only one of several air cleaning processes.

Removing Particulates

Most of the 'dust and dirt' in dirty air is made up of inert particles. These particles may be as small as 0.01 micron to as large as insects and leaves. No one type of air cleaning device could efficiently remove all of these contaminants. Equipment that removed the smallest particles would quickly become overloaded with the larger ones. In addition, not every cleaning application needs to remove the entire range of particles.

Air filters remove airborne dust in several ways. The air filter designer must consider the sizes and shapes, specific gravities, concentrations and electrical properties of the particulates. The most important characteristic is particle size.

Mechanics of Air Filtration

Air filters remove airborne contaminates by several different mechanisms. The three most significant are the following:

- Straining
- Impingement
- Electrostatic Attraction

Straining When looking at a dirty air filter the dust, lint and large particles collected on the filter surface are the most noticeable. These particles were **strained** from the air stream flowing through the filter. See Figure 13-1. While this surface dirt is eye catching, it is only a small portion of the matter removed from the air. Most of the airborne contamination is removed by the next method.

Impingement As air moves through a filter, it must bend or change direction to flow around the filter fibers. Particles in the air resist these direction changes because of their inertial mass. Because they cannot change direction, the particles collide with or **impinge** upon the filter fibers and stick to them. This method is shown in Figure 13-2.

As you might expect, impingement is more effective in a rapidly moving airstream. It is more difficult for rapidly moving particles to change direction than it is for slower moving ones. For this reason, most filters have a minimum, as well as a maximum, air velocity rating.

Electrostatic Attraction Dust develops an electrical charge as it passes through the air just as a person develops a static charge by walking on a rug. The statically charged particles then stick to the oppositely charged filter fibers by **electrostatic attraction**.

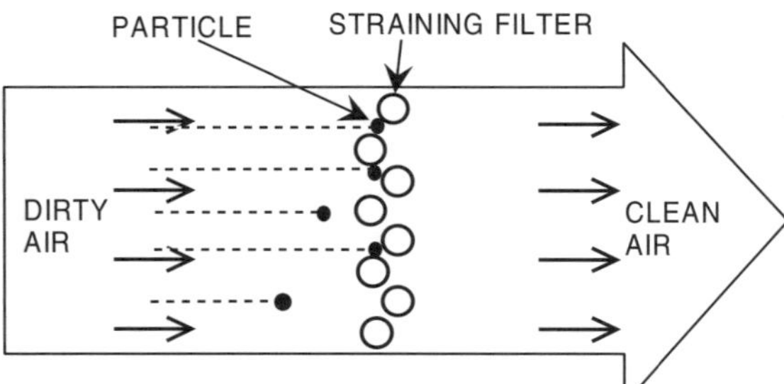

Fig. 13-1 Straining Mechanism

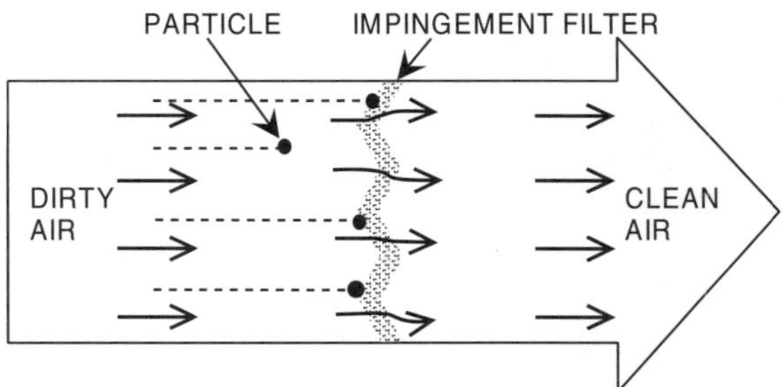

Fig. 13-2 Impingement Mechanism

Electronic air filters, with their electrically charged plates, very effectively use this removal mechanism. The airborne particles are charged with a power supply in the first section of the filter. In the second section they are collected on plates with the opposite charge. See Figure 13-3.

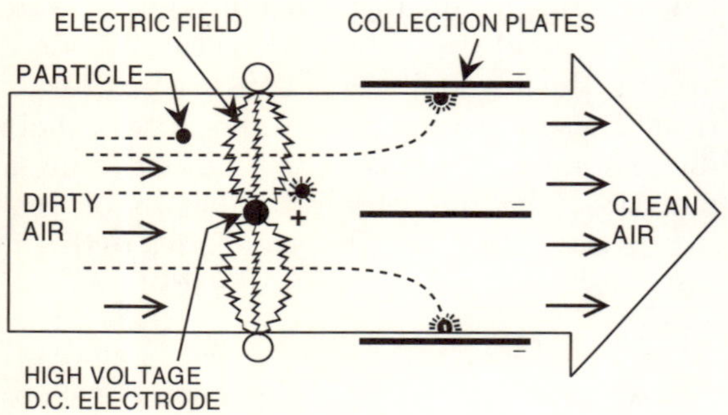

Fig. 13-3 Electrostatic Attraction Mechanism

Types of Air Filters

Filters may be classified in several ways. First, they may be described by their medium. Most filters are either:

- Dry, fibrous media types, or
- Viscous impingement types.

Dry, fibrous media filters are made of glass fibers, cellulose fibers, or synthetic fibers. The media may be formed into blankets of varying thicknesses, fiber sizes, and densities, or into fiber mats of random fiber size and density. The medium may be supported by a wire frame, it may be self-supporting because of its inherent rigidity, or it may be self-supporting because the air flowing through it inflates it into its expanded form.

Some dry filter media have enormous surface areas capable of filtering extremely small particles. To maintain proper airflow and minimize the amount of energy required to move air through filters, medium efficiency (30-60% ASHRAE Dust Spot Rating) **extended surface filters** are used. **Pleated-type, pocket or bag filters**, see figure 13-4, are often recommended by designers. These filters have a

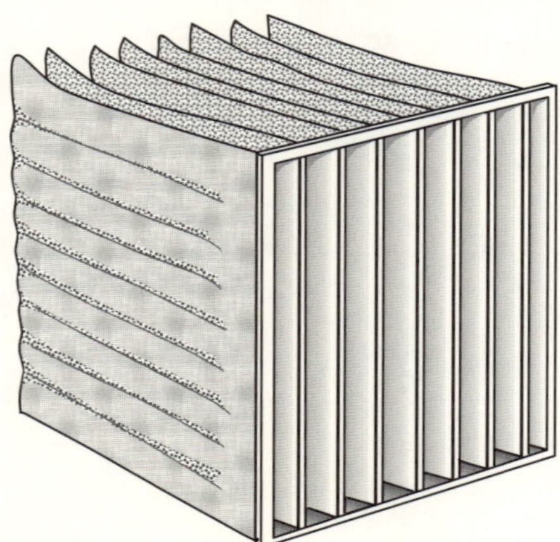

Fig. 13-4 Pleated and Bag Filters

Illustration courtesy of Airguard Industries

higher removal efficiency than low efficiency filters, yet they will last without clogging longer than higher efficiency filters.

In buildings that are designed to be exceptionally clean, the designer may specify the equipment to use both medium efficiency as pre-filters with higher efficiency (85-95% ASHRAE Dust Spot Rating) extended surface filters. Figure 13-5 shows this arrangement of pre filters used in conjunction with high efficiency filters. Examples of high efficiency filters are the HEPA (High Efficiency Particulate Air) and the ULPA (Ultra Low Penetration Air) filters. These filters are routinely used for clean room, toxic containment and nuclear applications.

Viscous impingement filters are made of coarse fibers that are formed into mats of high porosity and then coated with a viscous substance. Glass fibers, synthetic fibers, synthetic open-cell foams, animal hair, vegetable fibers, metallic wools, foils, screens and expanded metals have been used as filter media. The media is then coated with a viscous substance such as oil or adhesive. The viscous material retains the particles that impinge on the fibers, hence the name viscous impingement. These filters may be used to pre-filter the air before it flows into high efficiency filter or in fan coil applications.

Next, filters may be classified by whether they are:

- Disposable or cleanable, or
- Stationary or renewable

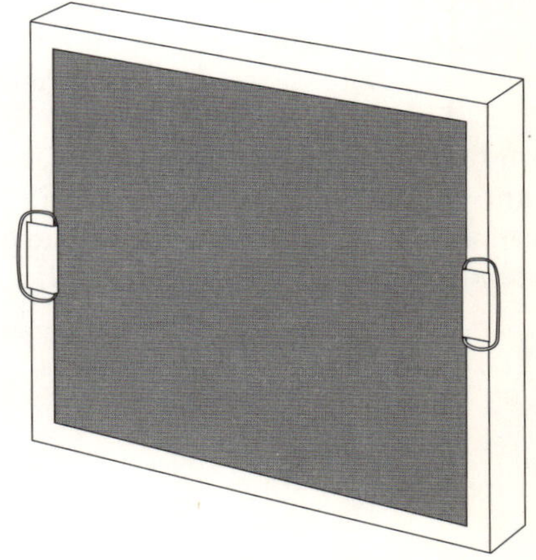

Fig. 13-6
Cleanable Viscous Impingement Filter
Illustration courtesy of Airguard Industries

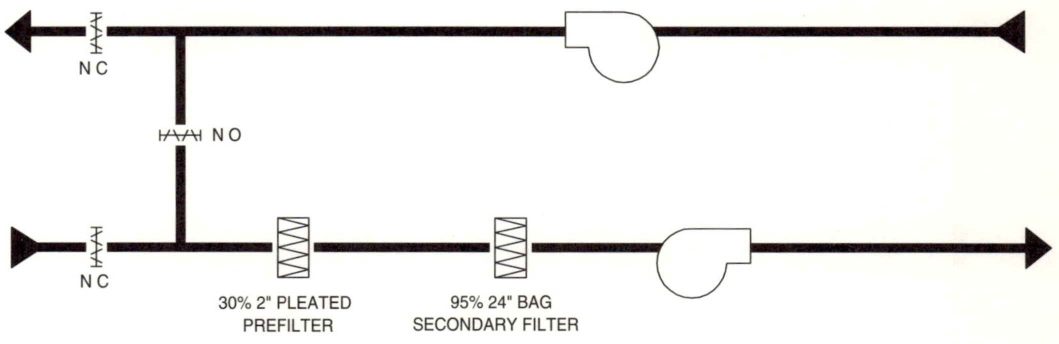

Fig. 13-5
Typical AHU Filter Arrangement

Disposable air filters are made of inexpensive materials and used until their performance declines. Then they are thrown away. **Cleanable air filters** are made from better materials that will stand up to repeated cleaning. Cleanable filters must be dismantled or taken off line for cleaning. See Figure 13-6.

Stationary air filters are made in rectangular panels that may be stacked together to fill the filter opening. When they become dirty, the panels are removed for cleaning or disposal. **Renewable media filters** are actually moving curtains of filter media. The medium may be the viscous impingement type or the dry fibrous type. The clean medium is unwound, manually or automatically, from its spool and moved across the face of the filter. See Figure 13-7 for an example. In the case of disposable, renewable filters, the dirty medium is wound onto a roll until the roll is filled. Then the entire roll is thrown away. In the case of cleanable, renewable filters, the moving curtain travels from the face of the filter into a reservoir, where it is automatically cleaned. The reservoir must be periodically cleaned of its collected dirt.

The last type of filter is the electronic air filter. Unlike the preceding filter types which rely on straining and impingement mechanisms, electronic air cleaners use only electrostatic attraction. **Electronic air filters** give a high voltage charge to incoming airborne dust particles and then capture the particles on oppositely charged collection plates. The parallel collection plates may or may not be coated with a viscous substance. In either case, the collection plates must be removed from the device for cleaning. See Figure 13-8.

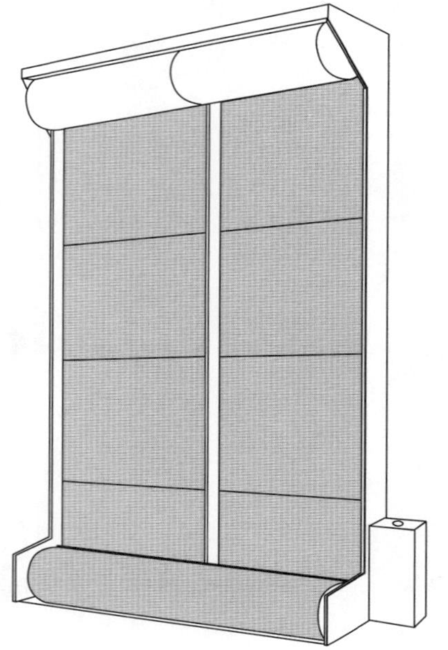

Fig. 13-7 Automatic Renewable Filter

Illustration Courtesy of Airguard Industries

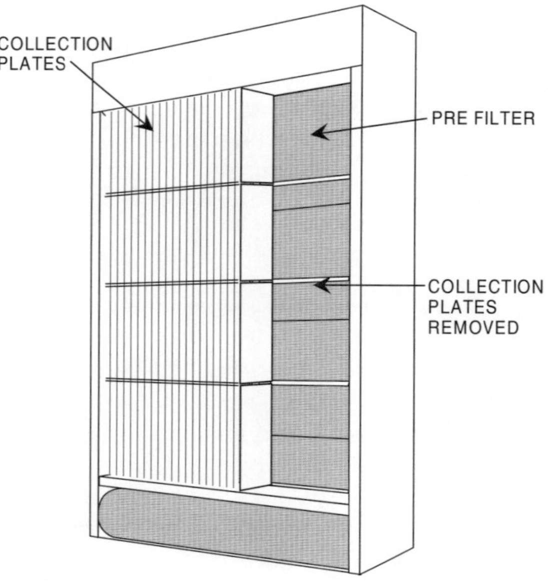

Fig. 13-8 Electronic Air Filter

Illustration courtesy of Airguard Industries

Evaluating Air Filter Performance

In general terms, the efficiency of an air filter is its ability to remove contaminants from the air.

Contaminant concentration is defined as the number, weight or surface area of the particles in the air. The efficiency of a specific filter therefore varies with the test procedure, concentration and the characteristics of the contaminating particles.

The atmospheric dust spot test is often used to rate medium efficiency air cleaners. ASHRAE Standard 52.1-1992 "Grauimeteric and Dust Spot Procedures for Testing Air Cleaning Devices Used in General Ventilation for Removing Particulate Matter" states that the removal rate is based on the cleaner's ability to reduce the soiling of a clean paper target. This ability is dependent on the cleaner removing very fine particles from the air. The measurement is made by comparing the light transmission of the stains made on paper target samples both upstream and downstream of the filter.

While measuring the change in contaminant concentration is a very precise way of evaluating filter effectiveness, the HVAC operator needs an easier method of checking operating performance. That method is to measure the air pressure drop across the filter.

Air cannot flow unimpeded through an air filter, even a clean one because of the interference of the fibers. This resistance is measured by an air pressure drop between the front and the back of the filter. The manufacturer rates a filter to clean a certain volume of air at a specified face velocity. The **face velocity** is the velocity of the air entering the filter. Since resistance increases as contaminants build up, the operator can measure its increase between the clean and dirty conditions. Then, when the resistance reaches a specified level, the operator knows it is time to clean or renew the filter.

This in-place test of filter efficiency is accomplished by installing manometers, draft gauges or differential pressure sensors on both sides of the filter. When these devices sense a specified pressure difference between the two sides, their transducers create an electrical signal. The electrical signal triggers an alarm to alert the operator. These devices work well with stationary, dry type filters and viscous impingement filters. However, for renewable media filters, timers and light transmission measuring devices work better.

The Problems of Gas and Odor Pollution

Removing particulate matter from the air is fairly straightforward, easy to do, and inexpensive. This is not always the case for gas and odor pollution. Cleaning the air of these contaminants may indeed be the HVAC operator's toughest challenge for several reasons.

1. Because they are difficult to predict ahead of time, gas and odor problems are often not known about or dealt with until after the facility and its HVAC system have been built.
2. Although many gaseous contaminants can be measured precisely, odors can seldom be measured.
3. While standards exist for many gases used in or produced by industrial processes, similar standards do not exist for non-industrial interiors.
4. Each gas behaves independently. Its behavior is affected by properties such as molecular weight, polarity, partial pres-

sure, diffusion, thermal conductivity, acidity and basicity.

5. HVAC operators have traditionally increased exhaust and ventilation rates to control gas and odor buildup. Since building air represents heating and cooling expense, operators have had to cut back on ventilation rates.

6. People's responses to gases and odors vary widely. What one finds irritating, another doesn't notice. Today, vague, difficult-to-locate problems are referred to as sick building syndrome. The annoying odor threshold is often barely measurable. And finally, the building occupants and test equipment will be the judges of whether a gas or odor has been adequately controlled.

Despite these difficulties, there are several positive actions for reducing gas and odor problems. ASHRAE suggests in standard 62-1989; Ventilation for Acceptable Indoor Air Quality two approaches. The first is the Ventilation Rate Procedure where a recommended quantity of ventilation air is provided based on the type of facility. The second is the Indoor Air Quality Procedure where acceptable air quality is achieved within the space by controlling known contaminants.

Controlling Gas and Odor Pollution

There are at least three methods for controlling gas and odor problems:

1. Remove or reduce the source of the problem.
2. Increase the HVAC ventilation rate.
3. Add a dry sorbent system of activated charcoal.

Removing or reducing the problem source and increasing the ventilation rate are the two most common means of dealing with gas and odor pollution. Increasing the use of scrubbers is usually not very practical because it adds too much moisture to the air and may become a source of microbiological contamination. And using odor modifiers or perfumes to mask pollutants that have penetrated carpeting, wall coverings and furniture only adds more gaseous contaminants to the air.

Activated carbon or activated charcoal is very effective at removing many odors and gaseous pollutants. It is produced by several different processes that greatly increase the surface area or number of pores in the material. After activation, the carbon will absorb many organic vapors.

Activated carbon is granular or pelletized in shape and quite small. Typically, it is held between perforated metal sheets that form a slim tray. Trays come in varying depths ranging from 1/2 inch to several inches thick. The trays install in the air handling equipment like particulate air filters.

Activated carbon can be highly efficient. Not all gas or odor problems require such efficiency, however. For this reason, most systems are set up to remove low concentrations of gaseous pollution. In these systems the sorbent tray filters do not treat all the mixed return air. Instead, only a portion of the return air is bypassed through them.

While activated carbon is very effective, it is expensive to buy and maintain. As a result, building owners are reluctant to use it. For most gas and odor problems, it is more practical to eliminate the cause(s) and increase the ventilation rate.

Summary

Despite its being desirable, not all building air is clean and odor free. As a result, air cleaning equipment has become a significant part of the modern HVAC system. In this chapter you learned that:

- Clean HVAC air helps protect people's health and productivity, sensitive equipment, furnishings and air conditioning equipment.
- Removing particulate matter is the most common reason for cleaning HVAC air.
- Air filters remove most particulate matter by the mechanisms of straining, impingement and electrostatic attraction.
- Air filters may be classified by their media (dry or viscous impingement), whether they are disposable or cleanable and whether they are stationary or renewable.
- A practical way to check filter performance is to measure the air pressure difference between the front and back of the filter.
- Correcting gas and odor pollution can be very difficult. There are several equipment options for solving the problem, but the most common approach is to reduce the cause and to increase the ventilation rate.

14

Air Moving Equipment: Fans And Ducts

Fans and ducts serve many roles in HVAC systems. They produce the pressure to distribute heated, chilled, humidified, dehumidified, air in all-air, and air-water systems. Together they collect return air, and exhaust air in these same systems while simultaneously controlling building pressurization. Fans are used for virtually every form of ventilation; in kitchens, clean rooms, drying facilities and industrial processes. Finally, fans may play an important role in a building's smoke control system. Upon completion of this chapter, you will be able to:

1. State how fans add kinetic and potential energy to air.
2. Identify the distinguishing features and drive arrangements of centrifugal fans.
3. Identify the distinguishing features of axial fans.
4. Explain fan performance characteristics, including pressure, flow rate, horsepower, efficiency and speed.
5. Use a fan's performance tables to check a fan's capability.
6. Evaluate pressure changes within a duct system.

Fan Basics

A fan is simply a number of blades on an axle or shaft supported within a frame or housing. The source of energy, typically an electric motor, drives the fan, typically through an arrangement of pulleys and belts.

There are two types of HVAC fans: centrifugal and axial. The names describe how the air flows through the fan impeller.

Air may enter one or both sides of the centrifugal fan. See Figure 14-1. The centrifugal force created by the rotating impeller or wheel of the **centrifugal fan** forces the air in a radial direction perpendicular to the fan shaft. This centrifugal force accelerates the air to a high velocity. As the air leaves the impeller, some of the velocity is converted into pressure known as static pressure.

In an **axial fan** air enters and exits the impeller along the same axis parallel to the fan's shaft. See Figure 14-2. The propeller-shaped impeller draws air through the impeller and moves it in a spiral motion. While the centrifugal fan adds energy to the air with centrifugal force, the axial fan uses only rotational force.

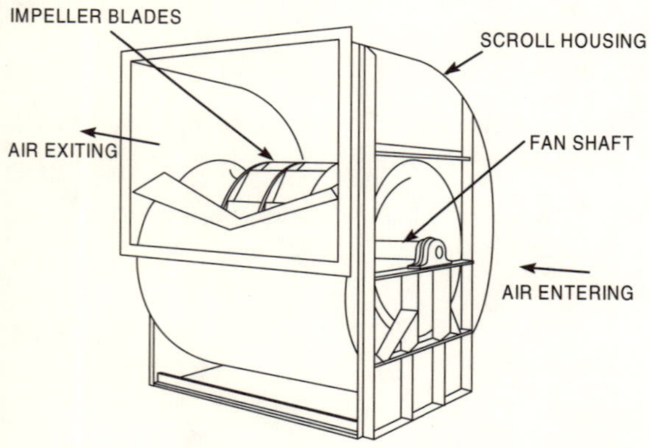

Fig. 14-1 Centrifugal Fan

Courtesy of the Trane Company

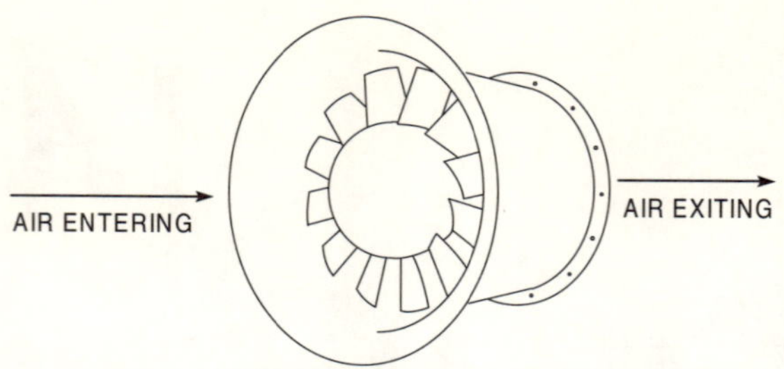

Fig. 14-2 Axial Fan

Courtesy of the Trane Company

Both centrifugal and axial fans add their mechanical energy to the air by raising the air's potential (or static) energy and its kinetic (or motion) energy. This energy transfer can be measured with a manometer as a pressure increase between the air entering and leaving the fan. In most HVAC systems, this pressure increase is less than 8½ in. wg.

Types of Centrifugal Fans

Centrifugal fans are available in four types, ten different drive arrangements and two special designs. The four types are based on the impeller design. The ten arrangements describe the air inlets and bearing arrangements. Lastly, the centrifugal fan has been adapted to two special housings and uses.

Centrifugal fans are named after the directions and shapes of their impeller blades or vanes. The four designs are:

1. Forward-curved
2. Backward-curved
3. Backward-inclined
4. Airfoil

Figure 14-3 shows the forward-curved impeller. The tip of the forward-curved blade is inclined forward in the direction of rotation.

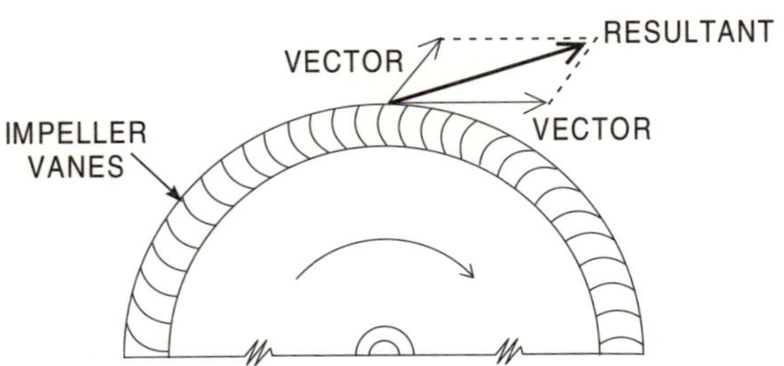

Fig. 14-3
Vanes of Forward-Curved Centrifugal Fans
Courtesy of the Trane Company

The vector arrows indicate both the direction and the amount of force that each impeller vane exerts on the air. A vector arrow points in the direction of a force. The length of a vector arrow is scaled to the amount of that force. The larger, darker arrow is the resultant arrow. The resultant arrow is the geometric sum of the direction and length of the vector arrows.

Look again at the forward-curved blade. The horizontal vector represents the forward rotational force the impeller applies to the air. The more vertically directed arrow represents the centrifugal, or outward directed force the impeller applies to the air. When added geometrically, these vectors produce a resultant pointing both forward and outward. This resultant line represents the direction of the air leaving the fan impeller.

Figure 14-4 shows the backward-inclined, backward-curved and airfoil blades. The vanes of all three types of impellers are pointed backward, in reverse to the direction of rotation. This orientation lets them glide through the air with less resistance and noise than the forward-curved types. Lessening their resistance improves their efficiency. Backward-inclined or airfoil fans will generally use less horsepower than a comparable forward curved fan. These blades can operate at higher speeds and move more air by virtue of those speeds. Additionally, when properly selected, airfoil and backward inclined fans are quieter than forward-curved fans.

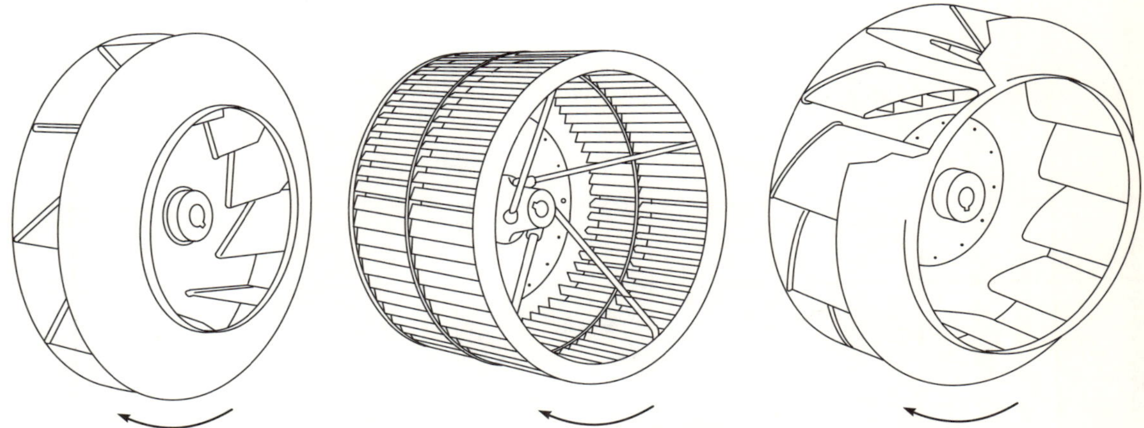

Fig. 14-4 Backward-Inclined, Backward-Curved and Airfoil Blades
Courtesy of the Trane Company

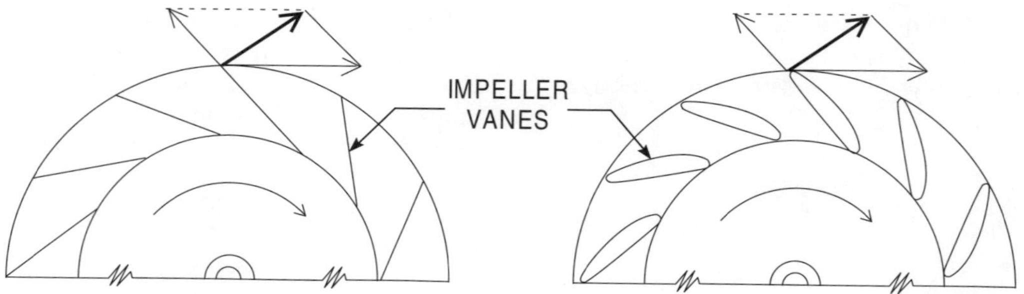

Fig. 14-5 Directional Components of Backward-Inclined and Airfoil Fan

Courtesy of the Trane Company

How does a backward-inclined blade move any air? Wouldn't its blades just slip through the air with little energy transfer? To answer these questions, look at the force vectors in Figure 14-5. The air flowing through these backward-inclined blades has a backward radial flow component *and* a forward rotational component. However, because the rotational force dominates, the resultant force is forward. And this forward resultant is the force that moves the air.

After you study the vectors of the three centrifugal impellers, the forward-curved design looks like it is the most effective. And it does create the most velocity. However, velocity is not the only consideration. High efficiency and low noise output are also important in fan selection. Backward-inclined blades, as shown in testing, are more energy efficient and quieter. Table 14-1 summarizes these impeller designs.

Centrifugal Drive Arrangements

Centrifugal fans are also described by their drive arrangements. They may be driven by a direct connection to their motors or by belts running over pulleys or sheaves. Except for fans and motors that are coupled, most HVAC fans are driven by belts. Pulleys and sheaves are more easily adjusted to meet different speed requirements than are direct drive connections. The Air Movement and Control Association (AMCA) has standardized the many centrifugal fan and drive arrangements. See Figure 14-6.

Impeller Type	Operating Characteristics			
	Speed	Power Use	Efficiency	Noise Level
Forward-Curved	Slow	Largest	Poor	Loudest
Backward-Inclined	Fast	Less	Good	Quiet
Airfoil	Fastest	Least	Best	Quietest

Table 14-1 Comparison of Centrifugal Fans

SW - Single Width
SI - Single Inlet

DW - Double Width
DI - Double Inlet

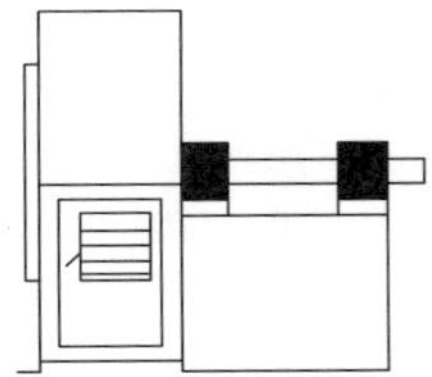

ARR. 1 SWSI For belt drive or direct connection. Impeller overhung. Two bearings base.

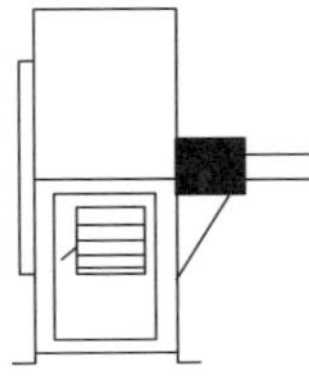

ARR. 2 SWSI For belt drive or direct connection. Impeller overhung Bearings in bracket supported by fan housing.

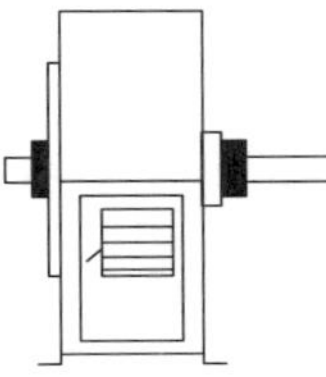

ARR. 3 SWSI For belt drive or direct connection. One bearing on each side and supported by fan housing.

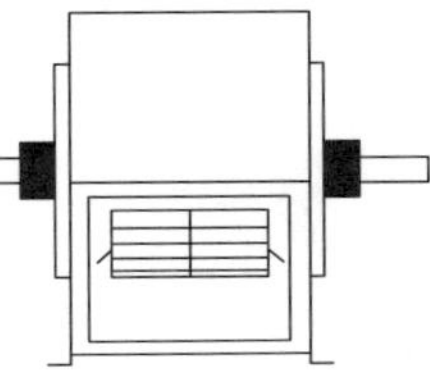

ARR. 3 DWSI For belt drive or direct connection. One bearing on each side and supported by fan housing.

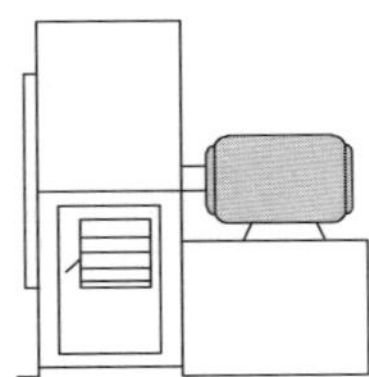

ARR. 4 SWSI For direct drive impeller overhung on prime mover shaft. No bearings on fan. Prime mover base mounted or integrally directly connected.

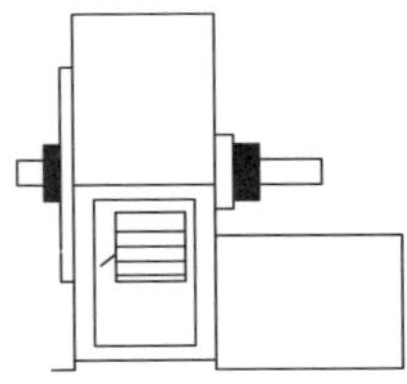

ARR. 7 SWSI For belt drive or direct connection. Arrangement 3 plus base for prime mover.

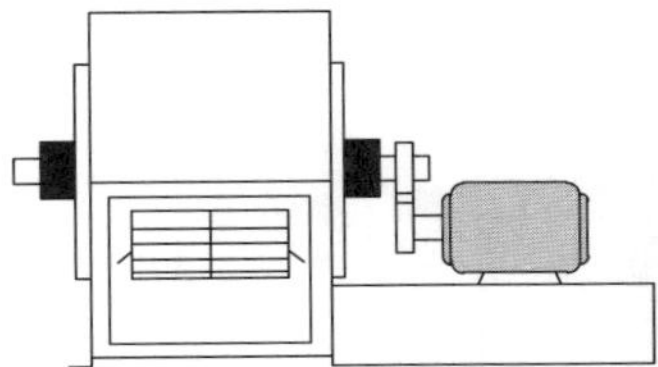

ARR. 7 DWDI For belt drive or direct connection plus base.

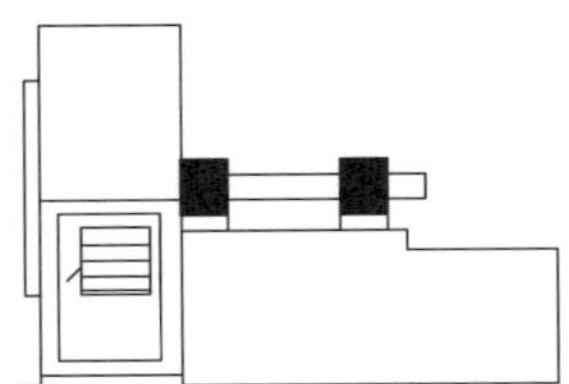

ARR. 8 SWSI For belt drive or direct connection. Arrangement 1 plus extended base for prime mover.

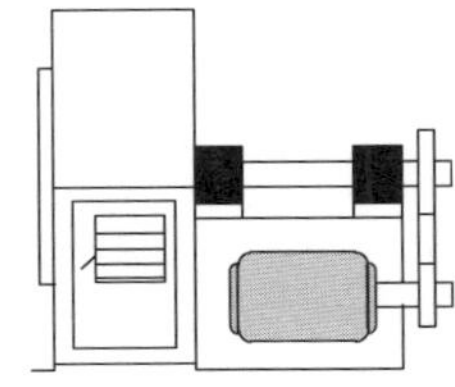

ARR. 9 SWSI For belt drive impeller overhung two bearings with prime mover outside base.

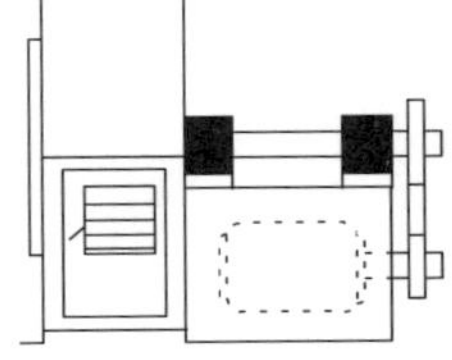

ARR. 10 SWSI For belt drive impeller overhung. Two bearings, with prime mover inside base.

Fig. 14-6 AMCA Drive Arrangements for Centrifugal Fans

Reprinted from AMCA Publication Standards Handbook 99-2404-78 with permission from the Air Movement and Control Association International, Inc.

Arrangements 1, 2, and 3 are the most common. The other arrangements are really variations of these three. Note the abbreviations used for fan widths and inlets. Fans set up to admit air in both sides are double inlet (DI) and are usually twice as wide as single inlet (SI) fans. Arrangements 3 and 7 are typical of double inlet, double width fans.

Purchase price and space needs are often the first considerations in selecting a fan arrangement. For small air volumes, single inlet fans are usually less expensive. Double inlet fans become more economical as the air volume requirement increases. For the same air capacity, a single inlet fan is about 30% taller than a double inlet, but only about 70% as wide.

Tubular and Roof Ventilator Centrifugal Fans

These are the two special centrifugal fan designs. The tubular centrifugal design creates straight line air flow by mounting the fan transversely in the housing. Straightening vanes built into the housing convert the radial air flow leaving the impeller into an axial air flow leaving the housing. See Figure 14-7. These fans are used where space or height may be limited

Centrifugal roof ventilators are packaged units. They are used wherever a low pressure, high volume fan is needed for roof ventilation. See Figure 14-8.

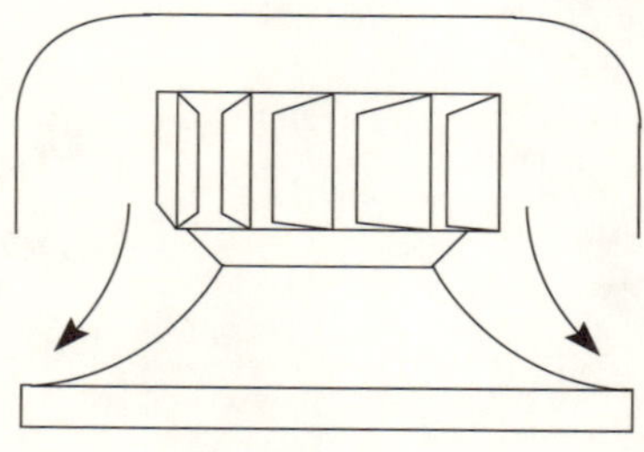

Fig. 14-8 Centrifugal Power Roof Ventilator

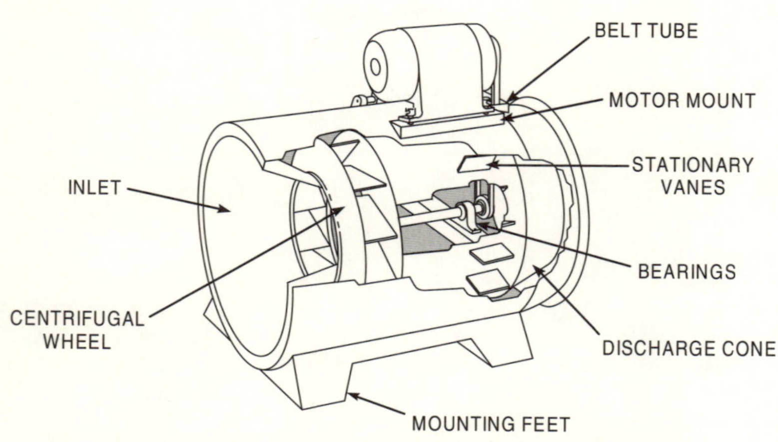

Fig. 14-7 Tubular Centrifugal Fan

Courtesy of the Trane Company

Types of Axial Fans

Axial fans are used in industrial HVAC applications where large volumes of air are needed, but fan noise is a minor concern. Axial fans are classified by how they are mounted in their housings. There are three types:

1. Propeller
2. Tubeaxial
3. Vaneaxial

Propeller fans are held in an open metal frame like the one shown in Figure 14-9. They are usually wall-mounted. The ordinary window fan is a propeller fan mounted in a free-standing cabinet. Propeller fans move large volumes of air with very little pressure increase. Typical pressures are less than 3/4 in. wg. Because they create little pressure, their horsepower requirement drops as their output increases. This relationship is exactly opposite that of centrifugal fans. As centrifugal fans develop more pressure they draw more horsepower.

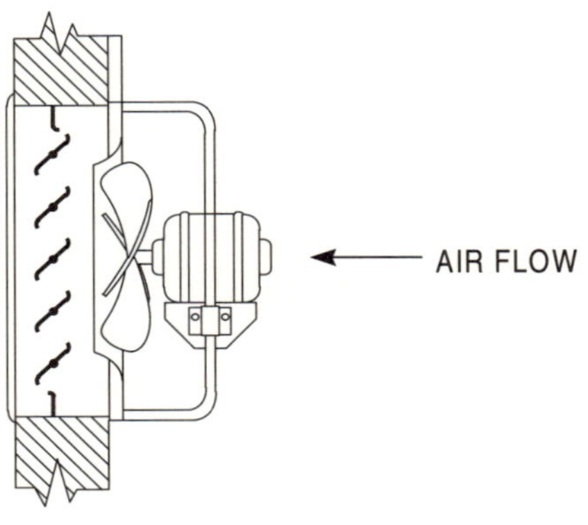

Fig. 14-9 Propeller Fan

Courtesy of the Trane Company

A tubeaxial fan is a propeller fan mounted in metal tube. See Figure 14-10. In this configuration, the propeller fan moves large volumes of air and develops 2-3 in. wg. The tubeaxial fan also adds a spiral motion to the air. This spiraling, turbulent flow creates large friction losses in ductwork. Consequently, the tubeaxial fan is not well suited to duct applications.

The vaneaxial fan is similar to the tubeaxial fan, with three significant changes. First, airflow straightening vanes are added behind the motor. See Figure 14-11. These vanes reduce the air turbulence enough to make the vaneaxial fan suitable for duct applications. Second, the diameter of the hub, or center of the fan, is more than half the blade diameter. Compare the end views in Figures 14-10 and 14-11. Third, the spacing between the blade tip and the housing is kept very close. The second and third differences allow the vaneaxial fan to develop greater pressures and operate more quietly than the tubeaxial type.

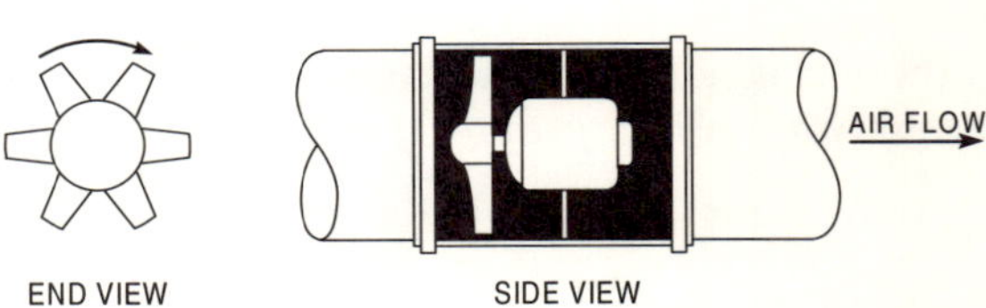

Fig. 14-10 Tubeaxial Fan

Courtesy of the Trane Company

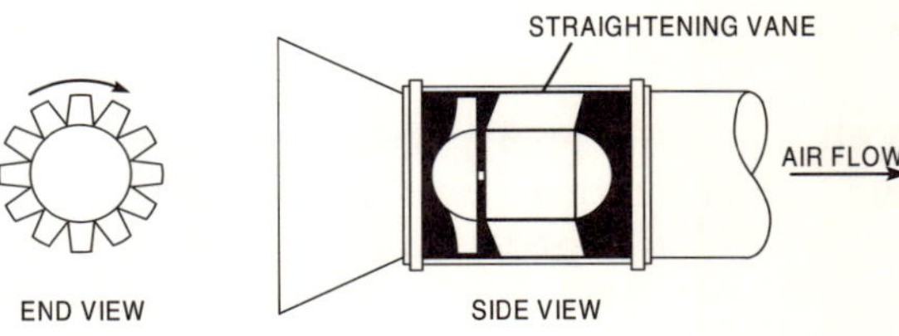

Fig. 14-11 Vaneaxial Fan

Courtesy of the Trane Company

Performance Characteristics of Fans

Fans are rated in laboratory tests that conform to standards set by the American National Standards Institute (ANSI), the Air Moving and Conditioning Association (AMCA) and ASHRAE. Fans are tested on a special test stand designed for the purpose. The testing setup is similar to the one depicted in Figure 14-18. The fan discharges into a duct fitted with air straighteners and pitot tube measuring locations. The outlet restriction is varied and the fan characteristics are measured at a series of points between shutoff and free delivery conditions. This data is then plotted to form the fan performance curve, or fan curve. These tests are conducted with dry air at standard conditions or **standard air** (70°F, 29.92 in. Hg barometric pressure and an air density of 0.075 lb/ft^3) to permit comparisons between different fans. This standard air, as it is called is identified as **scfm** or **standard cubic feet of air per minute**.

Fan performance is rated using five **fan performance characteristics**:

1. Volume
2. Pressure
3. Horsepower
4. Speed
5. Efficiency

Volume

Volume or **capacity** is a measure of the amount of air that flows through a fan. This capacity is expressed in **cubic feet per minute** or **cfm**. Fan volume varies, depending on the system resistances. A high duct resistance lessens fan volume; a low duct resistance permits greater fan volume.

Pressure

A fan imparts both potential (static) and kinetic (velocity) energy to the air. This energy input is measured as a pressure increase. Static pressure is the measure of potential energy in the system. Velocity pressure is the measure of kinetic energy in the system. Total pressure is a measure of both the static and velocity pressures. As an equation it is:

$$P_t = P_s + P_v$$

where: P_t = total pressure, in. wg

P_s = static pressure, in. wg

P_v = velocity pressure, in. wg

Pressure Sensing

The accurate sensing of pressures in all-air systems is important when controlling the pressures in a duct. However, to get accurate sensing, certain factors must be considered. First, which pressure is being sensed; static, velocity, or total pressure? Then, the proper sensing element must be selected, and properly located, to give the most accurate reading.

Static pressure (Ps) is defined as that pressure within a system exerted in all directions and measured at right angles to the direction of the air flow. Static pressure is the potential energy within an air system. Figure 14-12 illustrates the method of measuring static pressure.

At any cross section in a duct, the **total pressure** (Pt) is the sum of the static pressure (Ps) and the velocity pressure (Pv). Figure 14-13 illustrates the method of measuring total pressure.

Velocity pressure (Pv) is defined as that pressure within a system exerted in the direction of the air flow. It is sensed as the difference between total and static pressure. Velocity pressure is the kinetic energy within an air system. Figure 14-14 shows the method of measuring velocity pressure.

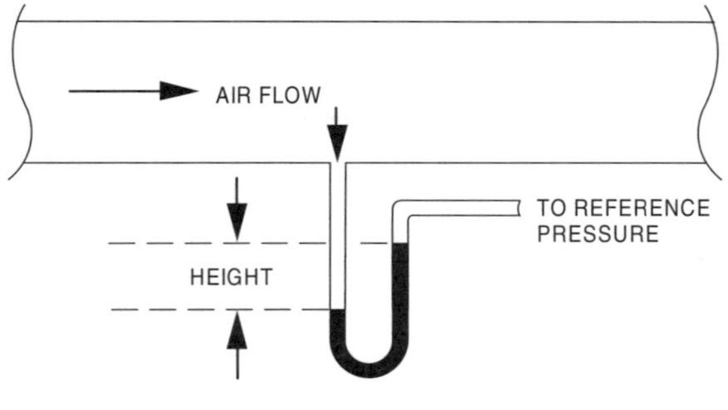

Fig. 14-12
Static Pressure

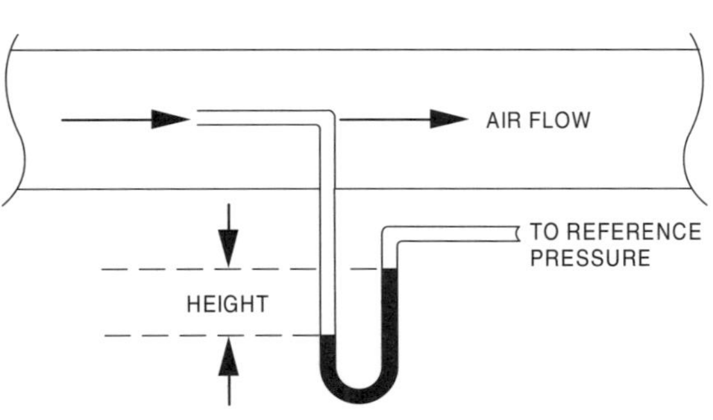

Fig. 14-13
Total Pressure

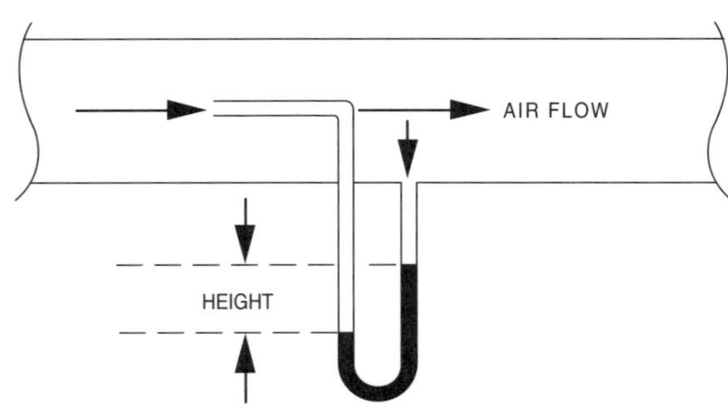

Fig. 14-14
Velocity Pressure

A static or total pressure sensing tip can be field-fabricated as shown in Fig. 14-15. For total pressure measurement, the tip is pointed directly into the air stream. For static measurement, the tip should be at right angles to air flow.

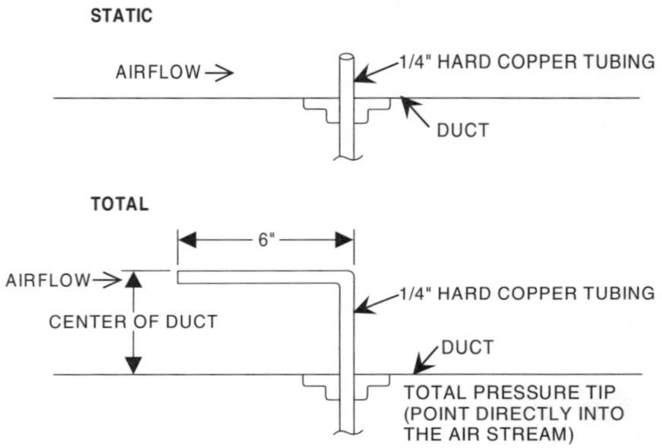

Fig. 14-15
Static or Total Pressure Sensing Tip

A standard pitot tube can be used to sense static, total, or velocity pressure. Figure 14-16 is a detailed drawing of a pitot tube. The **pitot tube** is a sensing instrument which combines both static and total pressure sensors in a single instrument. The pitot tube is inserted in the air stream, parallel to the direction of flow and with the opening always pointed into the air flow. The opening in the center of the tube is connected to the inner tubing and registers the total pressure on a manometer. The outer tubing has a number of small radial holes in its wall. The space between the two tubes conveys to the manometer the static pressure. The velocity pressure, as discussed, is the difference between the total pressure and the static pressure. These devices have proven to be the most accurate for duct static pressure sensing.

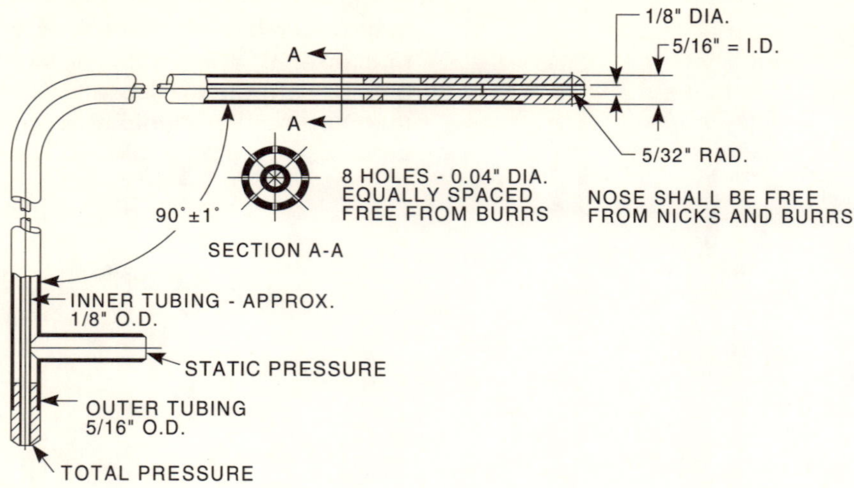

Fig. 14-16 Pitot Tube Detail

Figure 14-17 shows a common setup for measuring total pressure. The total pressure (Pt) is measured in the flow direction, to include the velocity pressure. The static pressure (Ps) is measured at right angles to the total pressure, so as to exclude the velocity pressure (Pv). Since this pitot tube measures both the total and static pressures, the pressure difference between the two columns of liquid is the velocity pressure, or Pv = Pt - Ps .

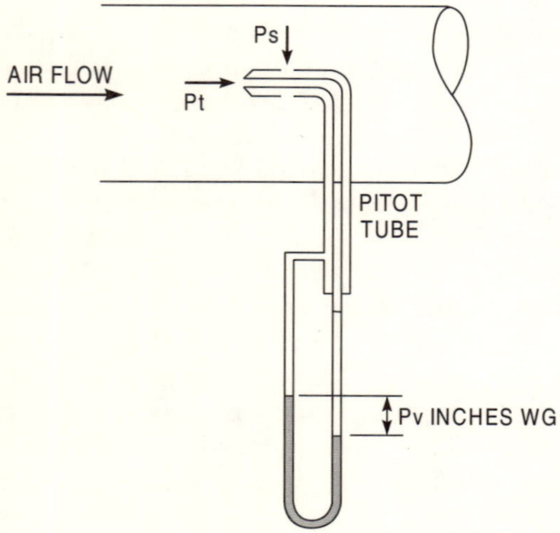

AIR FLOW
Ps
Pt
PITOT
TUBE
Pv INCHES WG

Fig. 14-17
Measuring Velocity Pressure by Difference

Fan Horsepower

The energy required to operate these fans is measured in **brake horsepower** or **bhp**. Rating power needs are important because of the cost of energy. Also, some fan designs are clearly more energy efficient than others.

Speed

Fan volume is also a function of rotational speed. In general, the faster a fan turns, the more air it can move. Fan speed is expressed in **revolutions per minute** or **rpm**. Fan speed is affected by the motor capacity, pulley diameter, air density and the duct pressure.

Efficiency

The power input to a fan is always greater than the fan output because of friction losses in the system. The ratio of output to input horsepower is called **efficiency**.

Performance Tables and Curves for Fans

Manufacturers test their fans for these performance characteristics and publish the results in **fan performance tables** or as **fan performance curves**. As you will see, each type of fan produces a unique set of performance data.

The performance of a fan is determined by strict laboratory tests following AMCA standards. Total, static pressure, flow rate and brake horsepower measurements are plotted from shutoff (blanked off) to free delivery (wide open). Figure 14-18 shows an example of the test equipment and the performance results.

Most fans can be operated over a range of speeds by adjusting the size of the pulleys or sheaves that drive them. So, rather than publish multiple performance curves, many manufacturers publish tables of performance data for the different speeds. These tables are often referred to as multi-rating tables. Since the condition of maximum efficiency is not readily apparent in tabular data, manufacturers usually underline those data or present them in boldface type. See Figure 14-19 for a fan performance table.

Performance curves help the designer picture changes in fan characteristics over a range of conditions. Figure 14-20 shows the static pressure, horsepower needs and static efficiency of a typical forward-curved centrifugal fan. Note that the peak pressure output and peak efficiency coincide at about the same CFM rating. Note that the horsepower requirements for a forward curved centrifugal fan continually increases with increasing volume delivery. Therefore generating high volume at a condition of low pressure can possibly overload the motor because of high current draw.

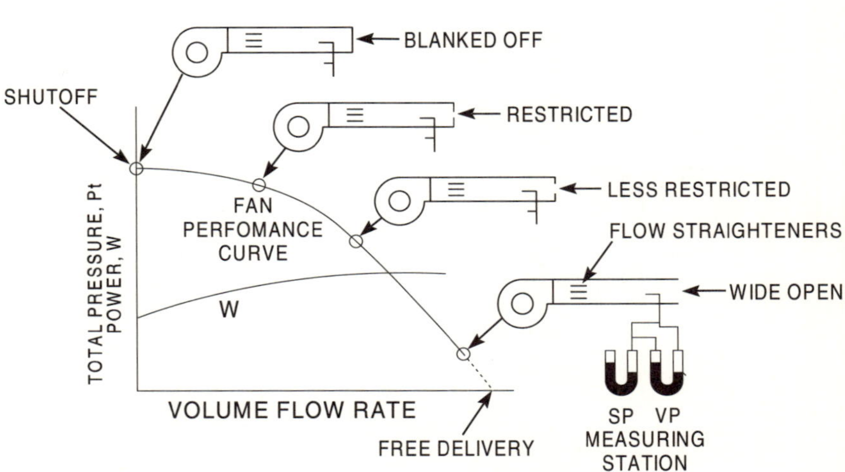

Fig. 14-18 Measuring Fan Performance at Several Points

	1/4" SP		3/8" SP		1/2" SP		5/8" SP		3/4" SP		1" SP		1-1/4" SP		1-1/2" SP		1-3/4" SP		2" SP	
CFM	RPM	BHP	RPM	BHP	RPM	BHP	RPM	BHP	RPM	BHP	RPM	BHP	RPM	BHP	RPM	BHP	RPM	BHP	RPM	BHP
2085	325	.10	376	.15	....	....	....	....	....	....	....	....	....	....	....	....	....	....	....	....
2502	351	.13	395	.18	438	.24	481	.30	....	....	....	....	....	....	....	....	....	....	....	....
2919	382	.16	421	.22	458	.28	495	.35	532	.42	....	....	....	....	....	....	....	....	....	....
3336	414	.19	450	.26	484	.33	516	.40	548	.47	613	.63	....	....	....	....	....	....	....	....
3753	447	.23	481	.31	513	.38	542	.46	572	.54	629	.70	686	.89	744	1.09	....	....	....	....
4178	482	.28	514	.36	543	.45	571	.53	598	.62	650	.79	702	.98	753	1.18	805	1.40	....	....
4587	518	.34	547	.42	575	.52	602	.61	627	.70	676	.89	723	1.08	770	1.29	816	1.51	863	1.74
5004	555	.40	582	.50	609	.59	634	.69	658	.79	703	1.00	748	1.20	791	1.42	834	1.64	876	1.87
5421	592	.47	618	.57	642	.68	667	.79	689	.89	733	1.11	775	1.33	815	1.56	855	1.79	895	2.03
5838	629	.56	654	.66	677	.78	700	.89	722	1.00	765	1.23	804	1.47	842	1.71	880	1.96	917	2.21

Fig. 14-19 Performance Data for a 27 in. Airfoil Fan

Figure 14-21 shows the performance characteristics of a backward-inclined centrifugal fan. Compare this fan with the forward-curved blade in Figure 14-20. Note that the backward-inclined fan develops pressure above its shutoff (blanked off) pressure. Note too, that soon after the fan's pressure capability begins to fall, so does its horsepower. This is called a non-overloading horsepower characteristic. The backward-inclined fan can operate more safely at a low pressure/high volume condition.

The performance curves for axial fans point out their advantage over centrifugal fans. See Figure 14-22 for typical propeller and vaneaxial fans. (Vaneaxial and tubeaxial fans have similar horsepower characteristics.) Note that the horsepower requirements of propeller and vaneaxial fans are greatest at shutoff pressures and decrease as volume delivery increases. This diminishing horsepower requirement makes axial fans popular for moving large volumes of air. Also note that the propeller fan has a broad range of high efficiency corresponding to its lower horsepower demand.

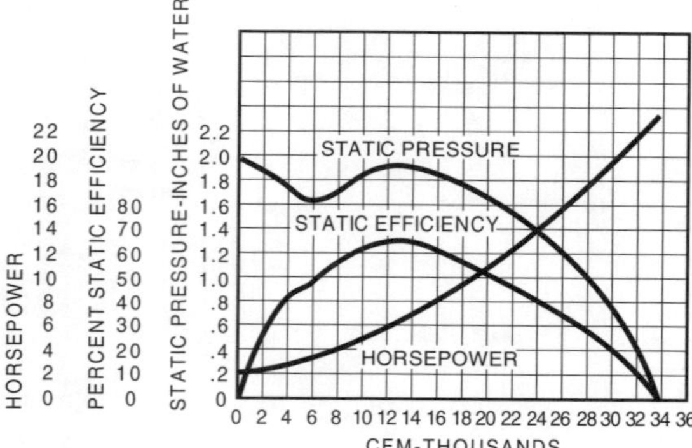

Fig. 14-20 Centrifugal Fan with Forward-Curved Blades

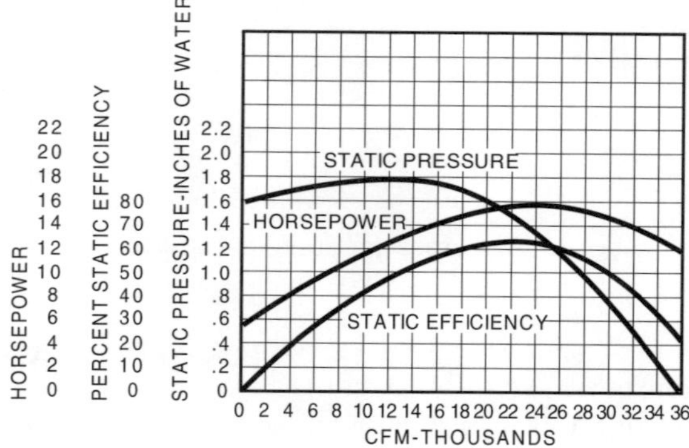

Fig. 14-21 Centrifugal Fan with Backward-Inclined Blades

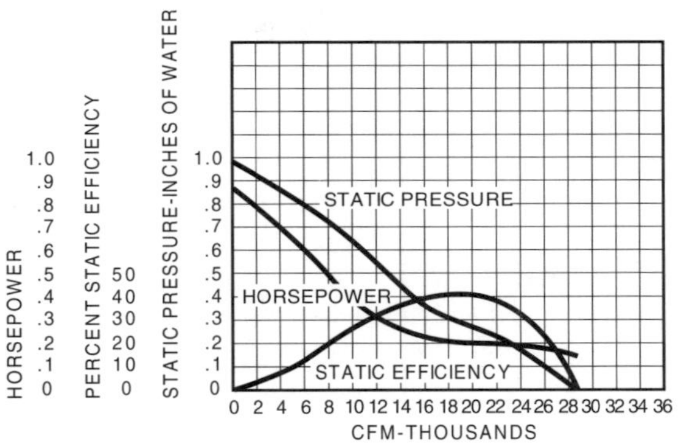

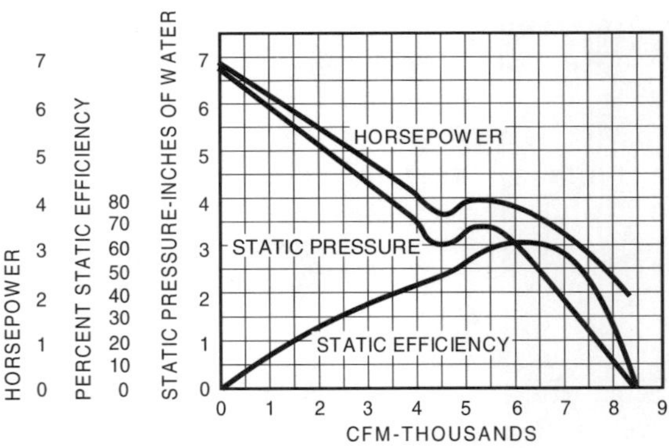

Fig. 14-22 Propeller and Vaneaxial Fans

Describing Air Flow in Ducts

The flow of air in ducts is described by the continuity principle and the Bernoulli equation. These concepts also describe the flow of water in pipes.

The **continuity principle** states that, at a given instant, the same quantity of air is flowing in every section of a duct. For example, if the output of a centrifugal fan is 1000 cfm, the separate flows in all the duct sections supplied by that fan would also total 1000 cfm. This conservation of flow is independent of duct shape or size. The continuity equation is:

$$Q = V \times A$$

where:
Q = amount of air, in CFM
V = air velocity, in FPM
A = cross-sectional area of the duct, in square feet.

The continuity equation is an expression of the conservation of the air mass within a duct. Continuity assumes that the density of the air does not change within the duct. Although air is readily compressed, most HVAC systems do not operate with enough pressure to do so.

The ducts in a system vary in size. Those nearest the air handling unit have large cross sections and those farther away are smaller. Lower velocities at coils and filters improve their efficiency. The continuity equation describes how these size changes affect the air velocity. The velocity in the larger ducts must be slower than the velocity in the smaller ducts.

The **Bernoulli equation** describes the conservation of energy within a fluid flowing in a duct or pipe. The Bernoulli equation expresses how the energy changes from potential to kinetic, and yet is conserved, as it flows from one point to another point within a duct.

The Bernoulli equation states the principle of energy conservation as the conservation of pressure within a fluid (in this case, air). The actual equation is too detailed to address in this text

Duct Characteristics

Now, let's relate the conservation of mass (continuity) and the conservation of energy (Bernoulli equation) to pressure variations in duct systems. Study the ideal duct portrayed in Figure 14-23. In an ideal air duct there are no frictional losses, directional changes that create turbulence, significant elevation changes, or momentary pressure shocks due to rapidly closing dampers. The total pressure remains constant throughout this ideal system and only the static and velocity pressures will vary. This is shown by the line of constant total pressure between points A through F in Figure 14-23.

Since the duct size does not change between points A and B, the air velocity, the static pressure and the velocity pressure are also constant. However, between points B and C, the cross-sectional area increases. Because of continuity, the velocity slows between these points. Since the velocity slows, the pressure due to velocity will also drop. However, since the total pressure remains constant, the dropping velocity pressure is balanced by an increasing static pressure. This phenomenon of velocity pressure being converted to static pressure is known as **static regain.** Static regain occurs only because of the continuity of flow and the conservation of total pressure and energy.

Between points C and D the pressures are again steady because the cross-section remains constant. Between points D and E the duct size decreases. Because of continuity, the velocity will increase in this region. And, because the velocity pressure rises with increasing velocity, the static pressure decreases.

The discussion of Figure 14-23 assumed that the air flow was ideal. Friction is a reality, however. Due to the viscosity of air, air drags along the interior surfaces of ducts. The rougher the duct, the greater is the frictional force. This frictional force is measured as a drop in pressure within the duct.

A more in-depth investigation would reveal several other facts about pressure loss due to friction. First, the loss caused by the roughness of the interior surface. Second, a duct

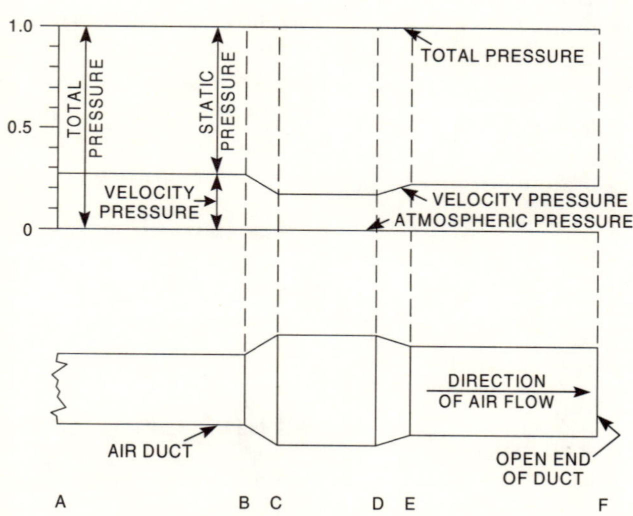

Fig. 14-23 Pressure Changes in Ideal Air Duct

Courtesy of the Trane Company

with a larger diameter would have less pressure drop. Third, the slower the velocity, the less the pressure drop. The results of Bernoulli's equation have been compiled for a variety of duct roughnesses, duct diameters and air velocities. They are expressed in tables and charts as loss in inches of water per 100 lineal feet of duct.

Ducts also experience dynamic losses due to directional changes in the air flow. Every joint, transition, fitting, damper and diffuser creates some turbulence in the air stream. Turbulent effects also show up as pressure losses. Scientists in laboratories have studied dynamic losses in all types of fittings and devices.

Dynamic losses are summarized in two ways. First, they may be expressed as the frictional loss in an **equivalent length** of straight duct. This is similar to the frictional equivalents discussed in the chapter on pumps. Or, manufacturers may publish tables of loss coefficients for their fittings and devices.

Now, study Figure 14-24 to see the effect of these friction and turbulent losses. In this example both losses are shown as losses in an equivalent length of duct. Between A and B the total pressure declines by the equivalent amount of these losses. Since continuity maintains a constant velocity and velocity pressure in this duct section, the losses affect only the static pressure. Note that at every size change there is both a total pressure loss and a static pressure loss.

Finally, by the time the air flow reaches the open end of the duct, the total pressure has fallen to zero. Since continuity has preserved the flow, the velocity component of pressure is still present at the outlet. As the air spills out of this duct and mixes with the larger space, even the velocity pressure will quickly fall to zero beyond the end of the duct.

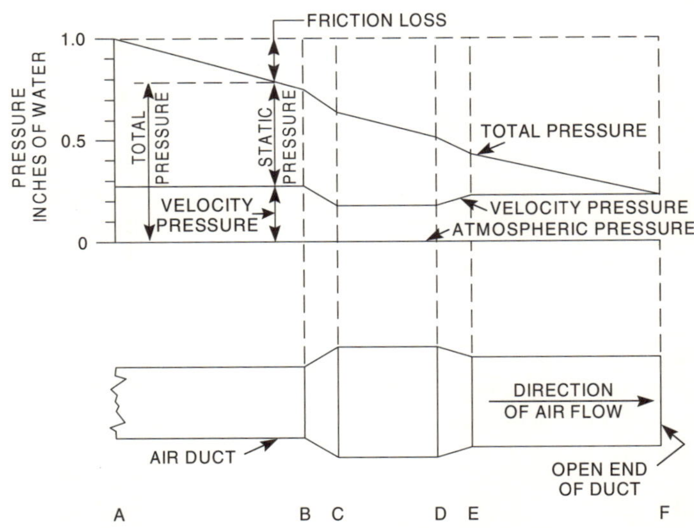

Fig. 14-24 Pressure Changes in Actual Air Duct

Courtesy of the Trane Company

This example points out that the static pressure loss is equivalent to the friction and turbulent losses. This example also points out how the measurement of either total pressure or static pressure can be used to design and evaluate duct systems.

Duct Design and Evaluation

The air volume necessary to condition a space can be calculated by the formula:

CFM = Qs / (1.08 x ΔT)

where: CFM = air flow in cubic feet per minute
Qs = amount of sensible heat added or removed, btu/hr
1.08 = constant (see chapter 6)
ΔT = difference between temperature of air entering the space, in °F and design temperature of space, in °F

A duct design that delivers this volume must also meet other criteria.

1. It must fit the available space.
2. It must not create too much transmission noise.
3. Its installation cost must be as low as possible.
4. Its operation cost must be as low as possible.
5. It must have minimal air leaks, heat gains and heat losses.

There are several ways to design ducts. The Bernoulli equation suggests that you could choose to control pressure, velocity or friction, and then solve for the other variables. Controlling one or two of the Bernoulli variables is the basis of several design methods:

1. Equal friction or constant pressure drop method
2. Velocity reduction method
3. Static regain method

The **equal friction or constant pressure drop** method is popular and works best for small, simple and symmetrical duct systems. It sizes all the ducts to have the same pressure drop per unit length. Generally a value of 0.1 in. wg is assigned to each 100 equivalent feet of ductwork. By carefully controlling the pressure drops throughout the system, the required fan capacity is exactly determined and fan selection is an easy matter. This design method is also very popular for return air systems.

For the **velocity reduction** method, velocity values are assigned to the various portions of the system and then the pressures are calculated. Velocities are selected at the fan discharge and progressively lower velocities selected for each change in air quantities caused by a branch takeoff or a duct split.

For the **static regain** method, the duct velocities are systematically converted into static pressure. Thus, by properly sizing the duct after a takeoff, the velocity conversion or regain of static pressure takes place. Controlling the static pressure determines the rate of discharge through each outlet. This is an advantage when the same size diffusers are used throughout a facility.

System Performance and Fan Selection

Now, you will learn how to match fan performance to duct system characteristics. Because of varying demand, throttling dampers and outlets, the air flow in a duct system is seldom constant. The range of this variation may be small, but it does vary. On the other hand, the duct system is a fixed system, and it responds predictably to air flow changes. The pressure loss due to friction in a duct system varies with flow.

Because friction increases at the *square* of the flow rate, the increase in friction loss is dramatic. You can develop a characteristic curve for a specific duct system by calculating this equation for a range of cfm changes. This curve is called the **system characteristic curve**. When plotted on a graph, the results look like the system curve in Figure 14-25.

Figure 14-25 also shows a typical fan performance curve intersecting the system curve at the **operation point**. Since the fan and duct systems will perform only at points on their respective curves, the point where these two curves cross is the operating point of the system. The fan is then selected to perform at the desired operating point for the system

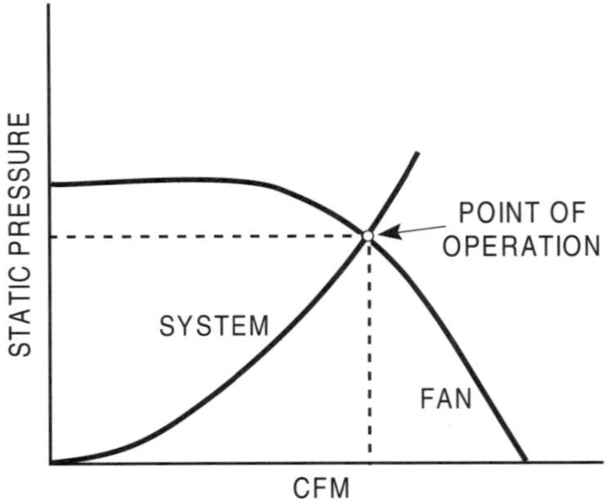

Fig. 14-25 System Operating Point

Summary

Fans and ducts are distribution system components. Together they deliver heated, chilled, humidified or dehumidified air to condition spaces.

- Fans add kinetic and potential energy to air. Ducts then distribute this air.
- Fans are available in centrifugal and axial arrangements. Arrangements are typically selected based upon system requirements of pressure and flow as well as efficiency and noise.
- Fans are selected by using fan performance curves or multi-rating tables.
- Air flow in ducts can be described by the continuity principle. Air flow in ducts will adhere to continuity principles of: the conservation of mass and the conservation of energy.
- Ducts are designed to deliver a required air volume using either of three methods: 1) equal friction, 2) velocity reduction and 3) static regain.

15

Humidifiers

The amount of moisture in the atmosphere varies from day to day and season to season. Since most HVAC systems draw in significant amounts of outdoor air, changes in atmospheric humidity may also affect a facility's interior weather. Because most people and systems work better within a fairly narrow range of humidity, controlling it is an essential element of HVAC systems. The topic of dehumidification was covered previously in Chapter 6. Upon completing your study, you will be able to:

1. Explain five reasons for controlling humidity.
2. Describe two mechanisms for adding moisture to air.
3. Recognize and explain the operation of steam, evaporative, atomizing and air washer humidifiers.
4. List the potential problems that result from adding humidity.

Why Control Humidity?

One of the most important reason for controlling humidity is to make people comfortable and efficient. However, human comfort isn't the only reason. Humidity also affects manufacturing processes and product storage, work place safety, public health and sound transmission.

Human Comfort Low humidity enhances the evaporation of moisture from the skin and makes a person feel cooler, regardless of temperature. Conversely, high humidity retards evaporation and makes a person feel warmer. Refer again to Figure 3-5. Note that the midpoint of the ASHRAE Comfort Zones, summer or winter, is a relative humidity of 45%. It is consistent with human comfort goals that most humidity control systems are designed to maintain relative humidity between 35% and 60%.

Production Efficiency and Product Storage Many materials such as paper, wood, leather and textiles are hygroscopic, that is, they absorb moisture. Changes in humidity can affect their size, volume, and consistency. Other material properties, such as static charge buildup, glue adhesion and cracking resistance, are also more manageable within narrow humidity ranges.

In general, most processes and products require relative humidity in the range of 40-60%. Ideally, these materials should be handled and stored in conditions of nearly constant humidity so that they won't shrink or stretch too much. Nearly constant humidity is also important for hygroscopic materials that are measured or sold by weight. However, some materials and operations require higher relative humidities. Textile and tobacco production facilities, for example, need 50% to 75% relative humidity and citrus fruit storage requires greater than 95%.

Work Place Safety As the concentration of dust or some gases in the air increases, it becomes potentially explosive. Dry air further raises this potential explosive because it supports static electricity better than wet air. However, relative humidities above 45% significantly reduce the space available in the air for dust particles and explosive gases to accumulate. Higher relative humidities also reduce static electricity.

One example of humidity control adding to work place safety is in hospitals. The life-support and anesthetic gases common to hospital operating rooms are potentially explosive. To combat the problem, the humidity in these rooms is often maintained above 50%.

Public Health Many microbes die faster at relative humidities between 45% and 55% than they do at humidities above 70% or below 20%. Physicians also believe that mid-range humidities of 35% to 55% reduce most people's susceptibility to colds and other respiratory disorders because moist nasal passages and throats are more resistant to infections. In addition, at humidities above 40%, dust particles settle out of the air more quickly, making the air cleaner and healthier.

Sound Transmission You have probably noticed how you can hear distant sounds easily in foggy weather or how relatively quiet the sounds become in desert or winter air. The reason for these differences is that moisture in the air absorbs the energy of sound waves. Although extreme outside humidity conditions cause unusual sound variations, most interiors have a narrower humidity range and their sounds are not affected as much. However, humidity is still a serious consideration in the acoustical performance of some theaters, music halls and auditoriums.

How to Add Humidity

There are several methods for adding moisture to the air. When considering these methods, keep in mind that most apply to all-air or air-water conditioning systems. The humidifier is centrally located in these systems, and the humidified air is distributed through the system ductwork. Figures 7-1 and 7-2 show the position of a humidifier in a central air system.

Hydronic or all-water systems do not have any built-in method of distributing humidified air. To overcome this limitation, several of the following humidification methods are sold as separate systems that can also include limited air distribution systems.

All humidifiers introduce vapor into the conditioned air stream or directly into the zone. Evaporative humidifiers rely on evaporating water. Those evaporating water from a water-filled pan are **pan humidifiers**, from a wetted medium are **wetted media humidifiers** or from minute water droplets in the air are **atomizers and air washers**. All evaporative humidifiers require sensible heat to change the water's state, whereas steam humidifiers create water vapor directly.

Steam Humidifiers

Because of their simplicity, steam humidifiers are the most common humidification method. Steam is already a gas and it is easily injected into an air stream, instantly adding water vapor. Because steam is already a vapor, the process doesn't rely on the sensible heat of the conditioned air to change the water's state. Since the temperature of the air stream does not change, the process is **isothermal**. If the facility is heated with steam, the same steam can be used for humidification. If not, the steam is produced within a separate humidifier.

The design of steam humidifiers must prevent the release of free water droplets into the air stream. If not trapped or eliminated, the ducts will become wet, leading to corrosion or bacterial growth.

Steam heated facilities may add steam directly to hot air ducts through an enclosed grid or an attached cup. See Figures 15-1 and 15-2. **Steam grid humidifiers** introduce steam through a series of orifices or openings in a dispersion manifold enclosed within a duct. The **steam**

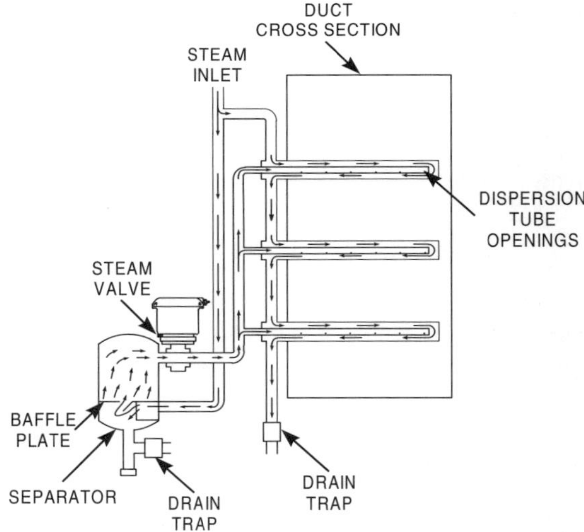

Fig. 15-1 Steam Grid Humidifier

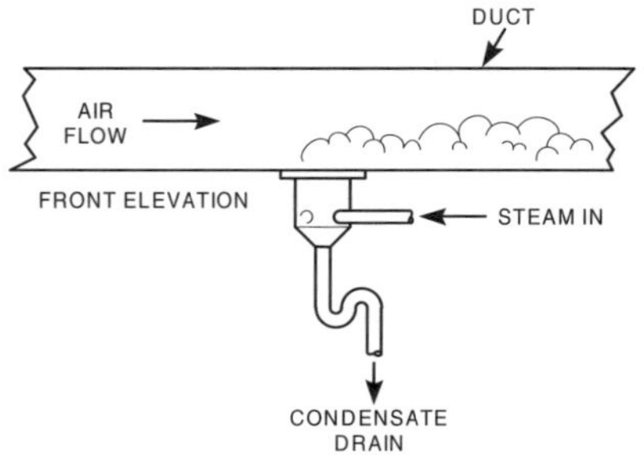

Fig. 15-2 Steam Cup Humidifier

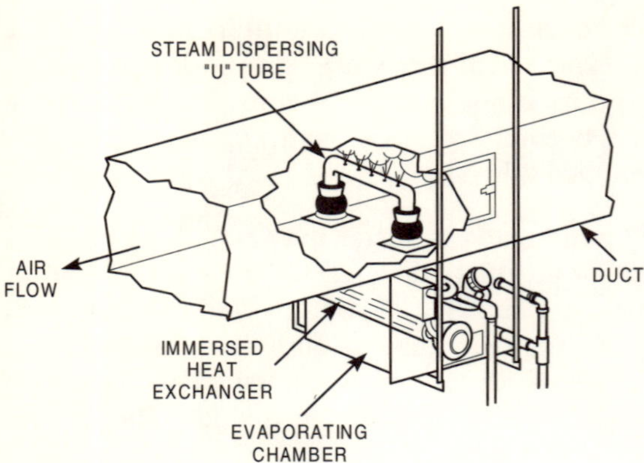

Fig. 15-3 Self-Contained Steam Humidifier

cup found in older buildings is usually attached to the bottom of the duct. Several cups may be used for broader dispersion.

For heating systems that do not have a steam supply, the self-contained steam humidifier in Figure 15-3 is an option. The **self-contained steam humidifier** creates steam by heating water within the unit and injecting the steam into an air duct through a dispersion manifold. These units usually use electrical resistance electrodes for heating the water. Self contained steam humidifiers are preferred from the standpoint of indoor air quality. The injected steam contains no boiler water treatment chemicals.

Evaporative Humidifiers

The two common evaporative humidifiers used in smaller commercial installations are the pan type and wetted media type (Figures 15-4 and 15-5). The **evaporative pan humidifier** relies on warm conditioned air to evaporate water from a water-filled pan and then to disperse the water vapor. The pan is attached directly to the warm air duct and the absorbent plates within the pan act as wicks to increase the wetted surface. An electric heater may be added to the pan to increase the evaporation rate.

The wetted drum humidifier is just one of several types of wetted media humidifiers. The **wetted media humidifier** uses a built-in fan or an induced air supply to evaporate water from a wetted medium. The medium is wetted by a spray or water stream flowing over it or the medium may be shaped into a paddle wheel, drum or belt that is rotated through a water reservoir.

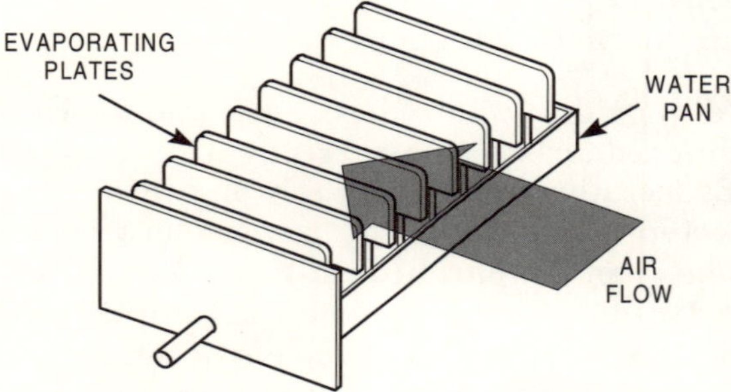

Fig. 15-4 Evaporative Pan Humidifier

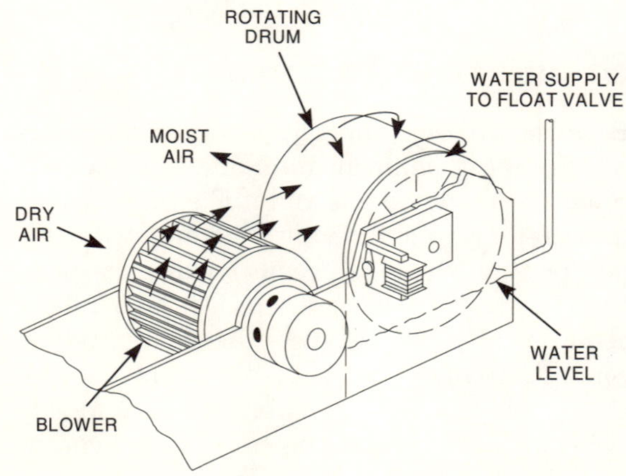

Fig. 15-5 Wetted Drum Humidifier

Atomizing Humidifiers

An **atomizing humidifier** creates extremely small water particles that are introduced directly into the air stream. Once in the air stream, the air's sensible heat evaporates these water droplets into water vapor. Atomizers create minute droplets by directing water onto a spinning disk or a cone similar to the one shown in Figure 15-6, or they force water through high pressure spray nozzles. Since atomization is a very efficient means of producing water particles, the process is limited only by the humidity content, sensible heat and volume of the air stream. The method does have a drawback, however. After vaporization, the minerals in the water supply will fall out of the air as dust particles causing a minor nuisance.

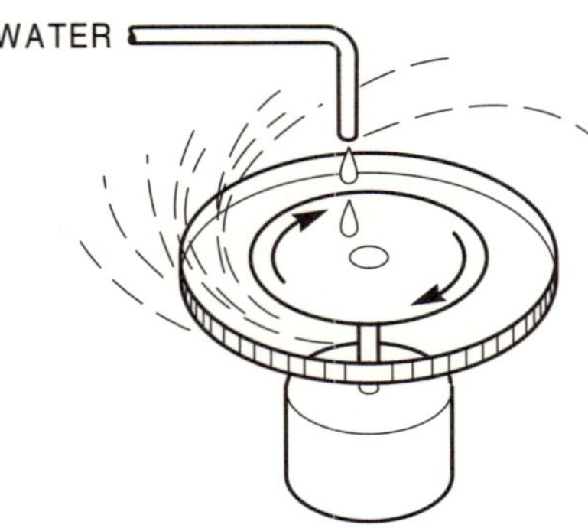

Fig. 15-6 Atomizing Humidifier

Air Washer Humidifiers

An **air washer humidifier** uses a heated water spray chamber to adiabatically add moisture to the air stream. See Figure 15-7.

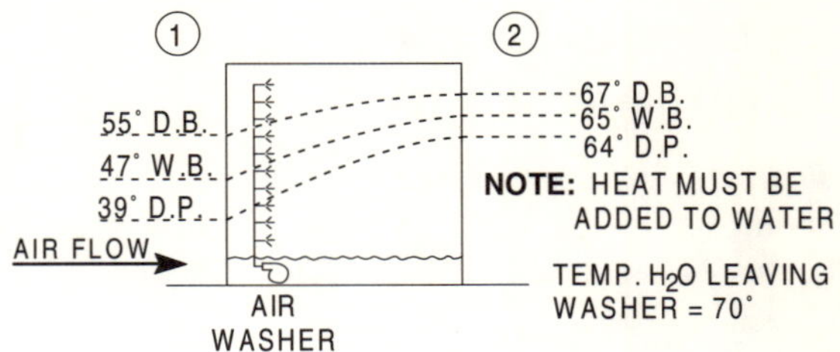

Fig. 15-7 Air Washer Humidifier

Most air washers are first selected for their cooling capacity. For example, air washers cool, in addition to humidifying, the large volumes of outside air used in textile mills and other industrial plants. However, by using heated water, the air washer can be an effective humidifier in these same facilities during the heating season. As a humidifier, the air washer still relies on the air's sensible heat to evaporate water droplets and atomized water particles. See the psychrometric process in Figure 15-8.

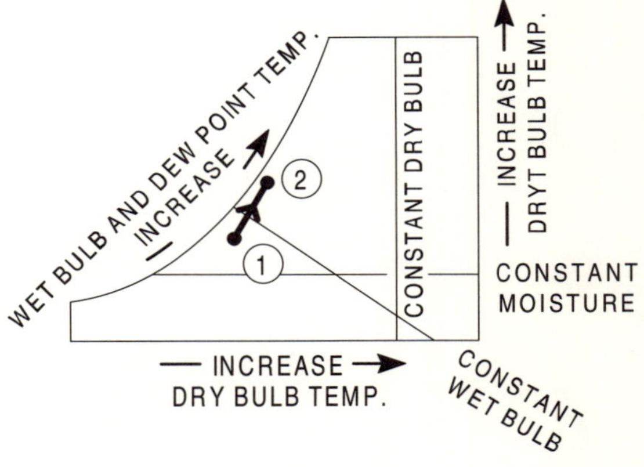

Fig. 15-8 Air Washer Humidification

Potential Drawbacks of Humidification

Despite its value and importance, adding moisture to the air also presents the system operator with several potential concerns.

1. Humidification equipment may require cleaning more frequently than most heating/cooling system components. Because it is nearly always wet, humidification equipment is an attractive breeding ground for many bacteria and fungi. In addition to the unpleasant odors they may produce, these organisms are potential threats to public health.

2. Poor water quality can cause rapid equipment deterioration. Hard water contains excessive amounts of dissolved calcium and magnesium. These minerals readily precipitate on steam boilers and evaporative surfaces, forming a crusty scale. On the other hand, water lacking these minerals can be quite corrosive to metallic parts. In evaporative humidifiers, the trick is to use water that has only a moderate amount of dissolved minerals.

3. Atomizers may require the use of expensive demineralized water. As atomized water evaporates, it leaves dissolved minerals behind. These mineral precipitates will fall out of the air as dust. If this dust is a nuisance, the operator may have to use demineralized water.

4. Visible and concealed condensation from too much humidity may cause structural damage. Condensation will form on any surface when the dew point of the surrounding air is greater than the surface temperature. This condensation can form on obvious surfaces such as windows, doors and walls or it can form deep within exterior walls or behind building facades.

Many characteristics of the facility determine whether condensation will form. They include: insulation quantity and placement, vapor barriers, infiltration and exfiltration rates, window decorations and placement, wall coverings, air diffuser and terminal unit design and placement, indoor relative humidity, and interior and exterior temperatures. It is not possible to completely discuss this topic here–just be aware that too much humidity in the wrong building can spell trouble.

Summary

In nearly every facility the best humidity is a moderate one–neither too dry nor too wet. Humidifiers are therefore an important part of the HVAC equipment for buildings in dry climates and northern latitudes. In this chapter you learned that:

- Human comfort, production efficiency and product storage, work place safety, public health and even sound transmission in theaters and music halls all rely on controlled humidity.
- Steam humidifiers are the most common. Because the process is isothermal it doesn't rely on the conditioned air's sensible heat.
- For small facilities, the evaporative pan humidifier and the wetted media humidifier are very popular because of their mechanical simplicity.
- Atomizing humidifiers are very efficient, but they require a nearly mineral-free water source.
- In many industrial plants, the air washer that cools the facility can also be used to humidify it during the dry air months.
- Last, humidification processes do have several drawbacks. They require a high degree of maintenance and they may cause structural damage to buildings with inadequate vapor barriers and insulation.

16

Control Systems For Occupant Comfort

Fundamentally a control system does four things: 1) establishes a final condition, 2) provides safe operation of equipment, 3) eliminates the need for ongoing human attention, and lastly, 4) assures economical operation. Hardware, software, installation materials (wire, tubing, etc.), HVAC processes and the final condition under control must all work together to insure occupant comfort.

After completion of this chapter you will be able to:

1. Describe control systems from a functional block perspective.
2. Define common terms used in the controls industry.
3. Explain the characteristics of each type of control response.
4. Give examples of temperature, humidity and pressure sensors.
5. Describe how loads affect the control system under heat gain and heat loss conditions.
6. Be able to state the operational differences between pneumatic and electric actuators.

The Fundamental Control Loop

Control systems can be simplified by breaking the system into functional blocks. When examined in smaller manageable pieces, any system becomes simpler to understand. For that reason, this chapter will begin with an overview of a control system from a functional block perspective. See Figure 16-1.

Final conditions at the end of the block diagram represent the results of the control system's efforts. They are the reason control systems exist. Most HVAC systems for occupant comfort control the final conditions of temperature, humidity and pressure.

System feedback is the actual condition under control, as monitored by the sensor. Whenever the sensor senses the variable under control, the system is referred to as a **closed loop.**

Sensors, at the beginning of the diagram, sense the controlled variable. Sensors measure the characteristics of the controlled variable and send an electrical signal to the controller (input).

Controllers analyze the sensor's input and make comparisons against a desired setting or **setpoint** value. Controllers then output either electric/electronic or pneumatic signals to the controlled device assembly to maintain the setpoint value.

Controlled devices regulate the flow of steam, water, or air. Valves or dampers regulate these fluids according to the controller's commands. The regulation of these fluids helps to balance the load, heat gain or heat loss taking place within the building. The **actuator** portion of the controlled device generally receives the source of energy to create movement, as well as the command signal from the controller. Typical actuators are pneumatically or electrically driven. The **manipulated device** is the portion of the controlled device which receives the actuator's movement, thus manipulating the flow of the fluid under control.

HVAC processes are varied, depending upon the building and the application. Generally the processes are temperature, humidity or pressure. These processes are controlled conditions in pipes, ducts, zones as well as primary equipment such as boilers, chillers or pumps. See Figure 16-2.

As described in the beginning of this section, the results of these processes affect the final condition in the space, duct, pipe, vessel or whatever is being controlled.

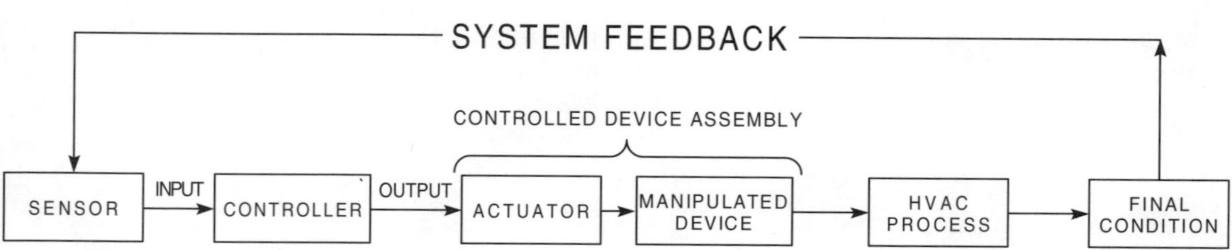

Fig. 16-1 Functional Block Diagram

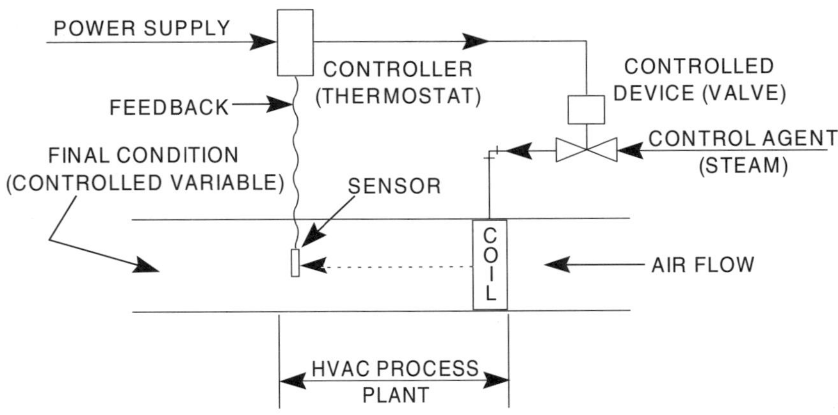

Fig. 16-2 Typical Control System

Reprinted with the permission of ASHRAE

Control System Types

Control systems can be categorized by energy source, as follows.

Self-contained control systems combine the controller and controlled device into one unit. An example of a self-contained control system is a self-contained valve with a vapor, gas, or liquid temperature sensing element that uses the displacement of the sensing fluid to position the valve. A steam or water pressure control valve that uses a slight pressure change in the sensing medium to actuate the controlled device is another example.

Pneumatic control systems use compressed air to modulate the controlled device. In this system, air is applied to the controller at a constant pressure, and the controller regulates the output pressure of the controlled device according to the rate of load change. Typically, compressed air at 20 psi is used. However, the controlled device can be operated by air pressures as high as 60 psi.

Electric control systems use electricity as the power source of a control device. This system may have two-position action (for example, off/on) in which the controller switches an electric motor, resistance heating element, or solenoid coil either directly or through electro-mechanical means controlled by a microprocessor. Or, the system can be proportional so that the controlled device is modulated by an electric motor.

Analog electronic control systems use solid state components in electronic circuits to create control signals in response to sensor information.

Digital electronic systems controllers utilize electronic technology to detect, amplify, and evaluate sensor information. The evaluation can include sophisticated logical operations and results in a output command signal. It is often necessary to convert this output command signal to an electrical or pneumatic signal capable of operating a controlled device. Figure 16-3 compares each type.

Control System Types		
Classification	**Sources of Power**	**Output Signal**
Self Contained	Vapor Liquid Filled	Expansion due to pressure resulting in Mechanical Movement.
Pneumatic	15 psi 20 psi 25 psi	0 - 15 psi 0 - 20 psi 0 - 25 psi
Electric	24 volt 120 volt other line voltage	0 - 24 volt 0 - 120 volt other line voltage
Analog Electronic	18 volts 21 volts 24 volts	0 - 16 volts 0 - 21 volts 0 - 24 volts other
Digital Electronic	24 volts 120 volts	0 - 20 milliamps 0 - 10 volts Digital other

Fig. 16-3 Control System Types

Sensors

Certain basic field hardware is necessary for a control system to function properly. **Sensors** provide information about the actual condition being monitored within the HVAC control system. Communication paths must be available to transmit sensor and control information. Often referred to as inputs, sensed signals convey either analog or binary information. **Analog Inputs** convey variable signals such as outdoor air temperature. **Binary Inputs** convey status signals such as fan status, ON or OFF.

The sensing component of field hardware must be calibrated and installed properly if the building control system is to be effective. See Figure 16-4.

Temperature Sensors

Temperature sensing elements are a critical part of any building control system. Various temperature sensors are available to meet a full range of applications.

Sensor configurations include either single point or averaging.

Single point insertion sensors are inserted directly into the fluid, air or water. Averaging sensors, a form of insertion sensors, are designed to produce a sensed signal which is the average of several temperatures. Averaging sensors are typically used in air ducts where different temperatures exist due to poor air mixing.

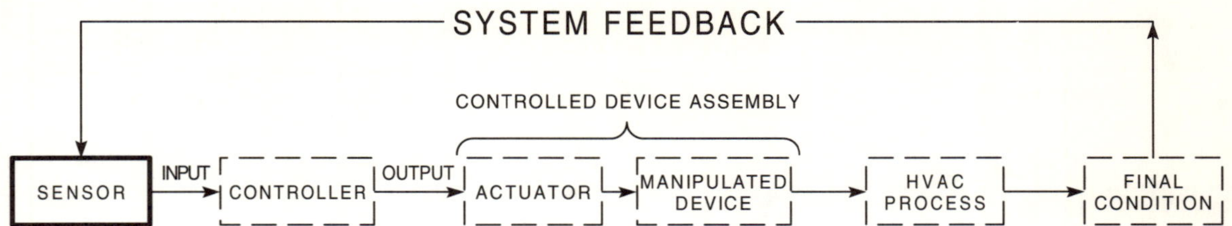

Fig. 16-4 Sensor

A **bulb and capillary** element contains thermally sensitive fluid (in the bulb) which expands through the capillary as the temperature increases. The familiar mercury thermometer is one example. Most control system applications connect a diaphragm to the capillary, so that fluid expansion changes the internal pressure and therefore the relative position of the diaphragm. This type of sensor is used in ducts **(insertion or averaging)** or piping **(immersion)** applications. Figure 16-5 shows an example of each.

A **bimetal** element consists of two dissimilar metals (such as brass and nickel) fused together, Each metal has a different rate of thermal expansion, so that temperature changes cause the metal strip to bend. The bending of the bimetal strip may cause a pneumatic signal to vary or an electric contact to open or close, actuating the manipulated device. The bimetallic element is simple and common, and is often used in room-mounted thermostats. An example of a bimetal element installed on a room thermostat is shown in Figure 16-6.

A **rod and tube** element consists of a high expansion tube inside of a low expansion rod, attached at one end. The high expansion tube changes length as the temperature changes, causing displacement of the rod. This type of sensor is sometimes used in immersion thermostats.

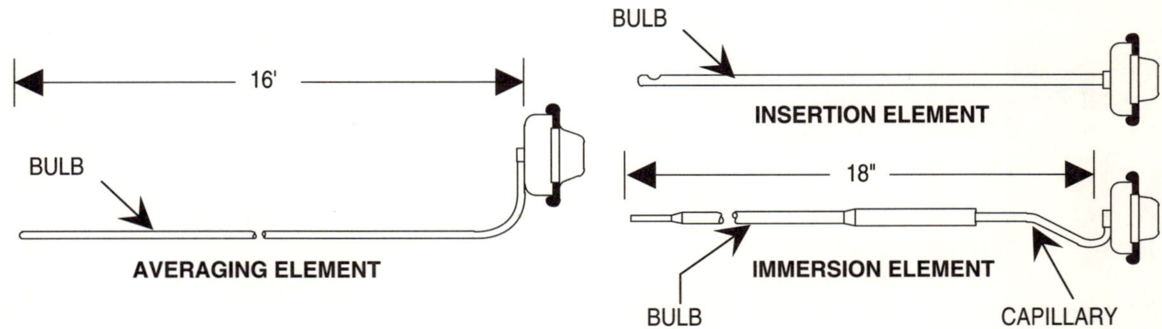

Fig. 16-5 Sensor Types

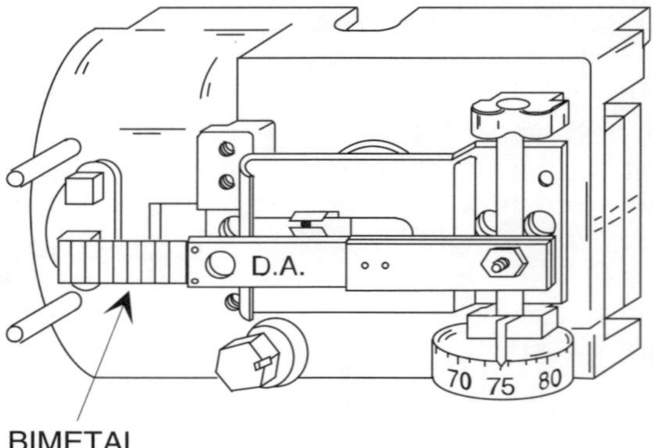

Fig. 16-6 Bimetal Element

A **sealed bellows** element is either a vapor, gas, or liquid filled element. The fluid changes in pressure and volume as temperature changes, forcing a movement that may make an electrical contact, adjust an orifice, or react against a constant spring pressure to activate a control. This type of sensing element is used in room thermostats and remote bulb sensing thermostats. An example of a vapor charged element is shown in Figure 16-7. Here, a corresponding temperature – pressure relationship will establish a given motion.

A **thermocouple** is a union of two dissimilar metals (e.g., copper and constantan) that generate an electrical voltage at the point of union. The voltage is a nonlinear function of temperature. This sensor can be sensitive to noise, as voltage levels are typically in the millivolt range, and changes per degree are in the microvolt range. The sensing element of a gas oven is typically of this type.

A **resistance** element or **resistance temperature detector** (RTD) is a metal wire winding of nickel, platinum or a silicon chip in which electrical resistance increases with rising temperature. Therefore, if a constant voltage is placed across the wire, the current would decrease. This change is then transformed by the controller to a suitable output signal. This type of sensing element is popularly used in electronic control systems.

A **thermistor** is a semiconductor device in which electrical resistance changes with temperature. It differs from a resistance element in that its resistance decreases as the temperature rises. It is usually used in an electronic circuit, and its output must be amplified and transmitted to provide a usable signal.

Pressure Sensors

Pressure sensing elements are designed to measure pressure in either low pressure or high pressure ranges. The device may measure pressure relative to atmospheric pressure, or the pressure difference between two points in a given medium.

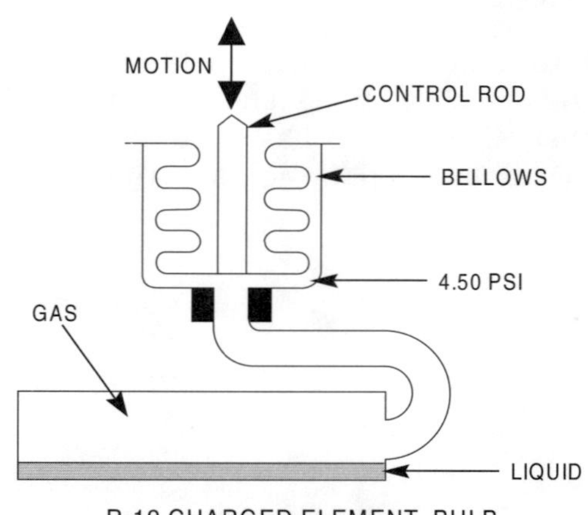

Fig. 16-7 Sealed Bellows

For higher pressures measured in pounds per square inch (psi), a **bellows**, **diaphragm**, or **Bourdon tube** may be used. See Figure 16-8. For lower pressures, usually expressed in inches of water column (in. wg), a large flexible diaphragm or flexible metal bellows is often used. The motion produced by the diaphragm or bellows is typically used with mechanical, pneumatic, or electric controls.

Newer digital electronic technologies are using **piezo-resistive sensing**. For this method, pressure sensitive micro-machined, silicon diaphragms are used. Silicon can be described as being a perfect spring which is ideal for this type of application. Unlike the large diaphragms used in mechanical controls, micro-machined diaphragms are small, perhaps 0.1 inch square. Resistors on the surface of the diaphragm change resistance when subjected to the stress caused by the defection of the diaphragm. This resistance is the sensed signal which communicates the pressure to the controller.

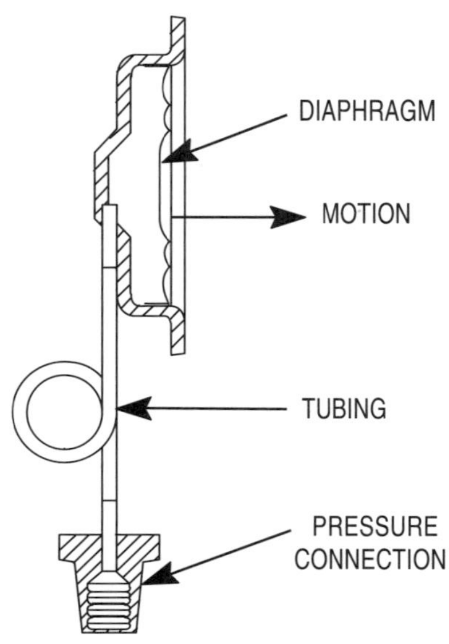

Fig. 16-8 Diaphragm Sensor

Humidity Sensors

Humidity sensing elements react to changes in relative humidity within a given temperature range. Two types of materials, organic and inorganic, are used. Whatever the type, most all humidity sensors are **hygroscopic**, or capable of retaining or giving up moisture. For many years organic materials such as hair, wood, paper, or animal membranes were used. Materials have been developed to eliminate the problems associated with these fragile materials. Cellulose acetate butyrate, or CAB, sensors have been effective for both pneumatic and electric controls.

More sophisticated, electronic sensors use capacitive sensing circuits. A typical material used for sensing relative humidity is a polymer whose dielectric constant changes with the number of water molecules present in the material. The polymer film is coated on both sides with a water penetrable carbon film forming a parallel plate capacitor whose capacitance (a function of the dielectric constant) changes with changes in relative humidity. Typical materials used for relative humidity applications include polymers coated with carbon.

Transducers and Communications

Transducers are devices which convert signals from one medium to another. Sensors and controllers frequently require interfaces between electronic, electric, pneumatic, and mechanical signals, necessitating the use of transducers. Transducers can be two-position (on/off) such as a PE (pneumatic-electric) or EP (electric-pneumatic) relay or a modulating EP transducer using a servomotor to position a control valve. An example of a modulating transducer is a signal device which converts a 3 – 15 psi into a 0 – 20 milliamp signal.

Communications are a necessary element of any control system. The most common choices are pneumatic tubing, shielded twisted pair (two wires foil wrapped to eliminate interference), coaxial cable (a conducting wire within a conducting cylinder) and triaxial cable (shielded coaxial cable), and fiber optics.

Current systems use digital transmission, with information encoded into binary digits. This allows transmission errors to be minimized, detected, and even corrected. Other systems employ analog transmission, where a parameter such as voltage or pressure is continuously modulated over some range, 0 to 5VDC or 3 to 15 psig, to represent a final condition. Such analog transmissions are susceptible to increased errors due to noise (partially removed by appropriate filtering) and attenuation (compensated for by amplifiers on long transmission paths). Another inconvenience of analog transmission is the need for individualized calibration of each sensor point to compensate for its physically unique transmission path.

Controllers

Controllers are devices which create changes, known as **system response** according to sensor information. Controllers play the critical role of maintaining the desired building conditions. See Figure 16-9.

Controllers produce various distinct types of control action to control a building's environment at desired settings. A few of these types of control action will be presented, beginning with the simplest and progressing through the more sophisticated. Other types of control action are available.

Two Position

Two position controls are simple devices that close or open a circuit under certain conditions and maintain the closed or open circuit until conditions change. Two position controls use a **differential** which is the change in the variable required to close or open a circuit. An example of a two position control is the home heating thermostat (either on or off). This two position thermostat incorporates a differential of 4°. For example, a home thermostat set at 75°F controlling a furnace may turn off at 77°F and turn on at 73°F, maintaining an average temperature of 75°F. An example is shown in Figure 16-10.

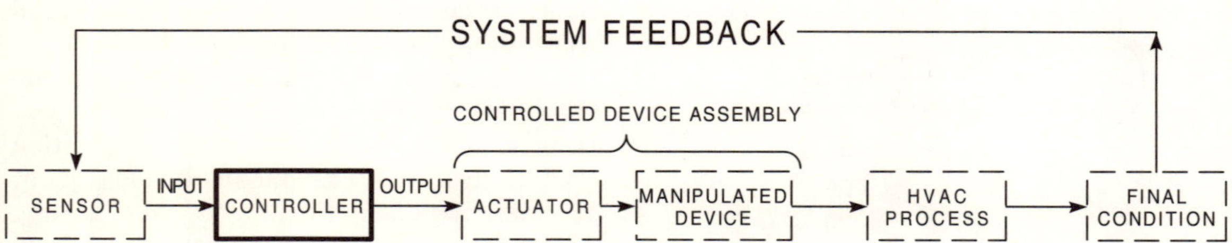

Fig. 16-9 Controller

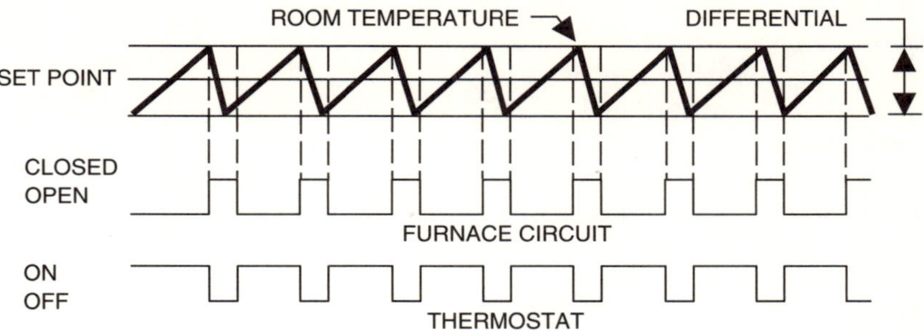

Fig. 16-10 Two Position Response

A characteristic of two position control systems is to **overshoot** the setpoint, the setting of the controller or thermostat. The overshoot is usually greater during low load conditions. For example, consider the home heating thermostat set at 75°F with a four degree operating differential. As the temperature falls to 73°F, the system heat input is less than the building heat losses because the furnace is off. Before the rate of heat input from the furnace can equal and surpass the heat losses, the system must operate for a period of time during which the temperature will continue to fall below the 73°F "turn on" temperature. Conversely, when the temperature rises to 77°F, the thermostat will "turn off" the heating system. Since the system heat input continues for a short time due to residual heat in the heat exchanger, more heat is provided. As a result space temperatures will continue to rise and overshoot.

Two Position with Anticipation

To combat the overshoot phenomenon just mentioned, two position controllers with **anticipation** were developed. This involves placing a heating element inside a heating thermostat which is activated when the thermostat activates. The heat from this element falsely loads the thermostat, causing it to deactivate before the controlled space overshoots the thermostat setting.

Floating

Floating control responses are similar to two position control responses, except that the controller produces a gradual continuous action in the controlled device. The controlled device is normally a reversible electric motor. This type of control action produces an output signal which causes a movement of the controlled device toward its open "stem down" or closed "stem up" position until the controller is satisfied, or until the controlled device reaches the end of its travel. Generally there is a **deadband** or a neutral zone in which no motion occurs. The controller for floating action systems normally has a deadband of 1° to 2° above and below the setpoint.

Assume a controller controlling a steam valve is set at 70°F and has a deadband of 2° (69° to 71°). As the discharge temperature falls below 69°, the controller will energize the steam valve motor to open the valve gradually until the temperature at the controller's sensing element has risen into the deadband. When the controlled temperature continues to rise above 71°, the controller will energize the steam valve motor to close the valve gradually. The motor's speed for this type of control is slow, and it may take a minute or two for the valve to move from fully closed to a fully open position. The motor speed must be compatible with the desired rate of temperature change in the controlled area. This type of control system will function satisfactorily in a heating or cooling system with slow changes in load. Figure 16-11 shows a graphic response.

Proportional Control

Proportional control describes the relationship between a controller and the controlled device. See Figure 16-12. In this system, the controller causes the controlled device to assume a position that is proportional to the

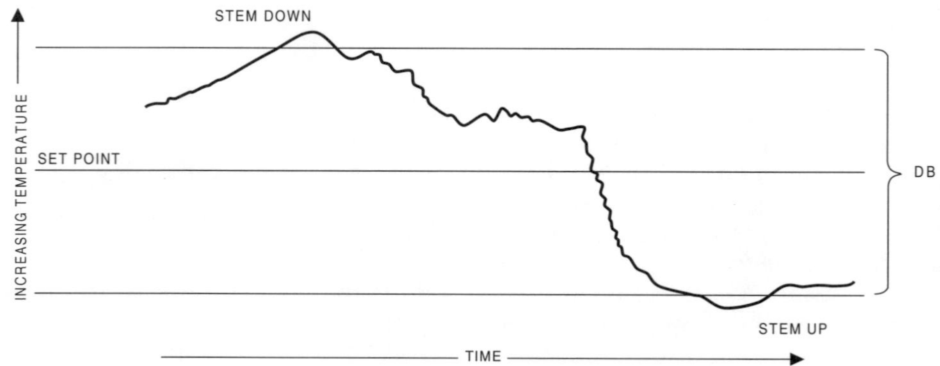

Fig. 16-11 Floating Control Response

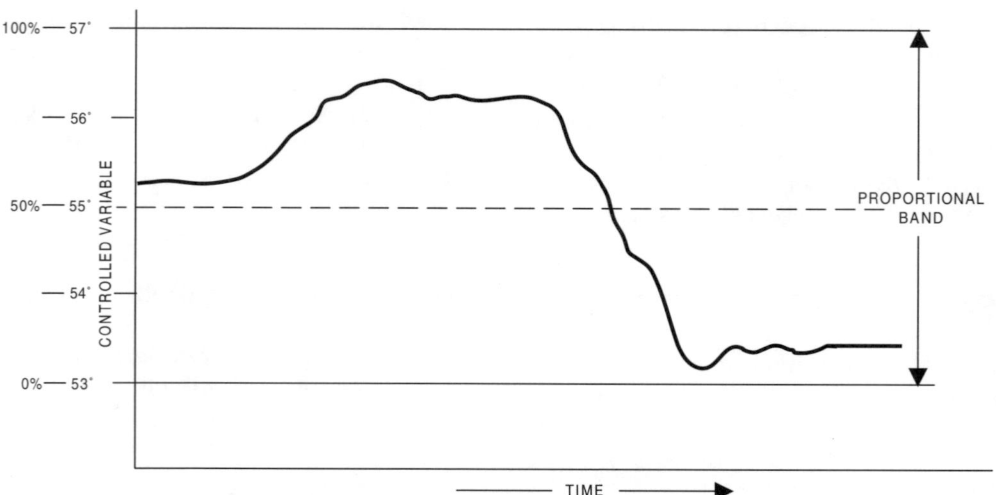

Fig. 16-12 Proportional

"magnitude of the load" as sensed by the controller. Proportional systems have an operating range called the "**throttling range**" or **proportional band**, which is the change in the controlled variable required to move the controlled device from open to closed. Such a system tends to reach a balance within its throttling range. The balance point is related to the magnitude of the load at a given time. If the throttling range of the control is 4°, and the setpoint is 55°F, the load and control system should balance at 57°F under maximum cooling requirements, and at 53°F under minimum cooling requirements.

For an example, compare a HVAC control system to the speed control system of an automobile. Proportional control systems, like your automobile, are subject to variations in the load much like the road. As the terrain changes, so does the **control point**, the value of the controlled variable for a given load. Going up hills, the speed drops off to 53 miles per hour. Going down hills the speed may increase to 57 miles per hour. The difference between setpoint and control point is known as **offset**. See Figure 16-13 for a temperature example.

Proportional Plus Integral Control

Many control strategies require a control response that will maintain setpoint. **Proportional plus integral control (PI)** provides this feature. Upon a load change, the control response will cause the controlled device to be positioned so that setpoint is maintained. Sometimes referred to as "automatic reset", this control scheme integrates the error (difference between setpoint and feedback) of the control response.

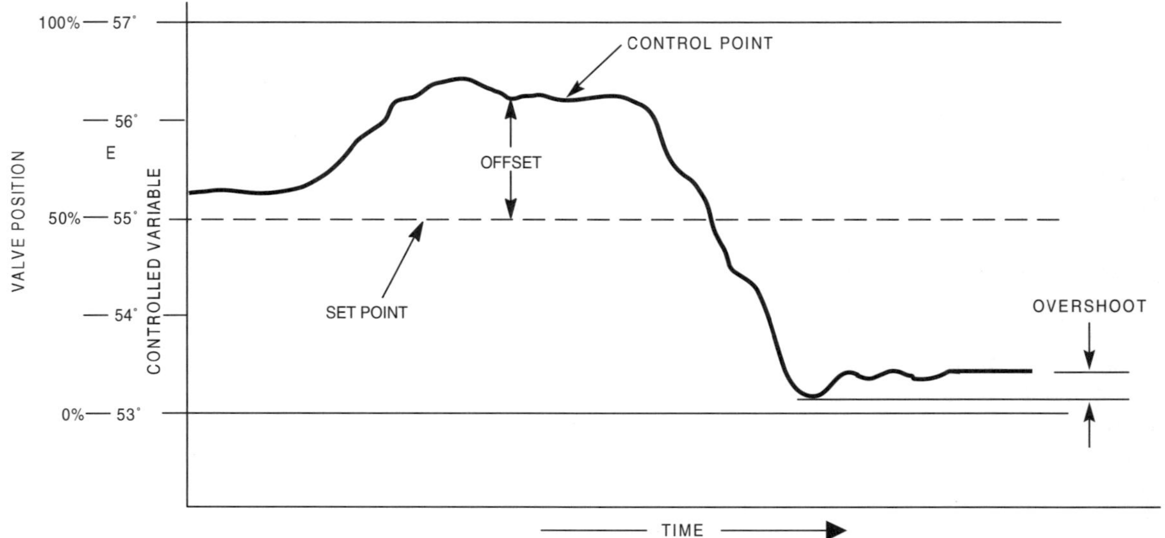

Fig. 16-13 Proportional Control Cooling Example

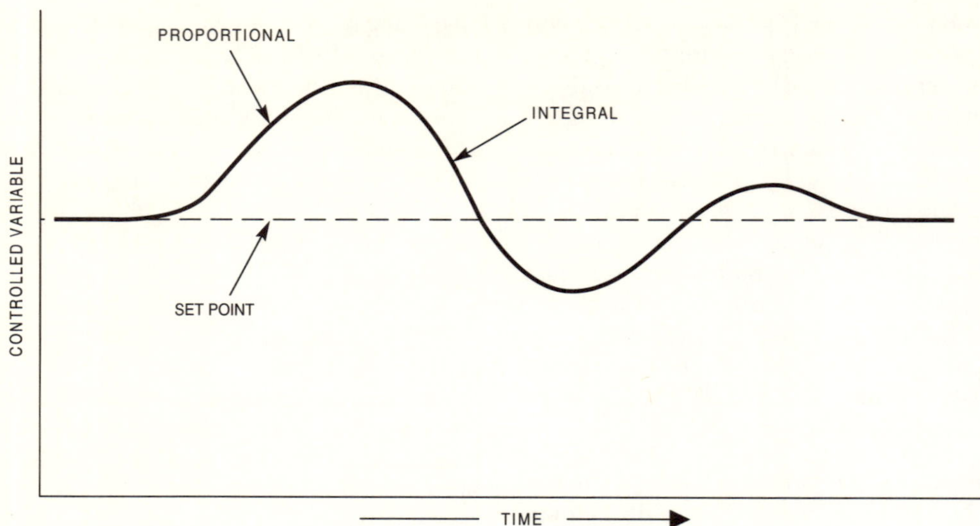

Fig. 16-14 Proportional Plus Integral Control

Have you ever noticed how the cruise control takes over and over time eventually restores the desired speed? Proportional plus integral action does exactly the same thing. The controller continues to drive the output signal until setpoint is established. Cruise control does this by taking over control of the throttle until the speed is back to the setpoint! See Figure 16-14.

Proportional Plus Derivative Control

This form of control, like proportional plus integral control maintains setpoint. **Proportional Plus Derivative Control (PID)** is especially effective in situations where overshoot cannot be tolerated. This control response is used where the system has rapidly changing loads. Since this is not the case with HVAC systems, this form of control is rarely required.

Controlled Devices

Most HVAC control systems require some type of **controlled device**. Water and steam flow devices are called **valves** while air flow controlled devices are called **dampers**. The **actuator** performs the function of receiving the controller's command output signal and produces a force or movement used to move the **manipulated device**, usually a valve or damper. Manipulated devices and their actuators make up the controlled device assembly known as the controlled device as shown in Figure 16-15.

Valves

Valves for automatic temperature control are classified in a number of ways. Valves are classified by body style, either **two-way** or **three-way**. Two-way valves control the flow rate to the heating or cooling equipment. Three-way valves control also the flow rate to

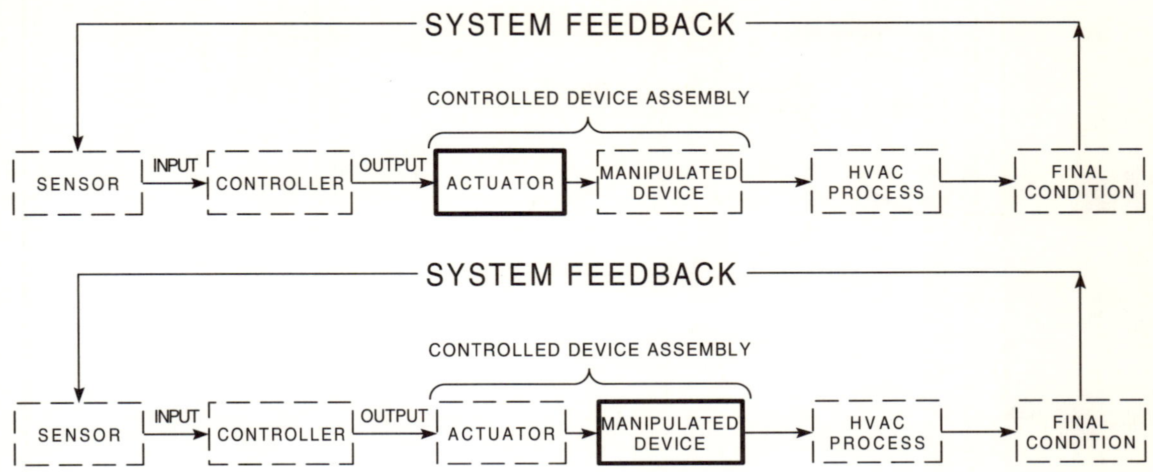

Fig. 16-15 Controlled Device

the heating or cooling equipment with the added advantage of maintaining a constant flow rate in the piping system. Figure 16-16 gives an example of each.

Other valve classifications exist, such as by flow characteristic, body pressure rating, as well as subtle internal differences. Aside from the two body styles discussed earlier, another classification is by normal position, **normally open** or **normally closed**. The normal position is the **fail-safe** which will occur upon loss of the control signal. This fail-safe is achieved by a spring on those valves so equipped. Given a loss in the control signal, the spring will return the plug (the part within the valve that controls the flow) to its normal position. Figure 16-17 gives an example of both normally open and normally closed two-way valves.

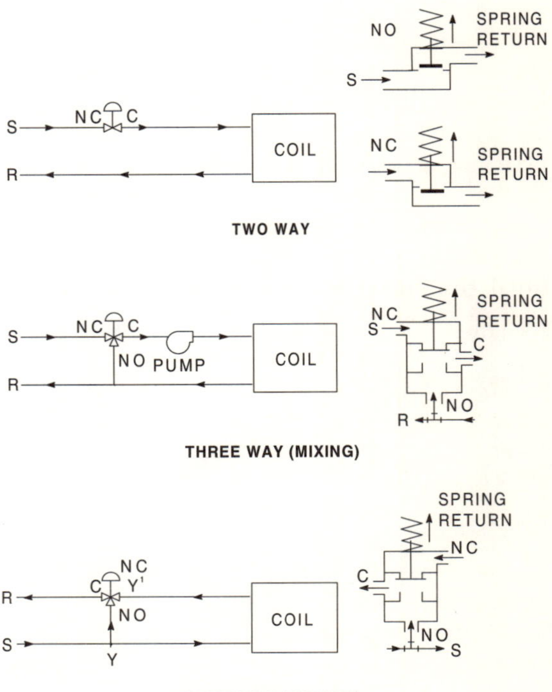

Fig. 16-16 Valve Body Styles

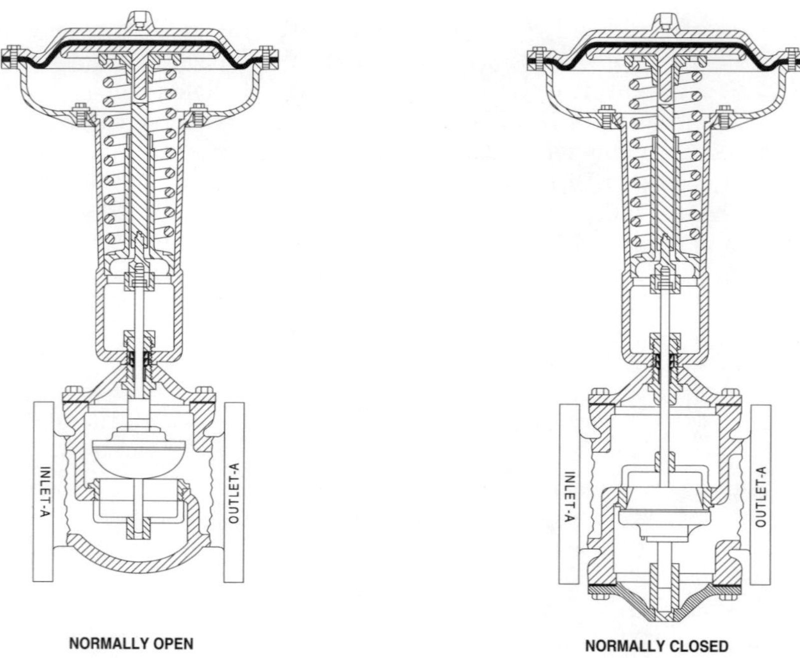

Fig. 16-17 Normally Open, Normally Closed Valves

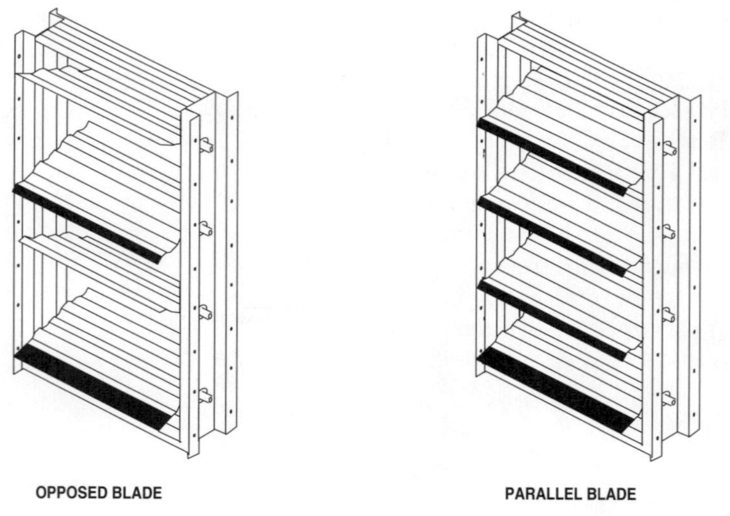

Fig. 16-18 Damper Blade Arrangement

Dampers

Dampers for automatic temperature control systems are much simpler than valves. Different classifications are based upon damper blade arrangement, leakage ratings, application, and flow characteristics. The two different blade arrangements are shown in Figure 16-18.

Like valves, dampers have a fail-safe. The fail-safe is established by the actuator and the way it is mounted to the damper. A typical fail-safe on an outdoor air damper is typically normally closed. Fail-safe on the adjoining return air damper would be normally open. The fail-safe positions of both dampers are arranged so as to prevent freezing outdoor air from entering the air handling system when the unit is off.

Actuators for Valves and Dampers

Actuators, sometimes called motors or operators, provide the force and movement required to stroke the manipulated device. The actuator must be powerful enough to overcome the fluid pressure differences against which the valve or damper must close. Also the actuator must be able to overcome any frictional forces from items such as valve packings, damper blade bearings and linkages.

Actuators may be pneumatic or electric. **Pneumatic actuators** consist of a pressure head, diaphragm, diaphragm plate or piston, and the associated connecting linkage. Pneumatic actuators produce a linear or straight line motion. Little has to be done to convert this motion into a form acceptable to valves or dampers. Since valve stems require linear movement the installation of pneumatic actuators are usually a direct mount. See Figure 16-19. Damper blades require a rotary motion which is easily accomplished by a crank arm.

Electric actuators are typically electric motors. Some actuators use slip-clutch mechanisms which engage or disengage to create or stop movement of the manipulated device. Other types of electric motors have a limited rotation of 270°. Figure 16-20 is one example.

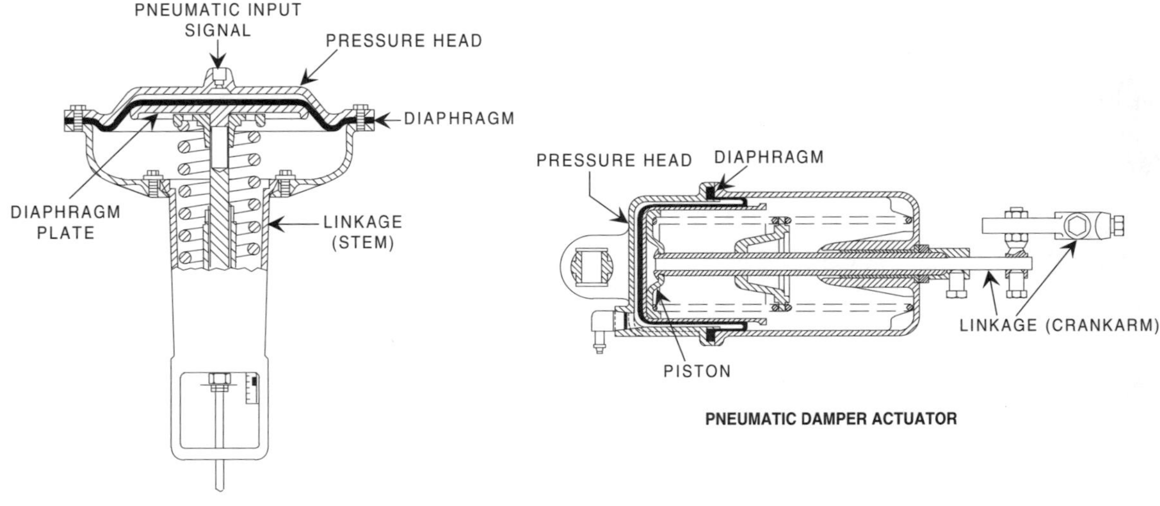

Fig. 16-19 Pneumatic Actuators

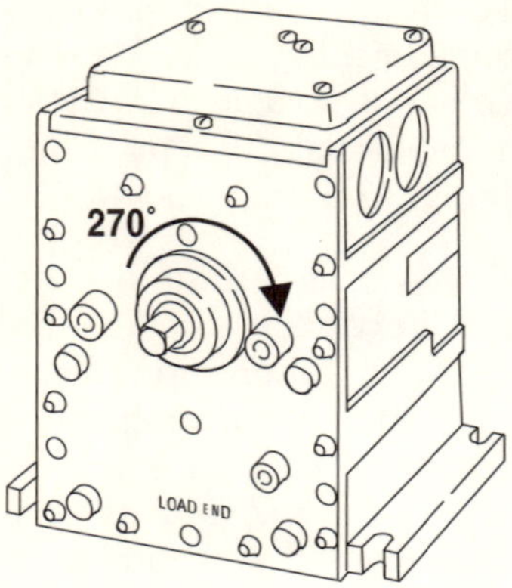

Fig. 16-20 Electric Actuator

HVAC Processes

Most HVAC systems have the capability for year-round air conditioning. Geography and climatic conditions greatly affect a system's design and the control strategies used. Regardless of the climate, most systems will have to provide control over both sensible and latent heat. Temperature, humidity and pressurization regulation or automatic controls ensure occupant comfort. Automatic controls regulate by controlling HVAC processes within predetermined limits. Figure 16-21 shows HVAC processes as part of the overall functional block diagram.

HEATING	COOLING
Hot Water	Chilled Water
Steam	Direct Expansion
Electric Heat	Outdoor Air
Heat Recovery	
Return Air	
Solar	

DEHUMIDIFICATION	HUMIDIFICATION
Chilled Water	Steam Injection
Chemical	Water Injection

Fig. 16-22 Control Agents

Control Agents

Control Agents are representative of the source under control: cooling, heating, dehumidification and humidification. Further breakdown of these processes reveals more detail on the exact type of process. For example, heating may be accomplished by hot water, steam, electric heat, heat recovery, solar energy or even return air. A detailed look at possible sources is shown in Figure 16-22.

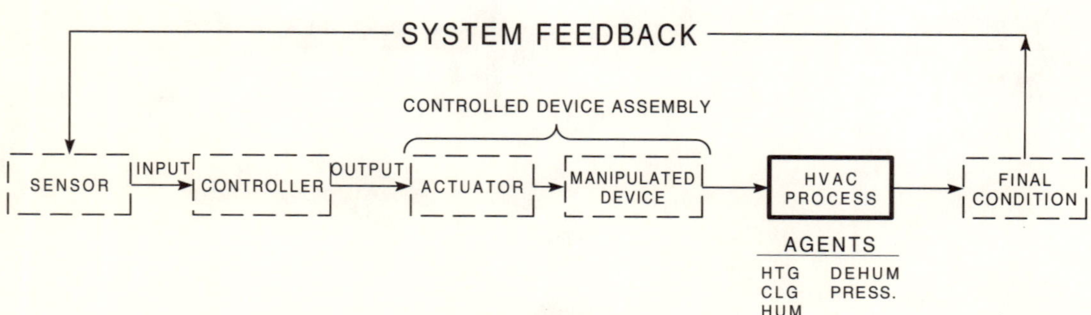

Fig. 16-21 HVAC Processes

	Outdoor Air Temp.	Boiler	Hot Water Pump	Exterior Zone Hot Water Temp.	Chiller	Chilled Water Pump	Chilled Water Temp.	Chilled Water Return T°	Air Handl. Unit Supply Air T°	Ext. Zone T°	Exterior Zone Valve Pos.	Interior Zone T°	VAV Box Damper Pos %
Severe Winter Day	-10°	ON	ON	180°	OFF	OFF	—	—	58°	68°	90%	71°	MIN 20%
Cool Spring Day	25°	ON	ON	160°	OFF	OFF	—	—	58°	70°	60%	71°	MIN 20%
Mild Spring or Fall Day	45°	ON	ON	140°	OFF	OFF	—	—	58°	71°	30%	71.5°	25% OPEN
Warm Spring or Fall Day	55°	*ON	ON	140°	ON	ON	45°	47°	58°	72°	0%	72°	50% OPEN
Mild Summer Day	75°	*ON	OFF	—	ON	ON	45°	50°	58°	73°	0%	72.5°	75% OPEN
Severe Summer Day	95°	*ON	OFF	—	ON	ON	45°	55°	58°	74°	0%	73°	100% OPEN

Fig. 16-23 Seasonal Scenario

*Boiler and hot water pumps may need to be on in order to provide reheat after dehumidification.

Operations

Operating the HVAC equipment is key to achieving optimal comfort levels. Equipment designed for worst case conditions will operate at maximum load less than 2% of the time. In order to "turn down" the HVAC equipment from full load design conditions, automatic controls are used to regulate flows, positions and temperatures.

Valves, dampers and variable speed devices are often used to regulate equipment. The equipment used is controlled at a level in response to the loads, on the system. Several seasonal scenarios for temperature control are presented in Figure 16-23. The system scenarios represent possible conditions for boilers, chillers, air handlers, interior zone VAV boxes, exterior zone heating valves and are based upon outdoor and zone conditions.

Final Conditions

Final conditions at the end of the functional block diagram, Figure 16-24, represent the results of the control systems efforts. The control strategies which achieve the final conditions produced by the control systems are varied, usually falling within certain tolerances established for the equipment under control.

Some very simple rules of thumb are given in Figure 16-25.

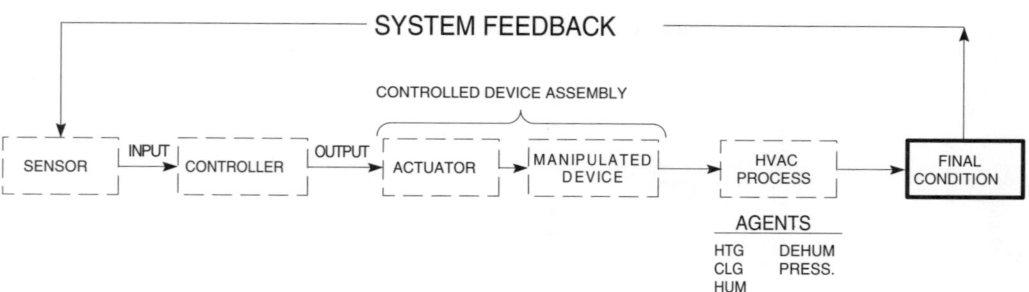

Fig. 16-24 Final Conditions

Temperature		Humidity	Pressure
Discharge Air	55-60° Cooling 75-105° Heating	Room (see Chapter 3)	Duct 1-3 in. wg
Hot Water	140-200°	Duct < 85%	Supply vs Return Piping Differential 20 psig
Chilled Water	43-45°		20 psig
Condenser Water	75-95°		
Room	(see Chapter 3)		

Fig. 16-25 Final Conditions

Feedback

Feedback, sometimes called **system feedback**, is transmitting the results of an action or operation back to its origin. See Figure 16-26.

Closed Loop Systems

Figure 16-27 shows a typical closed loop system controlling room air temperature. **Closed loop systems** use feedback for accurate control of HVAC processes. The controller in this system and its sensor measures the actual changes in the final conditions and actuates the controlled device to bring about a corrective change, which is again measured by the controller. Without feedback, this system would not control.

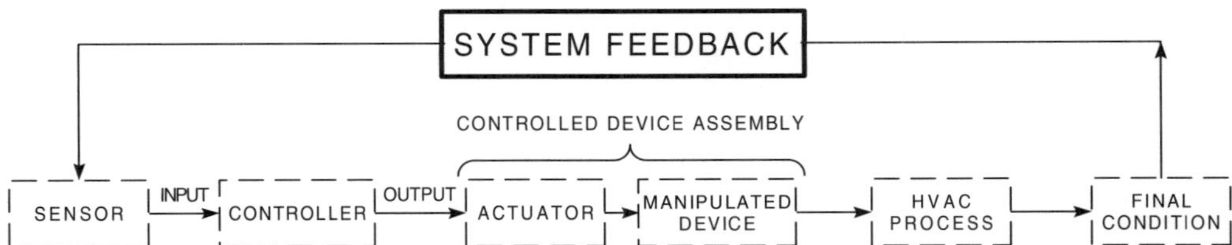

Fig. 16-26 System Feedback

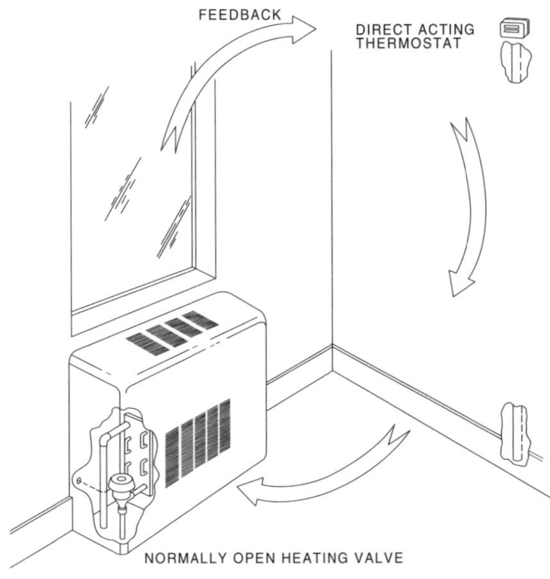

Fig. 16-27 Closed Loop System

Open Loop Systems

An **open loop** system is used to correct for load changes on final conditions. A typical example is shown in Figure 16-28.

Here an outdoor air sensor and its controller is arranged so as to cause an inverse relationship between outdoor temperature and hot water temperature. As the outdoor temperature decreases, the hot water gets progressively hotter. Notice that the outdoor sensor is in an open loop. Open loop systems simply sense; they do not control, as in the case of the outdoor air sensor. One can only sense outdoor air, it cannot be controlled. Yet sensing of outdoor air is critical in the proper function of this system.

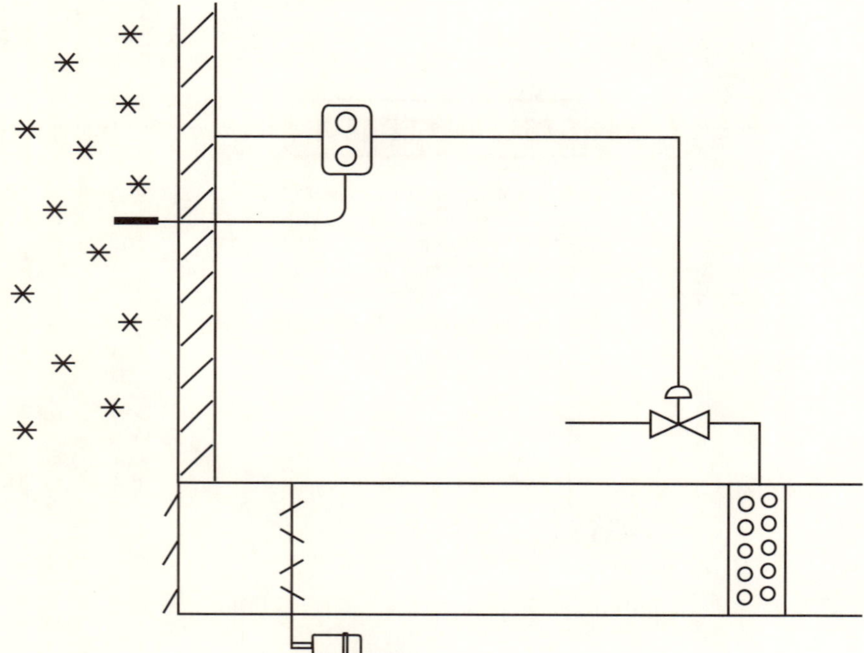

Fig. 16-28 Open Loop System

Summary

In Chapter 16 you learned the main components of a control system from a functional block perspective. By now, you are beginning to appreciate the wide and varied knowledge base of the control systems engineer. Some of the main points you investigated are these:

- The functional block approach to control systems simplifies the understanding of control systems.
- Sensors are available in many types depending on whether the control system is pneumatic, electric or digital, electronic or a hybrid and the type of variable being sensed. Temperature, humidity and pressure are typical sensed variables.
- Controllers receive inputs from sensors and compare these inputs against a setpoint. The controller then outputs to the controlled device assembly. Controller outputs may be two position, proportional or proportional plus integral.
- Controlled devices are typically valves and dampers. Actuators may be either pneumatic or electric. Actuators may come in spring return or non-spring return varieties. The fail-safe condition normally open or normally closed is a consideration for most control applications.
- HVAC processes; heating cooling, humidification, dehumidification and pressurization are facilitated by their control agents.
- The final condition under control eventually will produce a comfort level for the building occupants. A host of control strategies and control systems for boilers, chillers, pumps and distribution systems must work together to achieve occupant comfort.
- Feedback or system feedback is the communication of the final condition to the sensor and ultimately to the controller in control of the HVAC process.

17

Control Strategies for Occupant Comfort

Occupant comfort is the final outcome of the HVAC system. Recall from Chapter 3, Managing Human Comfort, that control of temperature, humidity, air flow, and air cleanliness helps to fulfill the comfort needs of occupants. Each thermal comfort requirement is ultimately a product of the HVAC system. The sensible and latent processes performed on air are accomplished via controlled preconditioning prior to the delivery of air to the space.

This chapter will address how HVAC control systems do their job of adding or removing BTU's and moisture to or from the air and distributing it throughout the building.

After studying this chapter, you will be able to:

1. Describe how zone control strategies for terminal equipment work for both temperature and humidity.
2. List four strategies used to control air handling units.
3. Identify one control strategy for each system: hot water, chilled water, and condenser water system.
4. Identify one control strategy for each type of distribution system; ducts and piping.

Zone Control

An advantage of **zone control** is that the actual load in the space is sensed and balanced by the controller's response. The only time the controller reacts is when it detects a load change. A load consists of heat transfer to or from the space. When this occurs, the space sensor and controller react by requesting heat or cooling to counterbalance the load.

Control Methods

Chapter 16, Control Systems for Occupant Comfort, discussed various control methods including the following:

- Two-position
- Floating
- Proportional
- Proportional plus integral
- Proportional plus integral plus derivative

Of these, the first four are most commonly applied to HVAC systems.

Two-Position

Two-position controllers turn the heating or cooling fully on and off as the temperature varies. The function is the same as that of most home thermostats. When heat is required, the thermostat turns on the furnace. The furnace continues to run until the temperature has risen a few degrees above the setpoint temperature when the furnace turned on. It then shuts off. This difference between the lower temperature (furnace on) and the higher temperature (furnace off) is referred to as the thermostat's temperature differential.

Two-position controls have limited use in commercial buildings because of the relatively crude form of control. Applications are usually fan coils in vestibules or unit heaters in shipping areas.

Proportional

This form of control uses a proportional controller. The controller's output signal varies in proportion to the change in the controlled variable measured throughout the controller's output range. This range may be from 0% (no value, closed) to 100% (full value, open), or design value. The controller's output variation is usually adjusted to produce a given number of units of change per degree change in temperature. For example, for each degree change in temperature, the units of change may be one or more pounds per square inch (psi), one or more volts, or one or more milliamps. Common terms for the controller settings which accomplish a unit of change per degree are **sensitivity** or **throttling range adjustment** for pneumatic controls and **bandwidth** for analog electronic controls. Another way to express the controller's response is by **proportional band,** which is the amount of change in the controlled variable required to cause the controlled device to travel from 0–100%.

The variation in output signal is either a direct or a reverse relationship to the variation in the temperature change measured. When modulating devices from open to closed, the controller does so because of a change in load. Many controllers operate with a setpoint aligned to the middle of the actuator's operating range. The setpoint of the controller is established by producing an output signal in the mid-range (50%) of the device or devices it is controlling. An example is shown in Figure 17-1a. Some controllers work from a 0% or closed position, opening only on a call for cooling or heating. See Figure 17-1b.

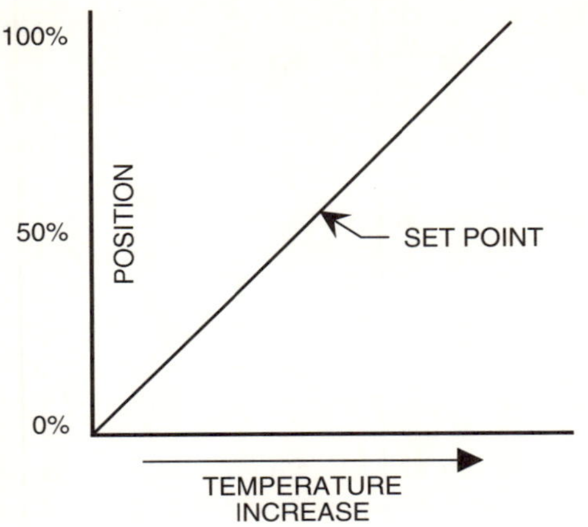

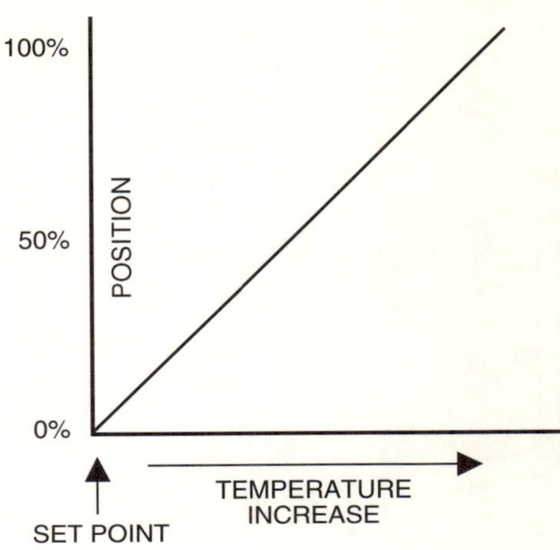

Fig. 17-1 a and b Temperature Output Signal Relationships

Proportional Plus Integral and Plus Derivative

When the load changes the controller will cause the controlled device to be positioned, or "drive," to whatever position is required to maintain setpoint. Many PI and PID controllers or sensor-controller arrangements are often applied in critical areas such as clean rooms where temperature or humidity must be controlled to tight tolerances. See Chapter 16 for the differences between PI and PID.

Summary of Control Methods

Of the types of zone control, two-position control is used primarily in electric control systems. Proportional control is used mostly in pneumatic control systems. Last, proportional plus integral control is used mostly in direct digital control (DDC) systems. The programmable intelligence, ability to perform complex control sequences, and networking capabilities make DDC the control system of choice.

Thermostat Placement and Tampering

Accurate placement of the thermostat and sensors is critical for proper sensing. Occasionally, the thermostat may measure the load inaccurately, for example, if it is next to a window or behind heavy drapes. Another disadvantage is that the setpoint on the thermostat may be tampered with by anyone passing through the room. Concealed adjustments will help prevent this tampering. A typical room thermostat with a concealed adjustment is shown in Figure 17-2.

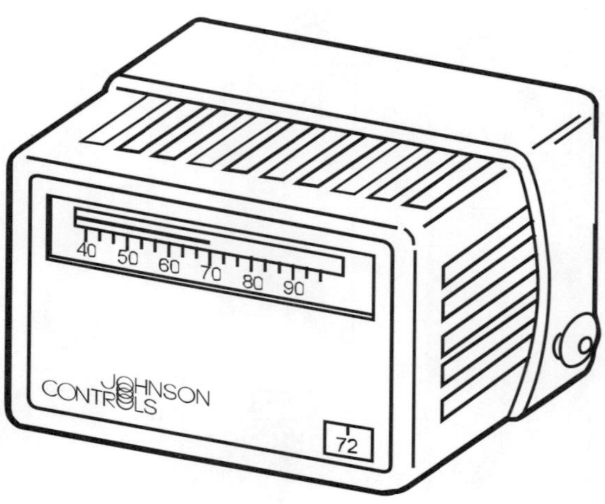

Fig. 17-2 Typical Room Thermostat (concealed setpoint dial)

Zoning

Zoning is a way of dedicating system components, including controls, to multiple spaces that have similar loads. Without zoning, achieving comfort each would be impossible.

A building may be zoned in various ways. Zones may be selected by interior, exterior, or by orientation; north, south, east or west. Each zone must have a uniform load. For example, consider an interior area. Interior zones don't have walls, roofs or floor area exposed to the outside environment. Variations in the number of people, the number and type of lights in use, and the operating status of machinery and office equipment in the zone can change the load. Fifteen people working in the zone with all lights on and all office equipment operating generate heat. These loads are actually heat and humidity gains that require only cooling.

Exterior zones, on the other hand, have at least one outside wall or roof exposure. The variations in temperature and humidity of the outside environment causes conductive transfer through the walls and windows. Infiltration losses contribute to the exterior convective heat gain or loss. Another exterior zoning consideration is the difference in the sun's radiation on the north, east, west or south sides of a building. Of course, the sun's heat radiation affects just the opposite sides of a building in the southern hemisphere.

Unlike interior spaces which are only heat gain, exterior zones are subject to both heat gains and losses.

A thermostat or humidistat in the zone senses zone conditions and depending on the deviation from setpoint, the control logic causes the heating and/or cooling apparatus to balance the zone requirement. If a zone is below setpoint, the thermostat will operate the heating apparatus. This may consist of a steam coil, hot water coil or electric heating coil and their valves or electric switches. Above setpoint, the heating apparatus would be closed or off and the cooling apparatus such as a chilled water valve or cold air damper would be opened.

Zone control is typically handled by a terminal unit such as the reheat coil shown in Figure 17-3, which may serve one or several rooms, or partitioned areas.

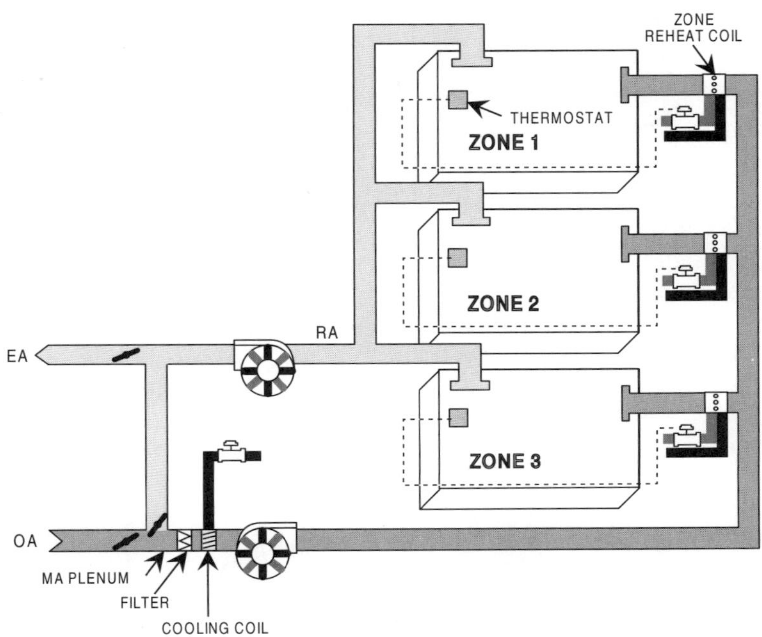

Fig. 17-3 Zone Control (each zone one room)

Terminal Equipment Controlled From the Zone

Each of the following terminal unit control strategies applies to a particular type of equipment. That equipment was previously discussed in Chapter 7, HVAC System Types. Review that section, if necessary, to reacquaint yourself with that equipment.

Baseboard Radiation

Baseboard radiation, or finned tube radiation as discussed in Chapter 7, HVAC System Types, provides a blanket of heat for exterior wall surfaces. The radiation system along the exterior walls radiates outward and convects upward, usually along the window areas, to replace the heat which flows to the outside. This radiation and convection prevents extreme variations of the existing heat in the space. Heat is required when the thermostat senses a drop in space temperature. The heat lost through the wall and windows must be balanced by an equal amount of heat input to maintain the space temperature. A valve controlled directly by a room thermostat modulates the flow of steam or hot water through the finned tube radiation. If electric heaters are used, a thermostat senses this heat loss and energizes one or more stages to balance the load.

Reheat Coils

Reheat coils are installed close to the zone in the ducts of either constant volume systems or as an integral part of a variable air volume box. In each case, a hot water valve is directly controlled by the zone thermostat. These coils receive a relatively constant air temperature of 55°F which may be heated or varied in volume.

The reason for the reheat coil is to prevent the space from being subcooled during low load conditions. Since the maximum design load does not always exist, 55° air can result in subcooling of the space. The reheat coil compensates for this by reheating the 55° air, applying a **false load** to the zone to prevent the space from subcooling. These coils are normally hidden from view in the ceiling of the controlled zone.

Unit Ventilator

Another popular type of terminal unit for exterior zones is the unit ventilator. The unit ventilator was originally developed for school classrooms, when ventilation control was first required by law. Here, the thermostat controls ventilation in addition to the heating and cooling. Several control strategies known as **ASHRAE cycles** are in use. These cycles use various combinations of ventilating and heating control strategies. Thermostats control dampers and heating and cooling valves to maintain space conditions.

Unit Heaters

Unit heaters are used where high output is required in a large space, such as a shipping and receiving area. During the winter, heat inside is rapidly released to the outside whenever the shipping and receiving doors are open. Generally, an electric thermostat senses this heat loss and turns on a fan which blows air through a steam or hot water coil to warm the space. The unit heater continues to run after the doors have closed because the space air temperature will be nearly equal to the outside air temperature.

Variable Air Volume Boxes

Variable air volume boxes can be used in either interior or exterior zones. When the zone needs cooling and the load changes, the thermostat's output varies to modulate the variable air volume box damper. The volume of air varies as a damper changes from its minimum to its maximum position. This increasing quantity of air, generally at 55°, balances the heat gain in the space. When the heat loss in an exterior space increases and the zone needs heating, the terminal units air volume decreases to a minimum, and heating is introduced by a reheat coil, baseboard radiation or both.

Control of Air Handling Units

Air handling units must be controlled to ensure that the air being made available to the zone is delivered at the right temperature and humidity. Economizer cycles are used to select the most energy efficient air streams for conditioning. Additionally, air quality is an increasing comfort concern. Evolving technologies for ventilation control are being introduced into the marketplace.

Temperature Control

Four methods commonly used to control the temperature of the air delivered by an air handling unit are as follows:

- Zone/room control
- Return air control
- Discharge air control
- Room reset of discharge

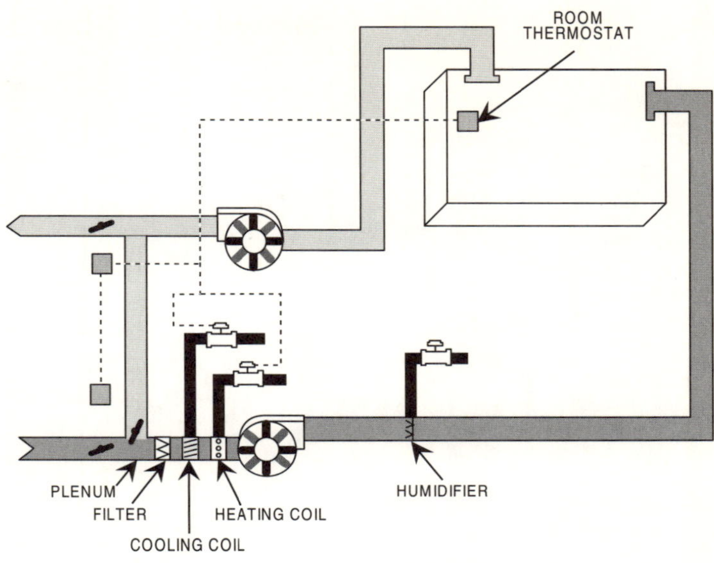

Fig. 17-4 Zone/Room Control

Zone/Room Control

In this configuration, the room thermostat located in a zone sends its control signal to the air handling unit. This control signal positions the heating valve, outside air damper, exhaust air damper, return air damper and cooling valve, so as to provide the desired air temperature to the zone. See Figure 17-4.

Return Air Control

This method can be described as a controller receiving its signal from the temperature sensor located in the return air stream. It is actually a control strategy which uses an average temperature of a large area or number of rooms. It is used when a single space sensor location is not representative of the entire area to be controlled. This strategy can be used to control heating, cooling or humidification apparatus. See Figure 17-5 for a temperature example.

Discharge Air Control

Discharge air control of entire unit is common to systems with multiple zones. In this configuration the sensor is located in the unit discharge and the controller signal positions the heating, outside air dampers and cooling apparatus to precondition the air so as to send a constant 55°F temperature to the zones. This air temperature is often the minimum temperature delivered to the zones. This strategy has been used with constant volume systems for many years. The zones have thermostats which control reheat coils which can add heat to prevent subcooling. Also, this strategy works well with the energy efficient variable air volume system, which is the all-air system of choice today. The zones have thermostats which control variable air volume boxes to modulate the quantity of 55° air delivered to the zone. See Figure 17-6.

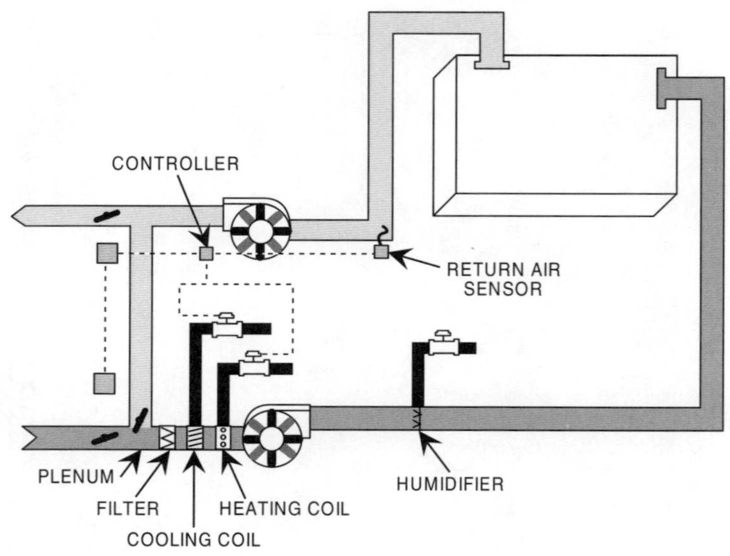

Fig. 17-5 Return Air Control

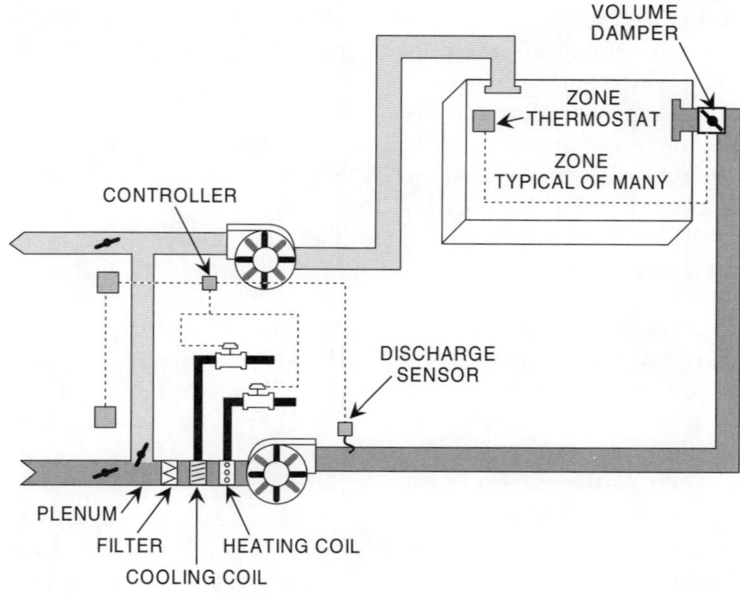

Fig. 17-6 Discharge Air Control

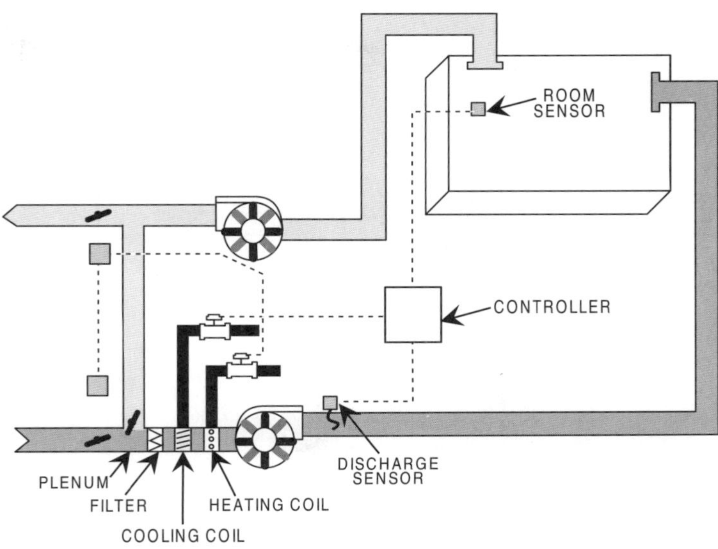

Fig. 17-7 Reset Control

Room Reset of Discharge

An example of this type of control for a single zone unit is shown in Figure 17-7. The use of two sensors, one in the zone and one in the unit discharge, allows for closer control of the temperature within the zone. The control of the unit is actually the combination of the room and the discharge control methods. Two feedback loops are utilized. The discharge control sensor's setpoint is adjusted higher or lower to compensate for changes in room temperature. The discharge sensor controls the unit directly so no undesirable variations in temperature reach the zone.

Humidity Control

Two methods for controlling the humidity of the air delivered by an air handling unit include the following:

- Room/return air humidity control
- Dew point control of discharge air

Room/Return Air Humidity Control

When a constant relative humidity (RH) is required in the zone, 50% RH for example, the humidifier is controlled to add moisture during the winter (low moisture content season) and cooling is controlled to dehumidify the air during the summer. This arrangement requires a continuous use of energy to maintain the humidity in the zone at setpoint. See Figure 17-8.

A more acceptable and energy saving concept of humidity control is the two setpoint method. Again whether sensed in the room or return air, the low moisture content season (usually midwinter) will be controlled at a low relative humidity, such at 35% RH. If the zone humid-

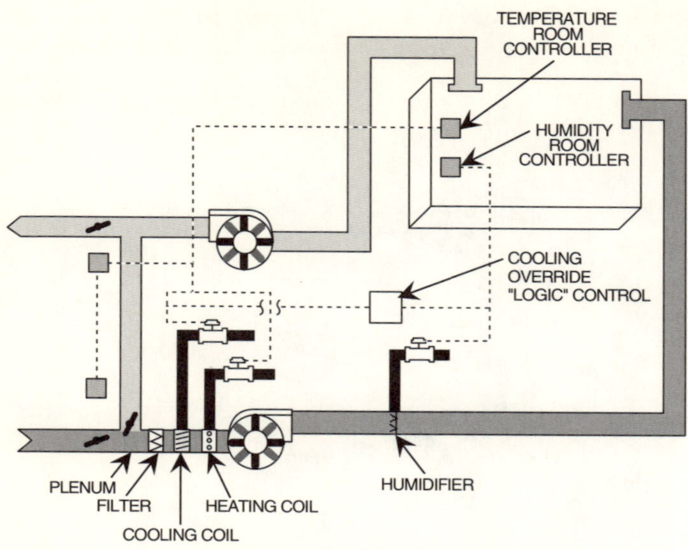

Fig. 17-8 Zone Control of Humidity

ity should go below 35%, the zone humidity controller adds moisture by modulating the humidifier. If the humidity increases above 65% RH, the zone humidity controller dehumidification signal overrides the temperature control signal to the cooling/dehumidifying apparatus to prevent the zone RH from rising too high. This is accomplished by the cooling override "logic" control. As a result of the dehumidification process, the space temperature may drop or be "subcooled." This would require the temperature control to reheat the subcooled air. Reheating subcooled air requires a heating coil on the downstream side of the cooling/dehumidification apparatus. The zone will fluctuate between 35% and 65% RH throughout the intermediate seasons of spring and fall. During this time, no energy to add or remove moisture will be used.

Dew Point Control

This strategy is specialized for certain process operations such as textile and tobacco production. In these types of manufacturing facilities, precise humidity control is required for quality control. Air can be dehumidified or humidified by controlling the cooling apparatus using a dry bulb temperature sensor downstream of a cooling coil. The coil which has a high efficiency rating, causes the leaving air of the cooling coil to be very close to saturation (100% RH). By sensing and controlling for dew point, achieved by dew forming on a dry bulb sensor, this arrangement will control temperature at saturation. This method provides an extremely predictable moisture content because the dew point can be equated to an exact humidity ratio. When this air is reheated to a required value, the relative humidity will be precise.

Air washers can also be used instead of high efficiency coils in dew point control applications. This is common in manufacturing facilities such as textile mills.

Economizer Cycles

The term **economizer cycle** is used to define the control strategy which allows **free cooling** from outside air, thus reducing mechanical cooling energy usage. This control strategy is applied to mixed air systems where the outside air or return air may be used to economize at the cooling coil. The reduced load on the refrigeration equipment results in tremendous savings in electrical energy.

There are three common types of economizer switchover cycles:

- Dry Bulb
- Enthalpy
- Floating switchover, adjustable differential

All three types of economizer switchover cycles choose between outdoor or return air streams. The way that economizer cycles choose which air stream to use is what distinguishes them from one another.

Dry Bulb Switchover

The dry bulb switchover cycle chooses whether the mixed air system should be using outdoor air for free cooling or return air based on the outdoor air dry bulb temperature.

It is a control strategy that saves energy year-round. In the winter mode of operation, it saves cooling energy by taking in free cooling from the outdoor air. In the summer mode of operation, it saves cooling energy by recirculating return air that has already passed through the cooling coil. The latent load on the cooling coil is reduced.

Winter Mode of Operation: When the outdoor air dry bulb temperature is below the switchover temperature (dependent upon locality), the temperature control system will have the ability to modulate open the outdoor damper upon a call for cooling.

Summer Mode of Operation: When the outdoor air dry bulb temperature is above the switchover temperature, the outdoor damper is placed to its minimum position providing minimum ventilation as required by code. In this mode of operation, the control system cannot modulate the outdoor damper beyond its minimum position. The primary source of air is the return air which has less total (sensible and latent) load.

Dry Bulb Switchpoint: Since this strategy senses only the sensible load (temperature only), designers must be careful when selecting the dry bulb economizer switchpoint. They must consider the latent loads at the dry bulb switchpoint. For example, Denver or similar "dry climates" have somewhat higher dry bulb switchover temperatures while coastal areas such as San Francisco have lower switchover temperatures due to the higher moisture content of the air.

Dry Bulb Switchover Logic:

**OA Temperature >
Switchover Temperature =
Summer Mode (Minimum Position)**

**OA Temperature <
Switchover Temperature =
Winter Mode (Free Cooling)**

Enthalpy Switchover

The enthalpy economizer switchover cycle chooses whether the mixed air system should be using outdoor air for free cooling or return air by measuring the total heat content or enthalpy of each air stream. The enthalpy economizer is sometimes referred to as the "true economizer" because it can sense both the sensible and latent components of the air. Dry bulb temperature and relative humidity are measured in both outdoor air and return air streams. This economizer will choose the air stream that imposes the least load on the cooling coil.

Enthalpy Logic:

OA Enthalpy >
RA Enthalpy OR OA Temp. >
RA Temp. =
Summer Mode (Min. Position)

OA Enthalpy <
RA Enthalpy AND OA Temp. <
RA Temp. =
Winter Mode (Free Cooling)

Enthalpy is a much more accurate measure of the load on the cooling coil. To maximize the efficient use of energy in a system, enthalpy should be used. However, all devices available today for sensing moisture content in air streams, especially those with wide temperature and humidity variations, require periodic maintenance. This must be considered in making the decision to use enthalpy switchover. If the potential for proper maintenance is not good, the best choice may be a dry bulb economizer or the floating switchover cycle, discussed next.

Floating Switchover–Adjustable Differential

The floating switchover–adjustable differential economizer switchover cycle chooses whether the mixed air system should be using outdoor air for free cooling or return air by measuring both outdoor and return air dry bulb temperatures. This economizer uses a differential temperature between outdoor and return air. This differential is the difference in temperature required to obtain free cooling from the outside air. The differential value is computer-generated from historical temperature and relative humidity data from the National Weather Service and is based on the system type.

Floating Differential Adjustable Logic:

OA + Differential >
RA = Summer Mode (Minimum Position)

OA + Differential <
RA = Winter Mode (Free Cooling)

Ventilation

ASHRAE Standard 62-1989 Ventilation for Acceptable Air Quality defines **ventilation** as the process of supplying and removing air by natural or mechanical means to or from a space. Air is provided at a specified volume known as the **ventilation rate**. The quality of the outdoor air used is subject to air quality standards for outdoor air as set by regulatory agencies such as the Environmental Protection Agency in the United States.

Ventilation rates may vary; a lobby area may require 15 CFM per person while a public rest room 50 CFM per toilet fixture. Local, state codes, ASHRAE Standard 62-1989 Ventilation for Acceptable Air Quality, or job specifications provide guidance to the designer and commissioning personnel.

Control of Primary Equipment

Each type of primary equipment—boilers, heat exchangers and converters, chillers and cooling towers—has a unique control strategy. Several common approaches to primary equipment control will be discussed here. There are many more which are beyond the scope of this text.

Hot Water Systems

Boilers are controlled by packaged controls installed at the point of manufacture. Hot water boilers operate at a fixed temperature, generally about 180°F. They are controlled at this fixed temperature either in a two position (on-off) manner or with some combination of low fire, modulating or high fire rates. This helps the boiler maintain a high efficiency and a long life.

Because overheating is a common problem with fixed temperature boilers, cooler return water is mixed with water leaving the boiler to obtain the desired hot water system temperature. This hot water system temperature is inversely reset from outdoor air temperature, commonly known as **outdoor reset of hot water**. See Figure 17-9. When outdoor air temperature goes down to 10°, the hot water temperature will be readjusted to a maximum heating value such as 180°. Conversely, when the outdoor air is 60°, the hot water tempera-

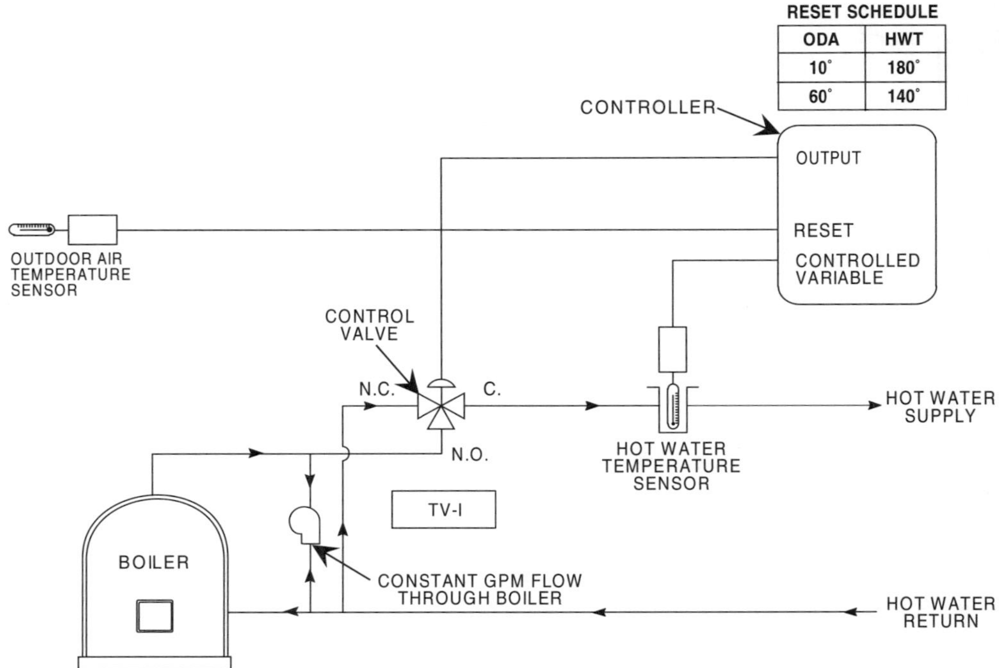

Fig. 17-9 Outdoor Reset of Hot Water

ture is readjusted to a light load condition, such as 140°.

Outdoor reset of hot water prevents overheating in the zone by providing a more controllable water temperature at the exterior zone valve. Colder outdoor air temperature increases the heat transfer through the walls and windows. The perimeter zone thermostat will sense this heat loss and proportionally open its valve. If the zone thermostat opens the valve too far to balance the heat loss, overheating might occur. By having the proper hot water temperature available to the terminal unit, the possibility of overheating is eliminated. Another benefit is the energy conservation which results from optimizing hot water temperatures to match the load.

Chilled Water Systems

Chillers, like boilers, generally come with control packages installed at the point of manufacture. Constant chilled water temperatures ranging from 42° to 45° are usually regulated by various capacity controls, such as inlet vanes or unloaders. Chilled water is required for two basic reasons:

1. To provide a minimum of 55° air temperatures to the zones.
2. To lower the dew point of the primary air in order to dehumidify the air in the zone.

When temperature rises in the zones, conditioned air lower than the zone's temperature and moisture must be fed to the zone. This cooler, drier air absorbs the excess heat and moisture. The warm return air carries the sensible heat from the zone and mixes with outdoor air being brought in for ventilation. This mixed air then rejects its heat to the cooling coil, causing the chilled water temperature to increase as much as 10 degrees. This chilled water returns to the chiller at 50 to 55° (worst case design load) to be cooled down to 43 to 45° before repeating the process once again. See Figure 17-10.

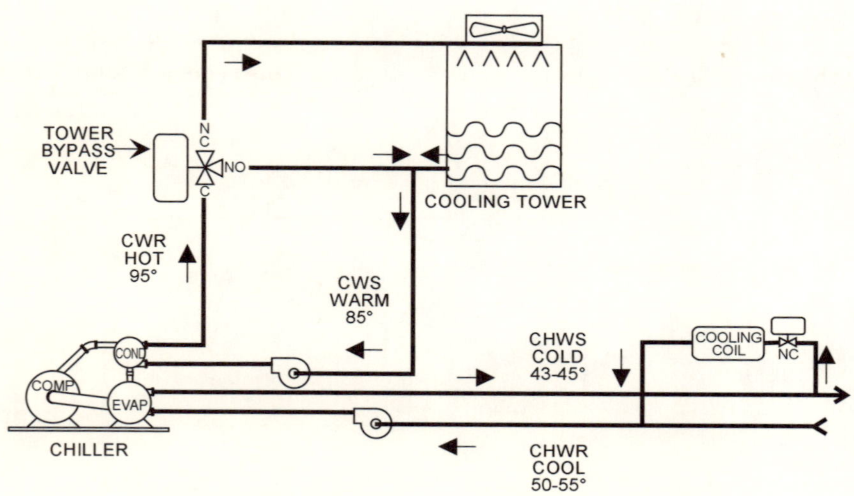

Fig. 17-10 Chilled Water System

Cooling towers have the important job of rejecting heat from the building. Whether there is one cooling tower or several, they are usually controlled using any of four methods discussed in Chapter 10: bypass valves, fan control, damper control, or a combination of any or all methods. Condenser water from the chiller is piped to the tower so that evaporative cooling may take place. Through this evaporative cooling process, heat is rejected to the outside air.

A bypass valve located close to the chiller or perhaps the tower, under proportional control, ensures that enough water is sent to the tower for evaporative cooling. Condenser water temperatures of 75° – 80° are desirable for most water-cooled chillers to achieve proper condensing temperatures. Given cold days, or "low-ambient" conditions, the bypass valve will bypass or re-route a portion of the condenser water to prevent overcooling the condenser water. In the summer as outside air wet bulb temperatures increase and as condenser water temperatures exceed the setpoint, the bypass valve will modulate sending full water to the tower. Next, stages of tower fans or low-high speed motor arrangements are energized to provide additional evaporative cooling of the condenser water, by increasing air flow through the tower. Dampers may also be employed to modulate air flows through the tower.

Control of Distribution Systems

Distribution systems supply and return the fluids, air and water used to transfer heat. Regulating the volume and pressure of these fluids helps ensure comfort for the occupants of a zone.

Fans –Volume and Pressure Control

Fans move air through ducts to zones at the volume required to transfer the heat and humidity needed for the comfort of the zones' occupants. Installing ducts of various sizes, larger at the fan or coil discharge and progressively smaller to the end, ensures an adequate volume of air for all the zones. Larger duct sizes at the coil section are required because the air stream must flow at a slower rate (500–700 ft per minute) to allow heat transfer from the coil to the air (heating) or air to the coil (cooling). The volume of air delivered to each zone is determined by the heat gains and losses of each zone due to heat transfer.

If an air stream flows at a constant volume through the ductwork, the temperature and humidity of that air varies as required by the zone sensors. This type of system is referred to as a **constant volume, variable temperature (CVVT) system**.

On the other hand, an air stream may be varied in volume as required by the zones. The airstream is usually controlled at a relatively constant temperature (55°F) and moisture content. This type of system is referred to as a **variable air volume (VAV), constant temperature (VVCT) system**.

Supply ducts which have variable flow rates have controls to ensure that the proper volume is delivered as required to ventilate and condition the zones. VAV supply fans are typically controlled by sensing static pressure. This control arrangement, **static pressure control**, is shown in Figure 17-11. Inlet vanes, discharge dampers, or variable speed drives modulate the fan's volume and hence control pressure within the duct.

The static pressure is the energy which pushes the volume of air through the duct and the VAV box to meet the zone's cooling requirement. As the temperature of the zone increases with a change in load, the zone's variable air volume box damper modulates open.

Pressures in the range of 0.75 to 1.5 in. wg are common. They are measured and controlled approximately 2/3 downstream of the longest duct run in the system. The reason for this location is to reduce the fluctuation of static pressure in the duct and also to make sure that the last variable air volume boxes have sufficient static pressure to operate.

Building pressurization control is required to maintain a slight positive pressure within the building to fend off unwanted infiltration into the building. Building pressure is sensed by an indoor sensor and often compared to outdoor atmospheric sensors. Pressures are typically maintained by modulating exhaust fans, return fans or relief dampers to control the building's pressure at desirable levels.

Pumps –Volume and Pressure Control

Much like fans, pumps move water at a volume required by the zones. The water must pass through boilers or heat exchangers, converters and chillers at a constant rate to allow heat transfer. Water piping then delivers the volume and temperature water for control by the thermostats in the zones as described in the previous section, Zone Control.

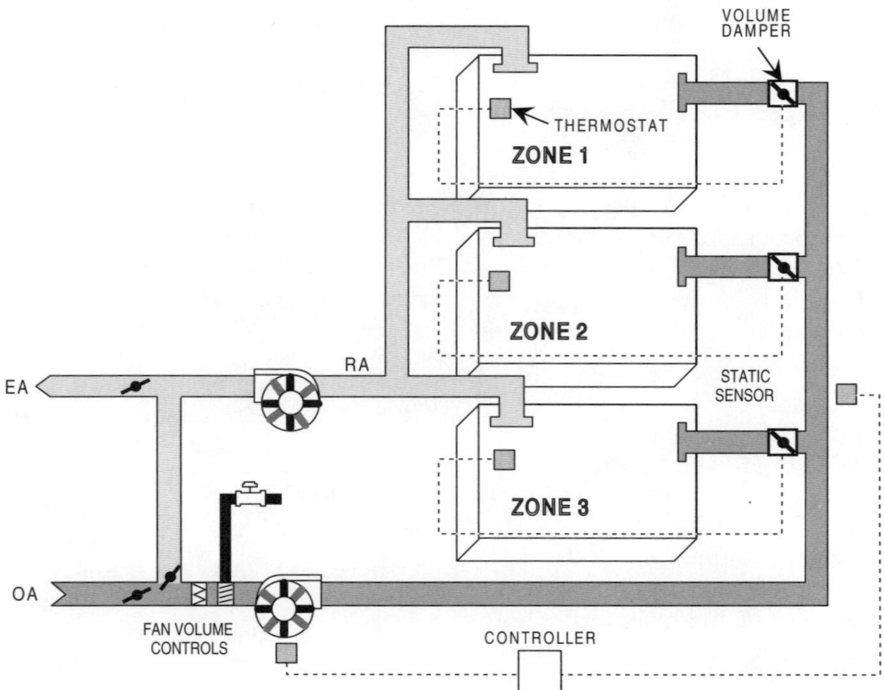

Fig. 17-11 Static Pressure Control

Much like air systems, pumping systems may be either constant volume or variable volume. Constant volume pumps are generally applied to primary loops through the primary equipment such as boilers and chillers. Flow in the secondary portion of the system, that which serves the terminal and air handling units, is variable. Constant volume pumping arrangements may be used as long as provisions are made to compensate for the varying flow through the units.

Because the flow of water varies in the piping, and along with it the pressure, an appropriate control strategy is differential pressure control. The **differential pressure control** system senses the differences in pressure between the supply and return pipe lines and the resultant control signal controls a differential control bypass valve which relieves excess supply pressure and volume to the return. See Figure 17-12.

Variable volume pumping arrangements are a preferred choice for the secondary portion of the system. A variable speed drive on the pump is often an even more efficient method of pressure control.

Whatever the control arrangement, differential bypass or variable speed pumping, the application helps maintain a constant inlet pressure to each valve. Doing this helps to prevent system pressures from overpowering of the zone control valves. The result is closer control of zone temperatures due to predictable inlet pressures at each valve.

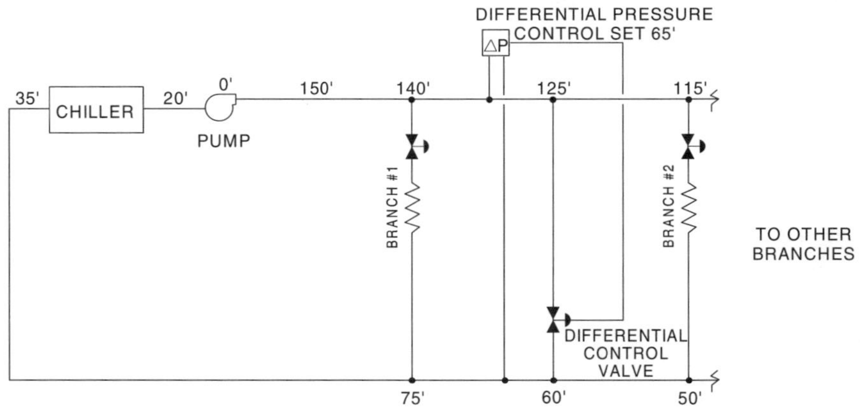

Fig. 17-12 Differential Pressure Control

Summary

- This chapter introduced control strategies which when properly implemented provide the required occupancy comfort.
- Zone control has the advantage of sensing the actual load in the space such as people, equipment and lights. A thermostat or humidistat in the zone senses zone conditions. Depending on the deviation from setpoint, the control logic generates the appropriate response.
- Air handling units can be controlled by any of four temperature control strategies: 1) Zone/room control, 2) Return air control, 3) Discharge air control and 4) Room reset of discharge. Two types of humidity strategies may be applied: 1) Room/return air control or 2) Dew point control of discharge air.
- Air handling units of the mixed air variety have economizer cycles to choose whether the system should be using outdoor air or return air. One of three methods is used to make this control decision: 1) measuring the outdoor air dry bulb temperature, 2) measuring the difference between outdoor air dry bulb temperature and return air, or 3) the enthalpy of the outdoor versus the return air stream.
- Control of primary equipment such as boilers, chillers and cooling towers is unique to each piece of equipment. Packaged control equipment generally control this equipment with standard control strategies.
- Control of distribution systems regulate the quantity and pressure of the heat transfer fluids; air and water.

18

Advanced Technology For Effective Facility Control

Facility management systems (FMS) have evolved dramatically over the years. When computerized building automation systems first appeared in the 1960's, they simply centralized monitoring and control functions by interfacing with pneumatic temperature control panels and motor control centers. Later, because of the energy crisis, energy management functions were added. After studying this chapter, you will:

1. Understand how facility management systems have evolved over the years.
2. State the differences between newer technology and older technology.
3. Relate numerous energy management features to energy conservation.
4. Understand in concept the wide range of computer software, hardware and firmware.
5. Describe how information management features benefit the facility manager and staff.

Advances in technology brought direct digital control, lighting control, fire management, security monitoring, distributed networks, personal computers, and sophisticated graphics. Electronic chips replaced pneumatic controllers. Personal computers (PCs) replaced minicomputers. Software programs replaced hardwired logic.

Each new advancement in the electronics and communications industries was eagerly adopted by facility management system (FMS) designers. Systems are now faster and more capable than ever before. Software programs, electronic components, sensors, actuators, hardware packaging, and communications networks are integrated, share information, and work together.

The overall purpose of a **Facility Management System** is to make the job of facilities people easier, to make a facility more efficient, and to keep a facility's occupants comfortable and safe.

An FMS can save money for building owners in several ways:

- Increasing the productivity of staff by having the system do mundane tasks for them.
- Reducing energy consumption (energy management programs).
- Identifying equipment needing maintenance and rotating the use of some equipment.
- Managing information.

When considering the use of any FMS, you must define the desired functions, make a realistic financial analysis, and determine the amount of time available for building personnel to learn and use the system.

The following discussion investigates many of the options available throughout the industry, although there may not be any single FMS which includes them all.

Integrated Control - Distributed Networks

In older systems, a dedicated computer known as a headend communicates with and controls remote field gear. The field gear reads the signal from a controller or sensor and sends it to the headend for evaluation and storage. If the headend determines control action is necessary, for example, a high temperature requires operating a cooling fan, the headend computer sends a signal to the field gear associated with the fan telling it to start the fan.

All of the programming for storage, analysis, and necessary actions is in the headend computer. The field devices, although possibly containing microprocessors largely for communications purposes, are primarily for converting and sending the signals from sensors, switches, and transducers so that the headend computer can monitor and control them.

Thousands of systems using the headend computer arrangement were installed in the 1970's and 1980's and are still being used successfully today. However, if installed today, more intelligent field panels would be available. The field panels are often direct digital control (DDC) panels compatible with most electronic and pneumatic sensors and actuators. They can control one large HVAC system or several smaller systems (including HVAC, boiler, chiller, and lighting systems), and they can interface to a headend computer for supervisory control and further data analysis, or they can work independently if they lose communications with the headend.

Figures 18-1a, 18-1b, and 18-1c compare how HVAC control has been traditionally done without a headend computer, how it can incorporate an FMS, and how it can be done by DDC field panels.

- Figure 18-1a shows an "all pneumatic" closed loop control.
- Pneumatic control of the setpoint.

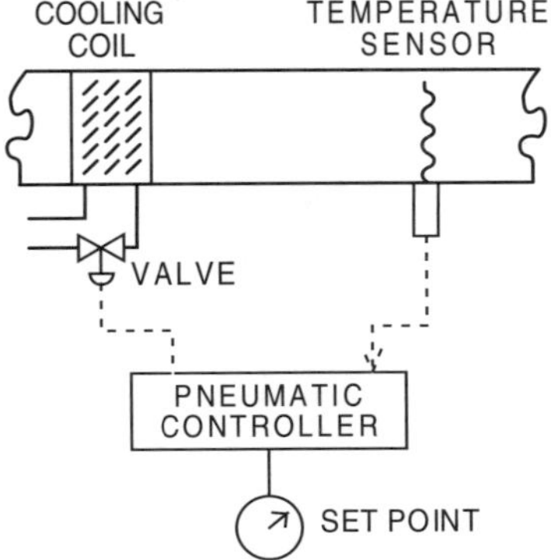

Fig. 18-1a Pneumatic Control Loop

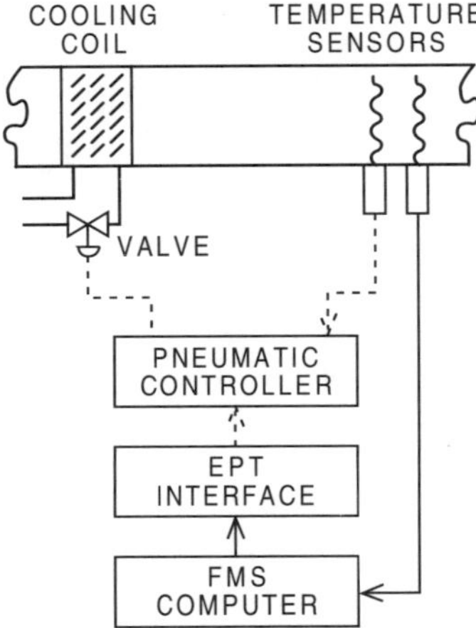

Fig. 18-1b FMS Loop

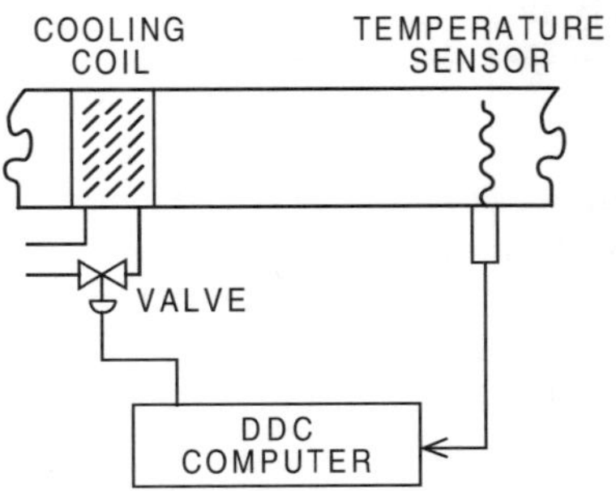

Fig. 18-1c DDC Loop

- Figure 18-1b shows a pneumatic controller in command.
- FMS computer setpoint controls through an electric to pressure transducer or EPT.
- Global (data access) across the system FMS control.

- Figure 18-1c shows that a DDC computer is the controller.
- Software control flexibility.
- Easy interface to the FMS network.

The DDC panels are often custom programmed or configured, individually, to suit a particular application. The DDC panels may be lighting control panels, fire panels or security panels. There might be a separate computer which can also interface to some of the same field devices for the purpose of maintenance management or security access control, such as monitoring card readers and door access.

With computer intelligence in the DDC field panels, the line between the responsibilities of the headend and the field devices becomes less well-defined. Certain programs, such as energy management functions, might best be programmed into the field panels, while others, such as centralized alarm reporting, are best done by the headend computer.

However, state-of-the-art control avoids using the headend computer. The trend is toward a network of microprocessor-based network control units (NCU's) or distributed intelligent controllers, each monitoring and controlling designated equipment and freely sharing data.

Network expansion units (NEU's) perform local control functions, such as starting and stopping fans and closed loop control of valves and dampers. A personal computer is an equal partner in the network providing a sophisticated operator workstation, storing network data, and storing and executing sophisticated monitoring and controlling applications.

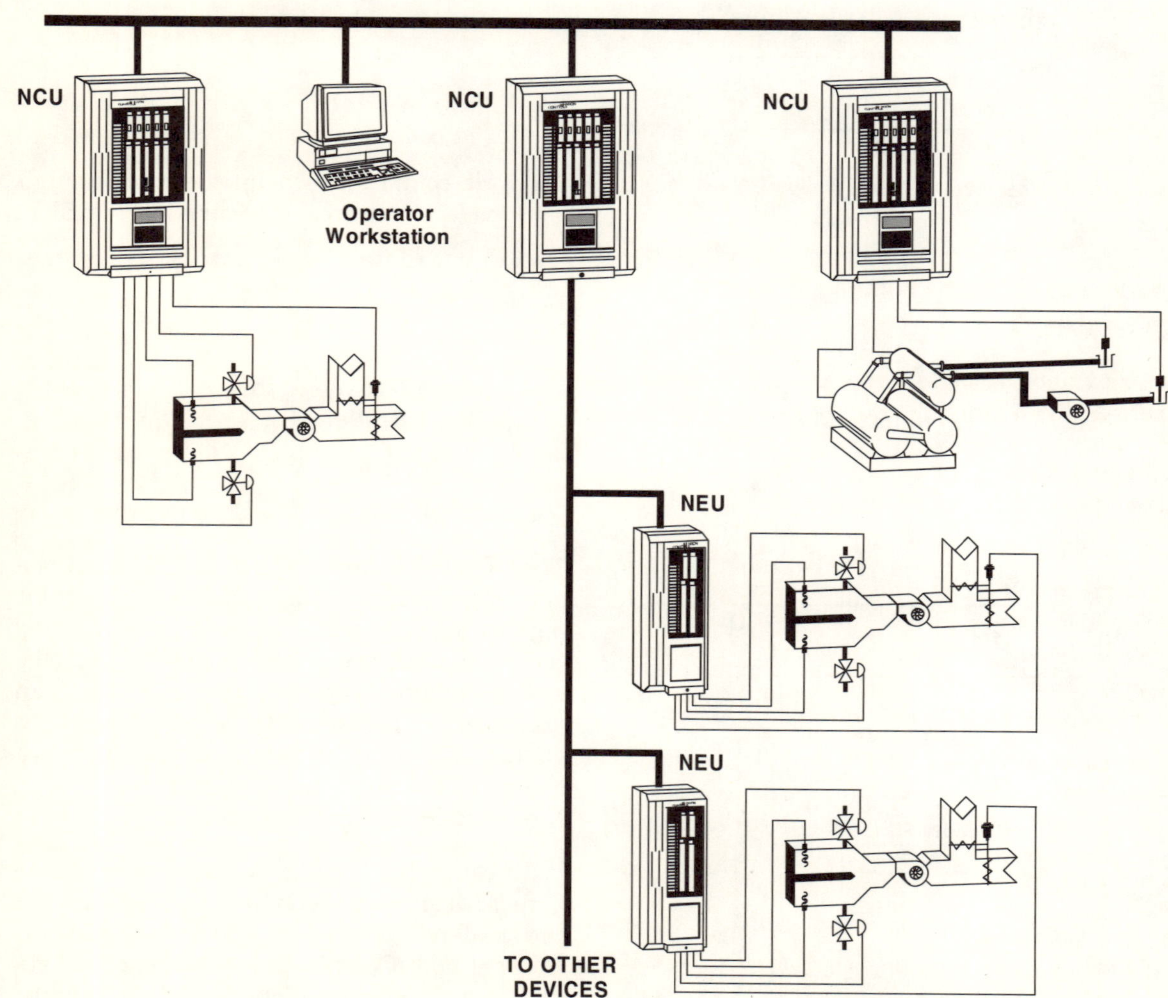

Fig. 18-2 Intelligent Network Devices in an FMS

Figure 18-2 shows one possible configuration.

The distribution of intelligent network devices provides complete stand-alone control capability when needed, providing maximum reliability: the network has no headend computer which would be the primary center for energy management programs. The most sophisticated distributed network devices use state-of-the-art communication modes which can rapidly share data, giving building operators complete and consistent information about the facility.

Even with the high degree of sophistication in the technology available, costs of such systems have not increased significantly, packing in more value for each dollar spent.

Communication between Network Devices

In addition to considering the type of configuration (headend centralizing communication for remote field panels or intelligent network devices), designers must consider the type of communication link between devices in the system.

The means of communication determines the speed, distance, and cost of communication. Some systems can mix various types of communication to accommodate a more complex network. Some of the most common types are:

- Coaxial cable and twisted-shielded pair may be least expensive and easiest to install, but they may be subject to electrical interference, such as lightning, which can cause component damage and loss of data.
- Fiber optic cable offers a high degree of quality in transmission, as well as protection against electrical interference.
- Dedicated leased telephone lines can transport data long distances to include numerous remote buildings into a single FMS network. They are expensive to use and are subject to the same failure as normal telephone lines. To reduce costs, many FMS systems offer the ability for the operator to dial up remote areas only when needed, or the program can dial up remote areas automatically to get data or to control equipment.
- Building electrical or telephone wiring can transport FMS network data, reducing the cost of labor and materials for wiring to implement the FMS. Systems of this type are called power line carrier systems.

Many networks can take advantage of existing communications links already installed in a facility. If the communications scheme is an industry standard link such as ARCNET™ or Ethernet™ that is used in thousands of office and industrial automation installations worldwide, installing and servicing the network should be easier and less expensive.

Small Building Systems

A smaller, less sophisticated (and less expensive) FMS might be appropriate:

- if the building is small,
- where potential energy savings are too limited to yield a large dollar savings,
- where the operating staff has little time to devote to using, or learning to use, the system.

Many small systems are available with a limited number of functions, including direct digital control, load control, time programming, and additional standard programs.

However, using the distributed network concept, the distinction between small building and large facility applications is not as clear. A smaller building simply uses fewer intelligent distributed devices on the appropriate communications network type.

Large Facility Systems

On a larger scale, an FMS can accommodate functions often associated with the needs of a larger facility:

- More equipment monitored and controlled
- Greater distance requirements
- Flexible programming
- Sophisticated data management and reporting schemes
- Simultaneous use of the system by multiple operators

Although a more sophisticated and complex system requires a greater initial investment, it can increase the return on investment by yielding significantly better energy savings and increasing productivity of workers.

Features For Optimal Control

Automatic equipment controls are designed to improve building efficiency and maintain occupant comfort while minimizing energy usage. These features often yield the most tangible and measurable energy and dollar savings for the building owner.

These features can be performed by the headend computer, a field panel, or traditional pneumatic controls. Therefore, our discussion will be general, not giving specifics of how the feature is programmed or configured.

Overall, the features reduce the amount of electricity a facility uses. The electric bill of a commercial building complex is a large part of the building's operating costs. Lights, HVAC, and computers are a few of the major consumers of electricity in a commercial building.

Like the electric bill received for your home, the electric bill for a commercial building is largely based on the total amount of electricity used, measured in KWH (kilowatt hours). The charge for each KWH varies from utility to utility, may vary from season to season (usually cheaper in winter), and may vary based on when it was used (usually cheapest at night). A rough estimate might be seven cents ($0.07) per KWH. The total amount used may also affect the rate, such as a "bulk rate" discount.

At any rate, the less electricity consumed, the lower that portion of the electric bill will be. This section discusses various features designed to reduce electrical consumption.

Optimal Run Time (ORT)

Optimal run time (ORT) refers to a single feature combining optimal start time with optimal stop time. Optimal stop time stops the building equipment before occupants leave the building at the end of the work day. State and municipal codes may prevent using optimal stop time because stopping equipment also means closing the outdoor air dampers. (Many areas have codes requiring minimum ventilation.)

Waiting until the last possible moment to start building equipment at the beginning of the work day will save electrical consumption (KWH). **Optimal Start Time** (OST) delays morning start-up without sacrificing occupant comfort when workers arrive.

Because fans are normally turned off at night and the temperature in the building is allowed to drift away from comfortable levels, the building temperature may drop to 50°F at night in winter and may be allowed to get up to 90°F at night in summer. However, by the time the building is occupied again in the morning, the temperature must be up to about 68°F in winter and down to 78°F in summer. If not, the occupants are likely to complain that they are uncomfortable.

Rather than starting at the same time every day, OST determines the latest time possible to start the fans in order to reach comfort levels by occupancy time. The amount of time necessary to reach comfort levels depends on many of factors:

- the outdoor temperature
- building insulation
- the building's ability to gain and lose heat
- how cold or warm the building got overnight
- how warm or cool the building must be by the time of occupancy to be considered comfortable

OST uses these factors and others to determine the latest possible equipment start time and still reach comfort at occupancy.

Load Rolling

Many fans, pumps, and HVAC systems in a building are operated continuously during occupied periods to provide the heating, cooling, and ventilation for which they were designed. However, since this equipment has enough capacity to maintain occupant comfort during the peak load conditions on the hottest and coldest days of the year, some may be turned off for short periods of time with no loss of occupant comfort.

The **load rolling** feature can significantly reduce overall KWH consumption by stopping (shedding) certain electrical loads (equipment) for predefined amounts of time periodically throughout the day. Load rolling is sometimes known as load cycling or duty cycling.

The program generally allows the user to define minimum on and minimum off times to avoid short cycling which could cause more cost in equipment maintenance than is saved by shedding loads. Similarly, maximum off times avoid discomfort caused by a single fan being off too long.

To maintain comfort, the program can automatically adjust the cycle times of loads to compensate for changes like space or outdoor air temperatures. In other words, before a load is shed, the controls can check a related temperature (such as space), and if too warm or too cool, the program can override (not shed) that load at that time, or it could just shed the load for a shorter time than usual.

In order that load rolling reduces KWH consumption, shed only loads like constant volume fans and pumps which do not need to make up for lost time when they are started again. For example, if you stop a constant volume fan for 15 minutes, you do not have to run it faster later to make up for the time it was off. Such loads are considered expendable loads. (The term *expendable* has nothing to do with relative importance.)

A deferrable load is the opposite of an expendable load. Compare the use of an expendable load like a constant volume fan or a pump to the use of a deferrable load like a chiller or a VAV fan. If you stop a VAV fan for 15 minutes, you would be reducing electrical consumption for that 15 minutes, but you would have to run the fan harder later to make up for the time it was off. The fan still uses the same amount of electrical power, but the usage is "deferred" to a later time. A load which must "make up" for the time it is off is known as a *deferrable* load.

Since turning off a deferrable load for a short time does not really save KWH, you should not use deferrable loads with the load rolling feature. However, you will find that it is appropriate to shed deferrable loads with the demand limiting feature.

Demand Limiting

Most residential electric bills are based primarily on total consumption. The more electricity (KWH) consumed, the higher the electric bill. In contrast, the electric bill of a commercial building is based on total consumption plus the rate of consumption (for example, KWH per each 15-minute period). The rate of consumption, known as electrical demand, is measured in KW (kilowatts). The electric utility monitors the total consumption (KWH) and the rate of consumption (KW) independently.

Each utility uses a slightly different method to calculate the highest demand it measures for the billing period. However, once it determines that peak demand, it adds an additional charge. And, as with the KWH consumption charge, the demand charge may vary by the time of day or by season. The demand peak often ratchets on a monthly or even on an annual basis.

The period of time that the utility company routinely uses to measure demand is called the demand interval. The demand interval is determined by the utility and is commonly 15, 20, or 30 minutes. For example, if a utility uses a 15-minute demand interval, it measures the KWH consumption in each 15-minute period or demand interval of the billing period, calculates the KW demand for each interval and, from that determines the demand portion of the bill.

Therefore, if the use of electricity in a facility could be kept spread out to maintain a relatively constant level throughout each day, instead of using a lot of power in one short period of time, the resulting dollar charge would be smaller.

The demand limiting feature monitors the rate of electrical consumption and starts shedding (turning off) loads when usage is exceeding a predefined demand limit demand target. Demand limiting is sometimes also known as peak shaving or load shedding.

Figure 18-3 shows how turning off some expendable loads or during peak times, even shedding a deferrable load which would need to be run longer later, can save on the demand charge by flattening the demand curve.

Economizer Switchover

The equipment in a commercial building conditions the air it supplies to keep its occupants cool. It can supply from one of two sources: outside air or return air. If return air is used, the control system closes the outdoor air to legal minimum limits and reconditions air already in the building.

Various methods can be used to determine the air source to use: simple dry bulb, enthalpy switchover, or floating switchover. (See chapter 17 for a discussion of these methods.) Which method to use depends on the building and system.

For the simple dry bulb, the program compares outdoor air temperature with a predetermined switchover setpoint. If the outdoor air temperature is lower than the setpoint and cooling is needed, the dampers are allowed to modulate, providing free cooling. If the outdoor air temperature is higher than setpoint, the program holds the dampers at minimum position to recycle as much return air as possible, using mechanical cooling to recondition it.

Enthalpy switchover is a more complex, but a more accurate, method of determining economizer switchover. It is based on these facts: 1) at the same temperature, humid air contains more heat energy than less humid air, 2) people feel warmer when the air is more humid, and 3) an HVAC system must use more energy to cool humid air. The total heat content of air (enthalpy) is calculated using dry bulb temperature and relative humidity (or dewpoint), among other values.

The program determines the enthalpy of the outdoor air and return air, and compares the results. If the outdoor air stream has less enthalpy than the return air stream, the dampers are allowed to modulate for free cooling. If the return air stream has lower enthalpy, the dampers are held at minimum position.

Supply Air Reset (SAR)

While a building is in occupied mode, it has heating and cooling requirements throughout the day. Supply air reset is a strategy which monitors the heating and cooling loads in the building spaces and adjusts the discharge air temperature to the most efficient levels that satisfy the measured load.

The cooling discharge temperature is raised to the highest possible value which still cools and dehumidifies the warmest room served by the fan system. Heating discharge temperatures are reduced to the lowest possible levels which still heat the coolest room.

SAR works best with a constant volume system in which the amount of air being supplied to the zones is always the same.

The system really has two control loops, as illustrated in Fig. 18-4. First, the room loop consists of the room temperature measuring element and the setpoint. As the room temperature varies around the setpoint, SAR calculates a new setpoint for the discharge loop. The second loop uses the new discharge set-

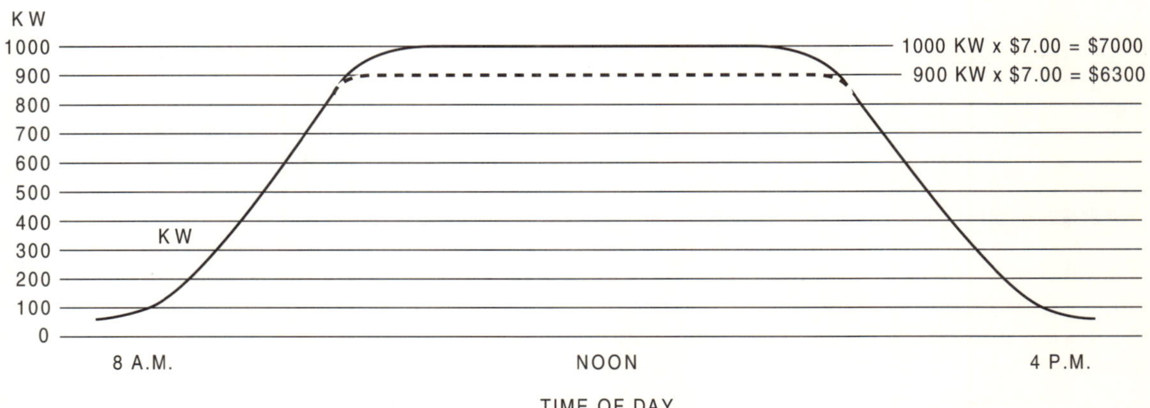

Fig. 18-3 Flattening Out the Demand Curve to Save Money

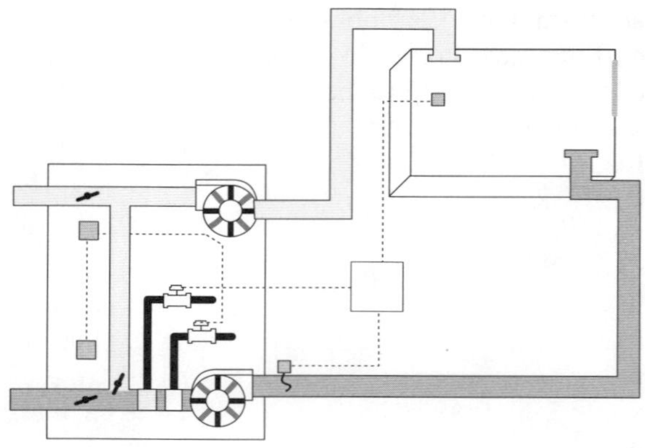

Fig. 18-4 Supply Air Reset

point and measures the discharge air temperature. As the discharge temperature varies around the setpoint, the program sends a new command to the valve or dampers of the mechanical heating or cooling equipment.

You can adjust various values, such as setpoints, proportional bands, and deadbands. Some computers/controllers allow you to choose either proportional control or proportional plus integral control (PI). PI Control is suitable for applications where the controlled variable must be right at setpoint, as in clean rooms or static pressure control.

Supply Water Reset (Chilled Water or Hot Water)

The supply water reset feature automatically changes the setpoint of the water supplied to the cooling or heating loop to the highest (for chilled water) or lowest (for hot water) temperature possible, while still satisfying the requirements of each zone it supplies.

To cool a commercial building, water or coolant is cycled throughout the building zones. As the water or coolant cools the zones, it picks up the heat. To get rid of the heat, the water is passed back through a chiller to cool it down again. The water is then recycled to the building zones for continued cooling.

If the chilled water setpoint is colder than necessary, the chiller wastes energy working to achieve the setpoint. Therefore, the **chilled water reset** feature adjusts the chilled water setpoint as high as it can, while still satisfying the zones.

Valves vary the amount of available chilled water supplied to each zone. The position of each valve varies as more or less cooling is required. For example, a gym where a basketball game is taking place may require more cooling (requires more chilled water, so the valve opens farther) than an office where several people are quietly working at their desks.

For the chilled water reset feature, the computer/controller checks if any of the valves are fully open, requiring all the chilled water they can get. If none of the valves is fully open (each is bypassing), each room is cool enough and does not require all the chilled water available to it. Therefore, the program determines that the temperature of the chilled water is colder than necessary to satisfy its current cooling requirements. The computer/controller increases the setpoint of the chilled water (supplying warmer chilled water). Warmer chilled water allows the chiller to use less energy.

Similarly, the hot water reset feature lowers the temperature of the hot water if all of its zones are bypassing.

Condenser Water Reset

Chiller plants are usually sized to reject their rated capacity through cooling towers sized to operate at design outdoor air wet bulb conditions. This sizing ensures that the plant will satisfy the design temperatures, but it wastes energy when conditions are not at the design wet bulb temperature.

As the chilled water removes heat from the building zones, it gets warmer. The chiller removes the heat from the chilled water, and condenser water removes the heat from the chiller itself. The chiller system pumps the condenser water to the cooling towers to reject heat to the outdoor air. The lower the condenser water temperature, the more heat it can remove from the chiller, reducing the energy necessary to cool the chilled water.

The **condenser water reset** feature saves energy by lowering the condenser water to the lowest possible temperature setpoint based on the ambient wet bulb temperature and the actual load being handled by the chiller.

Cooling the condenser water to the lower setpoint requires the cooling tower fans to expend more energy. However, when properly implemented, lowering the condenser water by, say, 1°F, can save more energy at the chiller than is used by the tower fans to lower it. The system invests some energy to save more energy.

The program can automatically reset the setpoint of the condenser water as low as possible to save energy. It takes into account the wet-bulb characteristics and uses either more or less of the water cooled at the cooling tower to achieve the new setpoint.

Chiller Sequencing

In many cases, more than one chiller is required to cool the chilled water to the desired setpoint. On a very hot day, all of the available chillers may be required. On a cooler day, only one or two chillers be needed. The **chiller sequencing** feature determines the most efficient combination of chillers required to run. It allows each chiller to run only within its efficiency range (for example, between 40% and 90% of its design capacity), and automatically starts or stops another chiller to keep all operating chillers within their range. Optionally, the program also checks the run time of various chillers to determine which to turn on or off next as cooling requirements change.

For example, in a building with three chillers, assume that the DDC controller determines that one chiller running at 98% capacity could sufficiently chill enough water to meet cooling requirements. Since 98% falls outside its efficiency range of 40-90%, the DDC controller would bring on another chiller. If both chillers online had equal capacity, each would run at 49% capacity to chill the water. Since 49% is in the most efficient operating range, using two chillers uses less energy than using one at 98% capacity. Whenever the DDC needs to decide which available chiller to bring on next, it can choose the one with less run time.

Information Management Features

The **information management** features are designed to help staff gather and analyze data to help them efficiently run the facility.

Once the FMS gathers and stores information about the facility and takes any actions as appropriate according to the program, that information is available to the user in various forms. The data may be available simply as summaries of existing conditions such as the current temperature and humidity of certain measured variables, and the on/off status of equipment. The more sophisticated and complex the capabilities of an FMS are, the more it can analyze, sort, and record greater amounts of data in increasingly different ways. The following discussion describes several features available in the FMS industry, although no single FMS necessarily includes them all. Similarly, other features exist which have been omitted here. If you already have an FMS or are in the process of choosing a new FMS, you must determine which features are most useful to you.

Many FMS vendors offer similar versions of each feature, but the features vary in the amount of detail they keep and in how easy they are to use.

Summaries

Summaries contain detailed information about specific aspects of the facility. For example, one summary might list all monitored and controlled equipment and variables with their current status. Another summary might just list the low/high temperature alarm limits. An FMS probably has many summaries, with at least one also associated with each information management feature such as runtime totalization or for energy management features like load rolling.

The data is probably stored in the headend computer of the system or, if the system has intelligent, independent network units, the data is stored in the network units and is also archived in a PC in the network. The summary data is gathered by the summary feature for output to the CRT, printer, or data file.

Depending on the system used, the names given to the controlled equipment and variables may simply be numbers, or they may be word-names assigned by the user for easy identification of the item.

In addition to selecting the summaries to be printed, most systems let you define when summaries should be automatically sent to a printer or data file. An example might be every weekday at 9 a.m.

Password

Password is a global security feature which prevents unauthorized people from using the FMS. It may also limit certain users to executing only certain FMS commands, and possibly to working with only certain areas of the facility.

Alarm Reporting

The FMS is usually programmed only to report abnormal alarm conditions. For example, a fan may normally cycle on and off during the day due to temperature changes or load rolling, and the FMS will not continually report each change though you can find out at any time what its current on/off status is. However, should the fan fail to turn on when the FMS commands it to, the FMS will issue an alarm message and probably sound a tone indicating the problem. The amount of description and detail contained in the alarm message sent to a screen or a printer; varies depending on the sophistication of the FMS.

Time Scheduling

The operator can easily program events to occur at a certain time on certain days. For example, you could schedule fans and lights to turn on and off at certain times, schedule various summaries to print, and schedule different temperature alarm limits for occupied and unoccupied times. The types of events you can schedule are determined by the FMS programming. Some allow you to program holiday or special event programs.

Trending

Trend records the status of certain variables at various intervals such as every 30 minutes and stores that data for your analysis. Trending is useful for HVAC system troubleshooting, giving you data about your facility, without taxing your staff to routinely go, monitor, and record the data.

For example, you may want to see what happens to a room temperature associated with a fan; you could use trending to sample the fan status and room temperature every 30 minutes and examine the trend summary at the end of the day to see how the temperature is affected by the fan.

The capabilities of the FMS determine whether you can choose the time interval, how many samples the FMS can store, how many different variables the FMS can trend at one time, and how the data appears on the trend summary. Many systems can even output the trend data in bar or line graph form.

Totalization

Totalization is a counting or summing process. It can often count:

- the units of time a particular status was in effect (for example, how long a fan was on or how long a temperature was high).
- the number of times an event occurred (how many times a motor started or a temperature was in alarm).
- the number of pulses recorded at a pulse sensor (for example, to determine the amount of chilled water used).
- the amount of a physical quantity used, calculated from the current rate of consumption (such as, how many gallons of chilled water, pounds of steam, or Btu's of cooling were used in a week, totalized from a different calculation determining the instantaneous rate of use).

Based on this easily accessible information, your staff can accurately schedule preventive maintenance, assess costs of running equipment, or even bill tenants.

Graphics

Although blueprints of the facility are useful to the operating staff, when changes are made, the blueprints are rarely updated. Furthermore, the blueprints do not indicate current operating characteristics of the facility. **Graphics** associated with the FMS can be a useful tool to help identify areas of concern, displaying in pictures, much of the same current information your staff could get from a typical summary of equipment status. However, a graphic has the advantage of visually associating one occurrence with another.

For example, to determine why a temperature is too high, a graphic could easily associate the temperature with whether a particular fan is on or with the temperature of the available chilled water or with the time of day when that area is occupied by 50 people. Without a graphic to visually pull all of the possible reasons together, the reason might never be realized as the explanation and possible solution.

Summary

The efficient operation on your facility ultimately depends on the efficient use of existing building equipment and the productivity of buildings people. The FMS is a tool to help the facility's staff make the most of their time on the job and to coordinate the equipment for optimal use.

When using an existing FMS, each feature should be used to maximize the benefit. Many existing FMS's have much power going unused, and the unused features could be implemented with a little investment of time and no investment of funds.

When investigating the implementation of a new FMS, first define what features are needed, how they will pay back, and how much money can be invested.

Typical ways that an FMS assists maintenance people from an energy management perspective are:

- Delaying the morning startup of equipment to the last possible moment (optimal start time).
- Using the source of air that is cheapest to cool (economizer switchover).
- Turning off selected equipment for short times during the day (load rolling).
- Conditioning the supply air only to the most efficient temperature (supply air Reset).
- Conditioning the supply water only to the most efficient temperature (supply water reset).
- Determining the most efficient combination of chillers to bring online (chiller sequencing).
- Limiting consumption (demand limiting) to reduce electrical costs by monitoring the rate of consumption.

List of Figures and Tables

Glossary

absolute zero — absence of all heat energy; temperature at which molecular motion stops.

absorber — in an absorption refrigeration cycle, a vessel which holds the liquid for the absorption of refrigerant vapor into solution.

absorption refrigeration cycle — process by which a circulating refrigerant is evaporated at low pressure, absorbed in an absorbing medium, concentrated by heating at high pressure, and condensed by cooling.

activated charcoal — powdered, granular, or pelleted carbon capable of adsorbing odors, gases, and vapors because of its fine pores.

actuator — mechanism which receives a controller's output signal and produces a force or movement to move a manipulated device such as a valve or a damper.

adiabatic process — heating or cooling without loss or gain in heat energy; a constant enthalpy process.

adiabatic saturation — theoretical achievement of 100% moisture saturation of air without gaining or losing heat energy.

ADP — see *apparatus dew point*.

air cock — see *boiler vent*.

air conditioning — process which removes enthalpy and water vapor from the air (partial air conditioning); process which controls temperature, relative humidity, air movement, and radiant heat energy, may also include removal of airborne particles and contaminants (full air conditioning).

air washer — device which uses liquid spray to wash particulates and impurities from air; evaporative device which uses a warm water spray in an enclosed chamber to cool or warm air by converting sensible heat to latent heat.

air washer humidifier — see *humidifier, air washer*.

air-off-the-coil temperature — actual temperature of air leaving a coil.

air-water system — HVAC system which distributes heated or chilled air and heated or chilled water to condition the air in a space.

all-air system — HVAC system which distributes heated or chilled air from a central plant through ducts to conditioned zones as opposed to systems which use air and another fluid or water.

all-water system — see *hydronic system*.

analog — variable which remains proportional to another variable over some specified range; for example, temperature may be represented by a voltage which is its analog.

analog input — data or signal indicating the condition of a physical variable within a minimum and a maximum range such as temperature, pressure, or voltage. When transferred to a controller, it is the initiating part of a control sequence.

apparatus dew point — temperature necessary for moisture in air to condense on a cooling coil.

approach — in a cooling tower, the difference between the wet bulb temperature of the entering air and the temperature of the leaving water.

ASHRAE — American Society of Heating, Refrigerating, and Air-Conditioning Engineers

ASME — American Society of Mechanical Engineers

atmospheric pressure — see *pressure, atmospheric*.

atomizer — device which sprays water as a fine mist.

atomizing humidifier — see *humidifier, atomizing*.

automatic reset — see *control, integral*.

axial fan — fan having its inlet and outlet on an axis parallel to the fan's shaft.

backflow — reverse flow in a water system from the normal or intended direction.

baffle — device for regulating the flow of a fluid; in a boiler, refractory designed to prevent direct flow of combustion gas to the chimney.

bandwidth — amount of change in analog input required to cause a controlled device to move from 0-100%.

barometer — device used to measure atmospheric pressure.

baseboard radiation — see *fin-tube convector*.

Bernoulli equation — mathematical equation used to express the conservation of energy as it changes from potential to kinetic energy.

bimetal sensor — temperature sensor consisting of two fused, dissimilar metals having a different rate of thermal expansion; for example a thermostat. See also *thermocouple*.

bin data — method used to organize weather data. The bin method organizes this data by the number hours a given condition, such as temperature, may occur in a given period of time.

binary input data or signal indicating one of two conditions; for example, the operating status of a fan as off or on.

blast tube combustion chamber in a boiler.

blowdown relief of steam below setpoint pressure to discharge water having large quantities of dissolved solids.

BoHP boiler horsepower.

boiler closed vessel which transmits heat from an external source, usually the combustion of fuel, to a fluid, typically water.

boiler, cast iron sectional boiler consisting of individual hollow cast iron sections serving as water jackets surrounding the furnace area.

boiler, electric boiler which uses electric resistance heating elements to heat water.

boiler, fire tube boiler in which combustion gases pass through tubes immersed in water.

boiler, high-pressure boiler designed to produce steam at a pressure above 15 psig or hot water at a pressure exceeding 160 psig and temperature above 250°F.

boiler, low-pressure boiler designed to supply steam or hot water for space heating typically at pressures below 15 psig for steam or 160 psig for water at 250°F.

boiler, water tube boiler in which water circulates in tubes and heat is applied outside the tubes.

boiler capacity maximum rate of heat output from a boiler per hour.

boiler horsepower a measurement of boiler capacity; equivalent evaporation of 34.5 lb water per hour to steam at 212°F (equal to 33,475 Btu/h or approximately 9809.5 watts).

boiler vent valve which vents air to the atmosphere when a boiler is filled and prevents a vacuum when a boiler is cooled or drained.

boiling point the temperature at which a liquid changes to a gas at a fixed pressure.

Bourdon tube gauge manometer which measures pressure using a closed, oval tube bent to a curve (tube tends to straighten under internal pressure).

Boyle's Law law which states that the product of the volume of a gas times its pressure is a constant at a fixed temperature; demonstrates that the volume of a gas at a constant temperature decreases under increased pressure or increases under decreased pressure.

breeching in a boiler, an outlet or passage for controlling the flow of burner exhaust gas to the atmosphere through a chimney or vent.

British thermal unit amount of heat energy needed to raise or lower the temperature of 1 pound of water 1°F; abbreviated Btu.

Btu see *British thermal unit*.

built-up system HVAC system in which the designer selects components individually on the basis of system specifications and operating characteristics as opposed to a factory-assembled (packaged) system.

bulb and capillary sensor temperature sensor bulb which contains a thermally sensitive fluid which expands or retracts through the capillary; for example, a thermometer.

bypass pipe or duct, usually controlled by a valve or damper, which conveys a fluid around a component in a system; air which partially contacts or fails to contact a cooling coil as it passes through the coil chamber.

calorie amount of heat energy needed to raise or lower the temperature of 1 gram of water 1°C at a pressure of 1 standard atmosphere.

capillary tube metering device consisting of a tube having a specific length and bore size.

Celsius scale standard scale for measuring temperature under standard atmospheric pressure; boiling point of water is 100°, freezing point is 0° (also known as centigrade scale).

centigrade scale see *Celsius scale*.

central system HVAC system which conditions air at a central plant and distributes it to zones using one or more fans and a duct system.

centrifugal compressor see *compressor, centrifugal*.

centrifugal fan fan which uses centrifugal force to move or add pressure to air.

centrifugal pump pump which uses centrifugal force to move or pressurize a liquid.

Charles' Law law which states that a volume of a gas at a constant pressure decreases when cooled or increases when heated.

chiller machine designed to cool water by compressing vapor in a refrigeration cycle to reject the heat.

chiller sequencing energy-conserving control strategy which determines the most efficient combination of chillers required to operate to meet a cooling load.

closed piping system piping system which has no openings to the atmosphere.

closed-loop control control system consisting of a sensor, a controller, an actuator, and a manipulated device, set up so that the sensor can sense how control changes affect controlled variable and signal the controller to react.

coaxial cable — transmission line consisting of a conductor centered inside and insulated from an outer metal tube which serves as the second conductor.

coil process line — line imposed on a psychrometric chart defining the efficiency of a cooling coil, which is the sensible, latent, and total heat removal capacity of the unit.

combustion — the rapid fracturing of chemical bonds that hold atoms together in a large hydrocarbon fuel molecule. This splitting of chemical bonds creates an enormous release of useful heat, along with the by-products of combustion.

compression tank — a tank, closed to the atmosphere and partially filled with water, used as a cushioning device by accommodating the expansion and contraction of water caused by temperature changes. See also *expansion tank*.

compressor — mechanical device which increases the pressure of a gas or vapor.

compressor, centrifugal — continuous-flow machine, such as a fan or propeller, which uses centrifugal force to compress refrigerant vapor.

compressor, hermetic — in a refrigeration system, a motor and compressor assembly enclosed in a sealed housing designed to reduce the possibility of refrigerant leaking at a shaft seal.

compressor, open — in a refrigeration system, a compressor driven by an external motor using a driveshaft; motor and compressor are separate units.

compressor, reciprocating — positive displacement machine which uses a moving piston to reduce the volume of a compression cylinder in order to reduce the volume of refrigerant vapor.

compressor, refrigerant — component in a refrigeration system which reduces the volume of a refrigerant by compressing low pressure vapor into high pressure vapor.

compressor, rotary — machine in which fluid is compressed through rotation of a positive displacement component such as a blade or a vane.

compressor, screw — rotary compressor which uses two intermeshing rotors (screws) to compress refrigerant in a compressor housing.

compressor, scroll — positive displacement compressor which compresses a refrigerant using a stationary upper scroll and a rotating (involute spiral) scroll.

compressor, semi-hermetic — in a refrigeration system, a compressor directly coupled to a drive motor contained within a gas-tight, bolted housing.

concentrator — in an absorption refrigeration cycle, a vessel where steam or hot water is used to evaporate water (refrigerant) from the absorption solution.

condenser device designed to change a fluid from vapor to a liquid by removing heat; in a refrigeration system, a heat transfer device which removes heat from a refrigerant by transferring it to water or air having a lower temperature.

condenser, evaporative device in which vapor is condensed within tubes cooled by the evaporation of water flowing over the outside of the tubes.

condenser water reset energy-conserving control strategy which reduces the temperature of condenser water to the lowest economical temperature on the basis of the ambient wet bulb temperature and the load being handled by the chiller.

conduction transfer of heat energy through physical contact between molecules.

constant pressure expansion valve metering device which controls the flow of liquid refrigerant into an evaporator by keeping the pressure constant.

constant volume use of a steady air volume at varying temperatures to control the condition of the air in a zone in contrast to the use of variable air flow at a constant temperature.

continuity principle at a given time, the same quantity of air is flowing in every section of a duct.

control, constant volume, variable temperature system control which varies the temperature and humidity of the air stream while maintaining air flow at a constant volume.

control, dew-point method of moisture control which uses a dry bulb temperature sensor downstream from a cooling coil to maintain discharge air at saturation by controlling the discharge temperature.

control, discharge air method of zone control in which controller receives its input signal from a temperature sensor in the discharge air stream.

control, floating method of control in which the controller produces a gradual continuous action in the controlled device.

control, integral method of control in which the controller positions the controlled device to maintain setpoint without error upon a change in load.

control, proportional method of control in which the amount of corrective action applied to a controlled device is proportional to the amount of load sensed by the controller.

control, return air zone control under which controller receives its input signal from a temperature sensor in the room.

control, variable volume, constant temperature system control which varies the volume of air flow to a zone at a constant discharge temperature.

control loop — sequential control system, consisting of a sensor, a controller, an actuator, and a manipulated device, connected to a mechanical or other process.

control point — in proportional control, the value of a controlled variable (such as temperature or pressure) which an automatic controller maintains under a fixed set of load conditions. Setpoint plus offset equals control point. See also *setpoint* and *offset*.

control system — device or series of devices used to regulate the operation of a controlled system or component either manually or automatically. If automatic, system responds to changes in temperature, pressure, or other controlled variable. See also *controlled variable*.

control system, analog electronic — system which uses solid state devices to regulate a controlled component.

control system, electric — system which uses electricity, often at line voltage, to regulate a controlled device.

control system, digital electronic — system which utilizes electronic technology to detect, amplify and evaluate sensor information. The evaluation can include sophisticated logical operations and results in an output command signal.

control system, pneumatic — system which uses compressed air to regulate a controlled device.

control system, self-contained — system which combines controller and controlled device into one unit.

controlled device — a device used to modulate or divert energy flow such as a valve or damper. See also *actuator*.

controlled variable — quantity or condition directly measured or controlled in a controlled system; examples include temperature, pressure, flow, and moisture.

convection — heat transfer that is a combination of two mechanisms; conduction and energy transport due to fluid motion.

convector — device which transfers its heat to air or other fluid by convection; terminal unit consisting of a finned tube or cast iron heat exchanger within a sheet metal cabinet or enclosure.

convertor — see *heat exchanger*.

cooling tower — evaporative cooling device that cools air by transferring heat from water to the cooler air of the atmosphere (evaporation).

cooling tower approach — difference between wet bulb temperature of entering air and temperature of leaving water.

cooling tower range — change in temperature of water passing through a cooling tower.

CPU see *central processing unit.*

CVVT control see *control, constant volume, variable temperature.*

Dalton's Law of Partial Pressures law which states that the pressure of a gas mixture in a common space is equal to the sum of the partial pressures of the individual gases.

damper movable or adjustable plate in a duct which controls the flow of a gas through the duct.

DDC see *direct digital control.*

deadband neutral zone or range of values within which no motion of a controlled device is required by the controller.

degree day average of a day's high and low temperatures added to or subtracted from 65°F.

dehumidify remove water vapor from the air.

demand charge that part of an electric bill based on the rate of electric power consumption in kilowatt-hours for a specified demand (time) interval in addition to the cost for the total electricity consumed.

desiccant absorbent or adsorbent material that removes water or water vapor from air or other material.

design conditions conditions that help to determine the *worst case* heating or cooling requirement.

dew point temperature at which air is 100% saturated with water vapor.

dew-point control see *control, dew-point.*

differential in a control system, the difference between the turn-on and the turn-off temperature or pressure as for opening or closing a circuit.

diffusion device device which mixes incoming supply air with air already in the conditioned space.

direct digital control control of a process or element with a computer using digital inputs and outputs for storing, retrieving, or processing data.

direct expansion coil in a heat exchanger, a coil having an expansion valve which controls the flow of refrigerant into the coil on the basis of the amount of superheat in the refrigerant as it flows out of the coil; in a refrigeration system, an evaporator which allows the air stream to be cooled by coming into direct contact with the refrigeration coil.

discharge air control see *control, discharge air.*

disengaging area surface area at the waterline of a steam boiler.

distributed control system a collection of interconnected control units, each having a specific function, which acquire data and control equipment or processes.

distribution system ducts or pipes which carry air, water, or steam to a conditioned space.

double duct system HVAC system which distributes both hot and cold air streams to a conditioned zone and mixes the air at the zone as opposed to mixing multiple air streams at a central plant.

draft an air current in an enclosed area or confined space.

drum enclosed vessel in a water tube boiler which distributes water to the tubes.

dry bulb temperature temperature as indicated on a dry bulb thermometer.

dry bulb thermometer see *thermometer*.

dry steam superheated steam which has more heat energy and less moisture than saturated steam.

dual-path system an air system that splits the mixed supply air into two streams. One stream is chilled and the other is heated.

duty cycling method of control intended to reduce electrical consumption and demand by turning off electrical equipment for predetermined periods of time during operating hours.

economizer cycle control strategy which reduces cooling load by allowing free cooling from outside air. See also *free cooling*.

efficiency-capacity curve graph used to evaluate pump performance by plotting the pump's efficiency on the vertical axis and its flow rate on the horizontal axis.

electric boiler see *boiler, electric*.

electric resistance coil coil which produces heat by opposing (resisting) flow of electricity through the coil.

electronic filter see *electrostatic filter*

electrostatic filter filter which removes dust particles from air by charging them with an electric field and attracting them to highly charged collector plates.

elevation head loss see *static head loss*.

enthalpy sum of the sensible and latent heats of the air and its water vapor in a pound of air (Btu/lb dry air); also known as heat content or total heat.

error see *offset*.

evaporation change of state from liquid to vapor.

evaporative condenser see *condenser, evaporative*.

evaporative cooler device which cools indoor air by evaporating moisture to exchange heat; commonly known as a swamp cooler; see also *evaporative cooling.*

evaporative cooling process of removing heat from air by exchanging sensible heat for latent heat such as bringing non-saturated air into contact with water sprayed through a nozzle (direct) or bringing non-saturated air into contact with a wet surface (indirect).

evaporative pan humidifier see *humidifier, evaporative pan.*

evaporator heat transfer device in a refrigeration system which changes refrigerant into low pressure vapor while absorbing heat from a space or water circuit.

exfiltration uncontrolled leakage of air from a building or space through cracks and openings such as those around doors and windows; opposite of infiltration.

exhaust air air vented to the atmosphere to maintain air quality within an area or space.

expansion tank a tank, open to the atmosphere and partially filled with water, used as a cushioning device by accommodating the expansion and contraction of water caused by temperature changes. See also *compression tank.*

extended surface filter filter designed with enormous surface areas capable of filtering extremely small particles.

facility management system integrated control of a building environment using computers or electronic devices.

Fahrenheit scale scale for measuring temperature under standard atmospheric pressure; boiling point of water is 212° and its freezing point is 32°.

feedback loop see *closed-loop control.*

feedwater water supplied to a boiler.

fiber optic cable long, thin, flexible fibers of glass or plastic through which data or images are transmitted using light.

fin-tube convector terminal unit which uses a finned tube to exchange heat by convection.

finned tube heat transfer pipe having a series of fins, tubes, or discs.

fire box the area of a boiler where combustion takes place, similar to a furnace.

fire side in a boiler, the area exposed to flame and combustion gases, as opposed to the water side.

fire tube the primary heat transfer surface area in a boiler.

fire tube boiler	see *boiler, fire tube.*
firmware	software permanently stored on a computer chip in read-only memory (ROM).
floating control	see *control, floating.*
flue	a duct or passage for conveying combustion products from a combustion chamber to or through a chimney; see also *breeching.*
flue gas	all gases in a flue during combustion in a combustion chamber.
flue stack	chimney.
flue tube	see *fire tube.*
FMS	see *facility management system.*
foaming	failure of steam bubbles to break at the water surface area at the water line of a boiler.
forced draft	using a fan or blower to supply (push) inlet air to a furnace.
four-pipe system	a system where one pipe supplies chilled water, a second pipe supplies heated water, a third pipe returns chilled water and a forth pipe returns hot water.
free cooling	method of temperature control which uses the temperature of outdoor air to reduce the temperature of indoor air.
friction head loss	reduction in pump head caused by fittings and devices in a piping system. See also *static head loss.*
full-hermetic compressor	see *compressor, hermetic.*
furnace	an enclosed structure in which heat is produced; a direct-fired heat exchanger.
furnace tube	see *blast tube*
grain	1/7000 of a pound; 7000 grains/lb (English measurement system).
gross output	for a boiler, the total amount of heat available per hour at a boiler outlet.
head	pressure exerted by a column of fluid; pressure a pump is capable of developing, usually measured in feet or pounds per square inch (psi).
head-capacity curve	graph used to evaluate pump performance by plotting the pump's head is plotted on the vertical axis and its flow rate on the horizontal axis.
heat	energy transferred from a source of higher temperature in the direction of a lower temperature.

heat, latent heat required to change the state of a substance without changing its temperature or pressure. Latent heat of fusion is amount of heat required to change ice to water at 32°F (144 Btu/per pound). Latent heat of vaporization is amount of heat required to change water to steam at 212°F (970 Btu/pound).

heat loss calculation see *peak heating load*.

heat, specific ratio of (a) the quantity of heat needed to raise the temperature of a given mass of a substance one degree to (b) the quantity of heat needed to raise the temperature of an equal mass of a standard reference substance, usually water, one degree.

heat content total heat of a substance; heat needed to raise the temperature of a substance to its present temperature. See also *enthalpy*.

heat exchanger device which transfers heat from a fluid in one chamber through the chamber walls to a fluid in another chamber.

heat load, total the heat required to maintain temperature, add humidity, and warm outside air (ventilation or infiltration).

heat recovery process of capturing heat that would otherwise be wasted.

heat wheel heat recovery device which uses loosely packed wire mesh or ceramic material in a rotating wheel to absorb heat from one air stream and give it up to an adjacent air stream.

heat transfer movement of heat energy caused by a temperature difference.

heat transfer coefficient heat flow; amount of heat transferred over a unit of time; rate at which heat energy moves through an area for each degree of temperature difference between the air on the warm side and the air on the cool side. Also known as U-factor, a reciprocal of the R-value.

heating degree day see *degree day*.

heating surface a boiler's total surface area where heat is transferred, including tubes, tube sheets and fire tube.

helical rotary screw compressor see *compressor, screw*.

hermetic compressor see *compressor, hermetic*.

high side float valve in a refrigeration system, a metering device which maintains the flow of liquid refrigerant so that the rate of flow into an evaporator is the same rate as the flow leaving the evaporator.

horsepower-capacity curve graph used to evaluate pump performance by plotting the pump's horsepower requirements on the vertical axis and its flow rate on the horizontal axis.

humidifier, air washer device which adds moisture to air adiabatically using a heated water spray chamber.

humidifier, atomizing device which adds moisture to air by spraying small water particles directly into the air stream.

humidifier, evaporative pan device which adds moisture to air by evaporating water from an open container.

humidifier, steam grid device which adds moisture to air by injecting steam through orifices in a dispersion manifold enclosed in a heating duct.

humidifier, wetted media device which adds moisture to air by evaporating water from a wetted medium using a fan or built-in air supply.

humidify to add moisture to air or other moisture-absorbing material.

humidistat instrument that measures and controls relative humidity.

humidity water vapor in air; airborne water vapor.

humidity, relative ratio of the actual partial pressure of water vapor in air compared to the saturation partial pressure measured at a specific dry bulb temperature; expressed as the percentage of saturation.

humidity, specific ratio of the mass of water to the total mass of a moist air; the weight of the water vapor in a pound of dry air expressed lb/lb dry air.

humidity ratio see *humidity, specific*.

hydronic system all-water heating system which in which the heat-conveying medium is hot water and the terminal units are radiators, convectors, or panel coils.

hygroscopic capable of retaining or giving up moisture.

impeller rotating part of a centrifugal compressor, pump or other machine which uses centrifugal force to move or compress a fluid.

inch of mercury pressure difference measured by a mercury-filled manometer, as in 1" Hg; the common method of measuring vacuum in an HVAC system.

inch water gauge common method of measuring atmospheric pressure; pressure difference measured by a water-filled manometer, as in 1" WG.

induced draft using a fan or blower inside the breeching to supply (draw) inlet air to a furnace.

induction unit a terminal unit in an air-water system consisting of a fin-tube heat exchanger, an air filter, and a high velocity air supply.

infiltration uncontrolled leakage of air into a building or space through cracks and openings such as those around doors and windows; opposite of exfiltration.

integral control see *control, integral*.

in-line pump centrifugal pump which has its inlet and its discharge in a straight line.

isothermal having a constant temperature.

Kelvin scale temperature scale on which the unit of measurement equals the Celsius degree and according to which absolute zero is 0°K, the equivalent of -273.15°C.

latent heat see *heat, latent*.

latent heat loss heat lost by evaporation. (see *evaporation*).

latent heat of fusion see *heat, latent*.

latent heat of vaporization see *heat, latent*.

ligament the portion of a boiler tubesheet between the tube holes.

load cycling see *duty cycling*.

load rolling see *duty cycling*.

load shedding see *duty cycling*.

low pressure boiler see *boiler, low pressure*.

manipulated device a control device such as a valve or damper excluding its actuator. See also *controlled device*.

manometer a double-legged, liquid column gauge which measures the difference between one pressure and a second pressure, commonly atmospheric pressure.

MAWP maximum allowable working pressure.

MBH 1000 Btus per hour.

mean radiant temperature average of the temperature of all surfaces in direct line of sight a human body; abbreviated MRT.

metabolism chemical changes in living cells which provide energy for body processes; the body process of converting food into heat energy.

metering device (refrigerant) device which controls the flow of liquid refrigerant to an evaporator.

mixed air — combination of air from two or more sources, for example outdoor air and return air.

mixing box — device which blends air streams having different characteristics, such as hot and cold air streams.

MRT — see *mean radiant temperature*.

mud drum — vessel in the lower part of a boiler where sludge and other impurities collect after settling out of water.

multi-zone system — HVAC system which mixes multiple air streams at a central plant and distributes the mixed air to the conditioned zones as opposed to a multiple zone system.

multiple zone system — HVAC system which distributes both hot and cold air streams to a conditioned zone and mixes the air at the zone as opposed to a multi-zone system.

NCU — see *network control unit*.

net load — the heat requirements of a system.

net output — for a boiler, the usable gross output per hour from a boiler after subtracting piping and pickup losses.

network control unit — device which monitors or controls designated equipment in a distributed control system and shares data with other control units.

network expansion unit — device which performs a local control function in a distributed control system, such as starting or stopping a fan.

NEU — see *network expansion unit*.

NOWL — normal operating water level.

offset — in proportional control, the difference between the setpoint and the control point; also called *error*. See also *setpoint* and *control point*.

one pipe — a system where a common supply and return pipe are shared by terminal units.

open compressor — see *compressor, open*.

open loop — a control system where the sensor senses a variable, but does not control the variable.

open piping system — piping system which has an opening to the atmosphere at one or more locations.

orifice plate — in a refrigeration system, a metering device used to balance refrigerant flow rate into the evaporator with the load in the evaporator.

outdoor air — air brought in from outdoors to compensate for exhaust air or other air lost in a process.

overshoot — a characteristic of two position controls where the actual condition created exceeds the setpoint value.

packaged system — HVAC system, usually factory assembled, pre-designed as a self-contained unitary system as opposed to a built-up or built-in-place system. A packaged system is usually assembled at the factory.

pass — horizontal run that flue gases take through a fire tubes in a boiler before exiting the breeching.

peak cooling load — worst case cooling requirement derived from the heat gain calculation.

peak heating load — worst case heating requirement derived from the heat loss calculation.

perfect washer — theoretical achievement of 100% moisture saturation in an air washer.

pickup load — see *pickup loss*.

pickup loss — additional heating capacity required to elevate a building's temperature after a cold start or recovery after a setback; also referred to as pickup load.

piezo-resistive sensing — a method of sensing pressure by measuring the change in volume (compression) of a material, such as silicon, in response to the pressure applied by a fluid.

piping loss — heat lost through pipe walls and insulation.

pleated-type, pocket or bag filters — see *extended surface filters*.

pneumatic-electric transducer — device that converts a pneumatic value to an equivalent electrical value such as volt or milliamp.

pop valve — see *relief valve*.

pound per square inch absolute — pressure exerted on a surface one inch square and equaling gauge pressure plus atmospheric pressure.

pound per square inch gauge — pressure exerted on a surface one inch square and measured using a gauge without considering atmospheric pressure.

pressure, atmospheric — pressure produced by the weight of gases in the outdoor atmosphere above the point where the pressure is measured.

pressure, static — pressure within a system exerted equally in all directions measured at right angles to the direction of air flow; the potential energy in a system.

primary air — air supplied to an induction unit.

primary system converts energy from electricity or fuel and is the source of hot or chilled water.

proportional band the change in the controlled variable required to move the controlled device from open to closed.

proportional control see *control, proportional.*

psia see *pounds per square inch absolute.*

psig see *pounds per square inch gauge.*

psychrometric chart graphic representation of the properties of moist air, each point representing a specific condition with respect to temperature and humidity, including the relationship of wet- and dry-bulb temperature, specific and relative humidity, enthalpy, moisture content, and dew-point temperature.

psychrometrics use of psychrometry and the psychrometric chart.

psychrometry branch of physics concerned with the study of air, temperature, and water vapor relationships, particularly the moisture in the air.

psychrometer device consisting of a wet bulb and a dry bulb thermometer used to measure humidity.

R-value see *thermal resistance.*

radiant panel terminal unit which exchanges heat by radiation through tubing grids installed in floors, walls, or ceilings.

radiation transfer of heat energy given off from molecules or atoms because of an internal temperature change.

radiator terminal unit in a hydronic or steam system used to deliver heat to a space by convection.

range in a cooling tower, the change in water temperature through the tower.

Rankine scale absolute temperature scale on which the unit of measurement equals a Fahrenheit degree and on which the freezing point of water is 491.67° and its boiling point is 671.67° (Fahrenheit temperature plus 459.67).

reciprocating cylinder see *compressor, reciprocating.*

refractory material having a high melting point, such as brick, which retains heat inside a boiler.

refrigerant fluid that absorbs heat by evaporation at a low temperature and pressure and releases heat by condensing at a higher temperature and pressure.

refrigeration process of extracting heat from a substance.

reheat heating supply air previously cooled below the temperature desired for maintaining the temperature of the conditioned space.

relative humidity see *humidity, relative*.

relief valve valve which opens to relieve overpressure in a vessel and close when pressure returns to setpoint.

resistance thermometer see *thermometer, resistance*.

return air air returning to a central unit from a conditioned space.

return air control control strategy in which a controller receives its signal from a temperature sensor in the return air stream from a large area.

rod and tube sensor temperature sensor consisting of a high expansion rod inside a low expansion tube, both attached at one end.

room separate partitioned area that may or may not require thermostatic control.

room reset of discharge method of zone control in which a controller receives its input signal from two temperature sensors, one in the zone and one in the discharge air stream.

rotary compressor see *compressor, rotary*.

safety valve see *relief valve*.

saturated steam steam at the boiling point corresponding to its pressure.

saturation temperature boiling point of a liquid at a fixed pressure.

scfm standard cubic feet (of air) per minute.

screw compressor see *compressor, screw*.

scroll compressor see *compressor, scroll*.

scrubber see *air washer*.

sealed bellows sensor temperature sensor filled with a vapor, gas, or fluid which changes in pressure and volume as temperature changes.

secondary system see *distribution system*.

secondary water hot or chilled water distributed to terminal units.

sedentary an activity between seated at rest and standing at ease.

semi-hermetic compressor see *compressor, semi-hermetic*.

sensible heat heat that changes the temperature of a substance without changing the substance's state; see also *enthalpy*.

sensible heat change addition or removal of sensible heat in air without changing its moisture content.

sensible heat factor ratio of air's sensible heat change to the air's total heat change.

sensible heat loss heat lost through the process of convection.

series loop all-water system design in which all water flows through each terminal unit.

setpoint in a control system, the desired value at which the controlled variable is set.

shell (boiler) metal welded to the tubesheets of a boiler to form its basic structure.

single-duct system an air system that uses one duct to distribute the air to each zone.

sling psychrometer see *psychrometer*.

software collection of programs and data used by a computer to process information; see also *firmware*.

specific heat see *heat, specific*.

specific humidity see *humidity, specific*.

specific volume volume of a substance per unit of mass (SV); volume of air (cu ft) per unit of dry air (lb), expressed as cu ft/lb dry air.

standard air air having a specific volume of 13.3 cu ft/lb dry air and a temperature of 70°F at an atmospheric pressure of 29.92 in. Hg.

standard cubic feet per minute measure of a fan's capacity using standard air.

static head loss reduction in pump head caused by an increase in elevation. See also *friction head loss*.

static pressure see *pressure, static*.

static regain conversion of the velocity pressure of flowing air to static pressure as the cross-section size of an air duct increases.

stays bolts or other devices designed to prevent tubesheet bulging when a boiler is under pressure.

steam generator boiler which produces steam.

steam grid grid of steam nozzles that release steam directly into the air.

steam grid humidifier see *humidifier, steam grid*.

steam quality the amount of vapor in a mixture of liquid and vapor, expressed as a percentage by weight.

superheat extra heat in a vapor when at a temperature higher than the saturation temperature corresponding to its pressure, for example the extra heat in steam which has not boiled although the temperature of the water is above its boiling point.

superheated steam steam at a temperature above its boiling point corresponding to its pressure.

supply air air entering a conditioned space.

supply air reset energy-conserving control strategy which monitors heating and cooling loads in building spaces and adjusts discharge air temperature to the most efficient level to satisfy the measured load.

supply water reset energy-conserving control strategy which automatically changes the setpoint of water supplied to a heating or cooling loop to the highest setpoint (chilled water) or lowest setpoint (hot water) to satisfy the heating or cooling load in a building space.

swamp cooler common name for an evaporative cooler; see *evaporative cooler*.

system characteristic curve graph used to evaluate the performance of a hydronic system by plotting the system's head in feet on the vertical axis and its flow rate on the horizontal axis.

temperature degree of heat or cold measured on a scale; the measure of the thermal state of adjacent substances which determines their ability to exchange heat.

terminal unit device by which the energy from a system is delivered to a zone (for example, a register or a radiator).

thermal comfort when a person is surrounded by an environment whose temperature and relative humidity permit the person to lose, without conscious effort, metabolic heat at the same rate he or she produces it.

thermal efficiency ratio of heat supplied from burning fuel to the heat absorbed by a fluid, typically water in a boiler.

thermal resistance a measure of the resistance of a substance which prevents heat from flowing through it; also known as R-value. The higher the R-value, the more inhibited is the transfer of heat energy. R-value is a reciprocal of the U-factor.

thermistor temperature sensor consisting of a semiconductor having a high negative coefficient of resistance; for example, the electrical resistance decreases as the temperature rises.

thermocouple junction of two dissimilar metals which generates, at the point of union, an electrical voltage proportional to the temperature sensed.

thermometer — instrument or device used to measure temperature. A dry bulb thermometer measures only sensible heat. A wet bulb thermometer measures latent heat content of the air; the bulb of a wet bulb thermometer is covered with a wetted wick, usually of muslin or cambric.

thermometer, resistance — temperature sensor consisting of a metal or silicon chip in which electrical resistance is a function of the temperature; that is, it increases or decreases as temperature rises or falls. See also *thermistor*.

thermostatic expansion valve — metering device which controls the flow of liquid refrigerant into an evaporator in response to the amount of superheat in the refrigerant leaving the evaporator.

three-pipe system — one pipe supplies chilled water, a second pipe supplies heated water and the third pipe carries the return flow.

throttling range — the change in the controlled variable required to move a controlled device from open to closed.

ton — in a refrigeration system, the amount of cooling 2000 pounds of ice can produce melting over a 24-hour period, 288,000 Btu/24hr, 12,000 Btu/hr.

total heat — see *enthalpy*.

total heat load — see *heat load, total*.

total pressure — the sum of static pressure and velocity pressure; see also *static pressure* and *velocity pressure*.

transducer — device which converts one type of signal to another or a device which receives data in one form and transmits it in another; for example, pressure as electrical voltage.

tubesheet — steel ends of a fire tube pressure vessel in which holes are drilled for inserting the tubes.

turbocompressor — see *compressor, centrifugal*.

two-pipe direct return — a piping design which uses a supply line and a return line.

two-pipe reverse return — a piping design where both supply and return pipe distances are the same. This keeps flow resistances equal for all terminal units minimizing balancing requirements.

U-factor — see *heat transfer coefficient*.

unit heater — enclosed terminal unit which consists of a heat exchanger such as fin-tube coils and a fan.

unit ventilator — enclosed terminal unit consisting of a coil, fan, and air filter capable of heating or cooling a space by exchanging heat from hot or chilled water in the coil; unit may also cool a space by bringing in outside air.

unitary system — see *packaged system*.

variable air volume — use of varying air volume to control the condition of the air in a zone in contrast to the use of constant flow with varying temperature

velocity pressure — pressure exerted in the direction of air flow; the kinetic energy within an air system.

ventilation — supplying or removing air from a space using natural or mechanical means.

ventilation rate — air provided for ventilation at a specified volume.

VVCT control — see *control, variable volume, constant temperature*.

water column — method of measuring the pressure a pump develops (one foot of water column is equivalent to 0.433 psi); fitting connected to a boiler drum designed to reduce turbulence and fluctuations in boiler water level.

water side — in a boiler, the tubes and vessels which contain water and steam, as opposed to the fire side.

water tube boiler — see *boiler, water tube*.

wet bulb temperature — temperature indicated when water evaporates from the bulb of a thermometer covered with a water-saturated wick and the heat of vaporization is supplied from the latent heat of the air.

wet bulb thermometer — see *thermometer*.

wetted media humidifier — see *humidifier, wetted media*.

zone — open area requiring separate thermostatic control. See also *room*.

zone reheat — a system where zone heaters temper, or reheat, air distributed from the centrally located system.

zoning — division of a building into separately controlled spaces for control of heating and cooling using a single thermostat.

Note: The definitions in this glossary are based on those found in this book, *Building Environments: HVAC Systems*, with supplementary information derived from *ASHRAE Terminology of Heating, Ventilation, Air Conditioning, & Refrigeration; McGraw-Hill Dictionary of Science and Engineering*; and *Webster's Ninth New Collegiate Dictionary*.

Index

A

B

C

D

E

F

G

H

I

K

L

M

N

O

P

R

S

T

U

V

W

Z

About the Author

Alan J. Zajac is a Lead Instructional Technologist, with the Johnson Controls Institute of Johnson Controls, Inc., Milwaukee, Wisconsin. He has performance development responsibilities for various business clients within the Controls Group of Johnson Controls, Inc.

Al is a lifelong technical trainer and performance technologist with wide ranging knowledge and skills. He develops training programs to support sales, estimators, engineers, technicians, project managers and installers. Al's technical knowledge and skill base include: HVAC systems, controls and applications, and contracting in the installation and service industry. Strong skills in the analysis, design, and delivery of appropriate instructional strategies enable Al to produce programs that get results.

An adjunct instructor at the Milwaukee School Of Engineering, Milwaukee, Wisconsin, Al developed and regularly conducts courses on HVAC systems.

Al received his BS and MS in Industrial Education from the University of Wisconsin – Stout, Menomonie, Wisconsin.

Al is a member of American Society of Heating, Refrigerating and Air Conditioning Engineers, Inc. (ASHRAE) and the International Society of Performance Improvement (ISPI).